I0037075

Evolution, Composition and Regulation of Supernumerary B Chromosomes

Evolution, Composition and Regulation of Supernumerary B Chromosomes

Special Issue Editors

Andreas Houben
Neil Jones
Cesar Martins
Vladimir Trifonov

MDPI • Basel • Beijing • Wuhan • Barcelona • Belgrade

MDPI

Special Issue Editors

Andreas Houben
Leibniz Institute of Plant Genetics and Crop Plant Research (IPK)
Germany

Neil Jones
Aberystwyth University
UK

Cesar Martins
UNESP—São Paulo State University
Brazil

Vladimir Trifonov
Institute of Molecular and Cellular Biology SB RAS
Russia

Editorial Office
MDPI
St. Alban-Anlage 66
4052 Basel, Switzerland

This is a reprint of articles from the Special Issue published online in the open access journal *Genes* (ISSN 2073-4425) from 2018 to 2019 (available at: https://www.mdpi.com/journal/genes/special_issues/B_Chromosomes)

For citation purposes, cite each article independently as indicated on the article page online and as indicated below:

LastName, A.A.; LastName, B.B.; LastName, C.C. Article Title. *Journal Name* **Year**, *Article Number*, Page Range.

ISBN 978-3-03897-786-5 (Pbk)
ISBN 978-3-03897-787-2 (PDF)

Cover image courtesy of Andreas Houben, Neil Jones, Vladimir A. Trifonov and Cesar Martins.

© 2019 by the authors. Articles in this book are Open Access and distributed under the Creative Commons Attribution (CC BY) license, which allows users to download, copy and build upon published articles, as long as the author and publisher are properly credited, which ensures maximum dissemination and a wider impact of our publications.

The book as a whole is distributed by MDPI under the terms and conditions of the Creative Commons license CC BY-NC-ND.

Contents

About the Special Issue Editors

Andreas Houben, Professor, senior research at the Leibniz Institute of Plant Genetics and Crop Plant Research (IPK), Gatersleben, Germany. Graduated in plant breeding from the Martin-Luther-University Halle-Wittenberg (Germany), and received a Ph.D. in the field of chromosome biology. His habilitation, obtained in 2010 from the Martin-Luther-University Halle-Wittenberg, was devoted to B chromosome research. He has published ca. 150 peer-reviewed journal articles on topics related to centromeres, chromosome evolution and B chromosomes. His current research interests focus on the function, regulation and evolution of plant chromosomes.

Neil Jones, Emeritus Professor at Aberystwyth University, UK. Graduated in Agricultural Botany at Aberystwyth in 1963, and then spent nine months in the Botany Department at Indiana University 1963–1964. Returned to Aberystwyth for his PhD in Chromosome Genetics 1964–1967, working on B chromosomes in rye. He designed, made, and analysed several experiments, before moving to Queens University Belfast 1967–1969. He returned again to Aberystwyth as a member of staff in the Agricultural Botany Department in 1969, gaining his DSc in 1967 and becoming Professor in 1991. Having become captivated by B chromosomes set about collecting hard copy the world literature on B chromosome from 1906 to 1980 as reprints and photocopies. After reading this mass of literature he published B Chromosomes, Jones and Rees, Academic Press 1982. This book gave him recognition in the B chromosome community, which he holds to this day. He will be keynote speaker at the fourth B chromosome conference in Brazil in July 2019. Other achievements include being visiting Professor at Kyoto University in 2004, Honorary professor in the Faculty of Biology in St. Petersburg University, Member of the Lithuanian Academy of Sciences, Senior Secretary and Vice President of the Genetical Society UK, and Dean of the Faculty of Science at Aberystwyth. He has published four books and 128 scientific papers.

Cesar Martins, PhD in Genetics and Evolution from Federal University of São Carlos (São Carlos) in 2000 and currently a Professor of Molecular and Cellular Biology and Genomics at São Paulo State University—UNESP (Botucatu). During most of his career Martins has worked with comparative cytogenetics exploring cichlids as a model group and his current research interest is directed to evolution of genes, chromosomes and genomes, with emphasis on B chromosome biology. Martins's lab explores cytogenetics, genomics and bioinformatics tools focused in a functional view of the cell environment under the effects of B chromosome presence. He has published 118 peer-reviewed papers, 4 books and 8 book chapters.

Vladimir Trifonov, Head of Comparative Genomics laboratory at the Institute of Molecular and Cellular Biology (Novosibirsk, Russia) and Lecturer in Genomics (Novosibirsk State University). Graduated in biology from Novosibirsk State University, received a PhD in the field of genetics (2002) from the Institute of Cytology and Genetics SB RAS (Novosibirsk). He did post-doctoral research at the Institute of Human Genetics (Jena, Germany) and University of Cambridge (UK). His habitation, obtained in 2019 from his institution, was devoted to the evolution of vertebrate sex chromosome. He has published ca. 130 articles in peer-viewed journals on molecular cytogenetics of vertebrates, clinical cytogenetics paleogenomics, and population genetics. His current research interests include: evolution of vertebrate karyotype; comparative genomics; development of new molecular cytogenetic approaches; structure, function and evolution of B-chromosomes and sex chromosomes; sex determination in vertebrates; human karyotype and chromosomal abnormalities, population genetics.

![genes](GCAT TACG GCAT *genes*)

MDPI

Editorial

Evolution, Composition and Regulation of Supernumerary B Chromosomes

Andreas Houben [1],*, Neil Jones [2], Cesar Martins [3] and Vladimir Trifonov [4]

1 Leibniz Institute of Plant Genetics and Crop Plant Research (IPK), Corrensstrasse 3, 06466 Gatersleben, Germany
2 Aberystwyth University, Institute of Biological, Environmental and Rural Sciences (IBERS), Edward Llwyd Building, Penglais Campus, Aberystwyth SY23 3DA, UK; rnj@aber.ac.uk
3 Institute of Bioscience at Botucatu, São Paulo State University—UNESP, Botucatu, SP 18618, Brazil; cesar.martins@unesp.br
4 Laboratory of Comparative Genomics, Department of the Diversity and Evolution of Genomes, Institute of Molecular and Cellular Biology SB RAS, Novosibirsk 630090, Russia; vlad@mcb.nsc.ru
* Correspondence: houben@ipk-gatersleben.de

Received: 9 January 2019; Accepted: 10 January 2019; Published: 20 February 2019

Supernumerary B chromosomes (Bs) are dispensable genetic elements found in thousands of species of plants and animals, and some fungi. Since their discovery more than a century ago, they have been a source of puzzlement, as they only occur in some members of a population and are absent from others. When they do occur, they are often harmful, and in the absence of "selfishness", based on mechanisms of mitotic and meiotic drive, there appears to be no obvious reason for their existence. Cytogeneticists have long wrestled with questions about the biological existence of these enigmatic elements, including their lack of any adaptive properties, apparent absence of functional genes, their origin, sequence organization, and co-evolution as nuclear parasites. Emerging new technologies are now enabling researchers to step up a gear, to look enthusiastically beyond the previous limits of the horizon, and to uncover the secrets of these "silent" chromosomes.

This volume, "Evolution, Composition and Regulation of Supernumerary B Chromosomes", consists of a series of new reviews and original research articles, and provides a comprehensive guide to theoretical advancements in the field of B chromosome research in both animal and plant systems. Beside "classical" B chromosomes, supernumerary A chromosomal segments and germ-line limited (termed L or "limited") chromosomes will be addressed. The topics include investigations into their DNA composition, transcriptional activity, and effects on the host transcriptome profile, drive, origin, and evolution.

Among animals, B chromosomes have been largely investigated in insects, fish, and mammals. Based on the study of nine grasshopper species from Europe, Asia, and Africa, Jetybayev et al. [1] state that the origin and evolution of Bs depend on particularities, such as the hotspots of chromosome rearrangements, the mobility of genomic elements, and the tendency of specific DNA fragments towards amplification. Milani et al. [2] use a combination of cytogenetics and bioinformatics to investigate Bs in three other grasshopper species from North and South America. Their data show the recurrent involvement of small A chromosomes, that are poor in genes and enriched in repetitive DNA, in the B chromosome origin. Hanlon and Hawley [3] present a review on B studies in *Drosophila*, and highlight *D. melanogaster* as a versatile model to advance our understanding of the complex B chromosome biology.

Another interesting animal group for B studies is represented by fish species. Komissarov et al. [4] investigate Bs of the Asian Seabass (*Lates calcarifer*), and give support to the view that Bs contribute to the variations in the genome among individuals and populations of the species. Among fish, cichlids have appeared as an interesting model for B studies. Coan and Martins [5] demonstrate that expanded

B transposable elements are under functional control and are not highly transcribed. Clark et al., [6] on the basis of both Illumina and PacBio sequence data of seven cichlid species, have accessed a "core" B chromosome dataset of genes and fragmented genes.

Among amniotes, Bs have been much better studied in mammals than in reptiles, birds, and amphibians. Although there are many reports on B chromosomes in amphibian species, only few studies have been done in lizards and no Bs were reported in avian genomes. This may reflect the fact that most amphibian and reptile genomes are poorly studied, and birds might lack Bs due to their characteristic genome evolution and karyotype structure. However, in songbirds, the Germline-Restricted Chromosome (GRC) has been reported [7,8].

Two review papers on B chromosomes in mammals—by Rubtsov and Borisov [9] and by Vujosevic et al. [10] —are devoted to B chromosome origin, content, activity, and evolution in mammals. These two articles complement each other, as they address Bs from different perspectives. Both reviews summarize recent data obtained from different mammalian species, and make several important conclusions: They suggest a considerable heterogeneity of mammalian Bs based on their origin and subsequent evolution. Bs found in modern species seem to be different in their origin and have undergone different evolutionary trajectories, although they might have been shaped by similar evolutionary mechanisms. Rubtsov and Borisov [9] suggest several models of B chromosome origin, while the article by Vujosevic et al. [10] provides an updated list of mammalian species with Bs, and gives a detail description of research experiments accomplished on these species.

The research article by Makunin et al. [11] provides novel data on B chromosome content and evolution in the red fox (the first mammalian species with Bs, whose genome has recently been sequenced and assembled (Kukekova et al.) [12], and in the raccoon dog, the carnivore species, where B-specific coding genes were discovered almost 13 years ago. Using new generation sequencing, the authors argue that the origin of B chromosome in these species is independent. Through the analysis of B content in different mammals they conclude a frequent and independent re-use of the same genomic regions in B chromosome formation. They suggest that such a re-use may be connected with gene functions.

Borisov and Zhigarev [13] consider another mammalian species—*Apodemus peninsulae*—a good model for B chromosome studies. They summarize their data, collected over 40 years in different geographical regions of species distribution, and conclude that the variability of B chromosome systems results from stochastic processes in populations.

Another important set of contributions come from plants. H. Su et al. [14] discuss the maize B alongside the latest progress of centromere activities, including centromere mis-division, inactivation, reactivation and de novo centromere formation. Drive is one of the most important B chromosome features. R. N. Jones [15] summarizes the mechanisms of drive, which enable B chromosomes to enhance their transmission rates by various processes of non-Mendelian inheritance in plants and animals. A. Marques et al. [16] demonstrate how new genomic approaches have shed light on the origin and accumulation of different classes of repetitive sequences in the process of B chromosome formation and evolution. M. Dhar et al. [17] summarize the characterization of a novel B which was discovered in *Plantago lagopus*. This B was found to be composed of mainly 5S rDNA-derived sequences and various types of repetitive elements. The transmission of the *Plantago* B through the female sex track does not follow Mendelian principles.

Supernumerary chromosomal segments represent additional chromosomal material that, unlike B chromosomes, is attached to the standard A chromosome complement. Using the *Prospero autumnale* complex (Hyacinthaceae) as a model, T-S. Jang et al. [18] decipher the possible origin of supernumerary chromosomal segments as by-products of the extensive genome restructuring within a putative ancestral *P. autumnale* genome, predating the complex diversification at the diploid level and perhaps linked to B chromosome evolution.

The germ-line limited (termed L or "limited") supernumerary chromosomes of the dipteran *Sciara coprophila* was revised by Singh and Belyakin, [19] adding knowledge to the imprinting phenomenon of such extra elements.

We believe that this volume will be an important resource for a wide variety of audiences, including junior graduate students and established investigators who are interested in chromosome biology and genome evolution.

Author Contributions: A.H., J.N., C.M. and V.T. designed and wrote the manuscript.

Funding: This research was funded by the DFG (1779/30-1) to A.H, FAPESP (2015/16661-1) and CNPq (305321/2015-3) to C.M., and RFBR (17-00-00146) to V.T.

Conflicts of Interest: The authors declare no conflict of interest.

References

1. Jetybayev, I.Y.; Bugrov, A.G.; Dzuybenko, V.V.; Rubtsov, N.B. B Chromosomes in Grasshoppers: Different Origins and Pathways to the Modern B_s. *Genes* **2018**, *9*, 509. [CrossRef] [PubMed]
2. Milani, D.; Bardella, V.B.; Ferretti, A.B.S.M.; Palacios-Gimenez, O.M.; Melo, A.S.; Moura, R.C.; Loreto, V.; Song, H.; Cabral-de-Mello, D.C. Satellite DNAs Unveil Clues about the Ancestry and Composition of B Chromosomes in Three Grasshopper Species. *Genes* **2018**, *9*, 523. [CrossRef]
3. Hanlon, S.L.; Hawley, R.S. B Chromosomes in the *Drosophila* Genus. *Genes* **2018**, *9*, 470. [CrossRef] [PubMed]
4. Komissarov, A.; Vij, S.; Yurchenko, A.; Trifonov, V.; Thevasagayam, N.; Saju, J.; Sridatta, P.S.R.; Purushothaman, K.; Graphodatsky, A.; Orbán, L.; Kuznetsova, I. B Chromosomes of the Asian Seabass (*Lates calcarifer*) Contribute to Genome Variations at the Level of Individuals and Populations. *Genes* **2018**, *9*, 464. [CrossRef] [PubMed]
5. Coan, R.L.B.; Martins, C. Landscape of Transposable Elements Focusing on the B Chromosome of the Cichlid Fish *Astatotilapia latifasciata*. *Genes* **2018**, *9*, 269. [CrossRef] [PubMed]
6. Clark, F.E.; Conte, M.A.; Kocher, T.D. Genomic Characterization of a B Chromosome in Lake Malawi Cichlid Fishes. *Genes* **2018**, *9*, 610. [CrossRef] [PubMed]
7. Pigozzi, M.I.; Solari, A.J. Germ cell restriction and regular transmission of an accessory chromosome that mimics a sex body in the zebra finch, *Taeniopygia guttata*. *Chromosome Res.* **1998**, *6*, 105–113. [CrossRef] [PubMed]
8. Torgasheva, A.A.; Malinovskaya, L.P.; Zadesenets, K.S.; Karamysheva, T.V.; Kizilova, E.A.; Pristyazhnyuk, I.E.; Shnaider, E.P.; Volodkina, V.A.; Saifutdinova, A.F.; Galkina, S.A.; et al. Germline-restricted chromosome (GRC) is widespread among Songbirds. *bioRxiv* **2018**. [CrossRef]
9. Rubtsov, N.B.; Borisov, Y.M. Sequence Composition and evolution of mammalian B chromosomes. *Genes* **2018**, *9*, 490. [CrossRef] [PubMed]
10. Vujošević, M.; Rajičić, M.; Blagojević, J. B Chromosomes in Populations of Mammals Revisited. *Genes* **2018**, *9*, 487. [CrossRef] [PubMed]
11. Makunin, A.I.; Romanenko, S.A.; Beklemisheva, V.R.; Perelman, P.L.; Druzhkova, A.S.; Petrova, K.O.; Prokopov, D.Y.; Chernyaeva, E.N.; Johnson, J.L.; Kukekova, A.V.; et al. Sequencing of supernumerary chromosomes of red fox and raccoon dog confirms a non-random gene acquisition by B chromosomes. *Genes* **2018**, *9*, 405. [CrossRef] [PubMed]
12. Kukekova, A.V.; Johnson, J.L.; Xiang, X.; Feng, S.; Liu, S.; Rando, H.M.; Kharlamova, A.V.; Herbeck, Y.; Serdyukova, N.A.; Xiong, Z.; et al. Red fox genome assembly identifies genomic regions associated with tame and aggressive behaviours. *Nat. Ecol. Evol.* **2018**, *2*, 1479–1491. [CrossRef] [PubMed]
13. Borisov, Y.M.; Zhigarev, I.A. B Chromosome system in the Korean field mouse *Apodemus peninsulae* Thomas 1907 (Rodentia, Muridae). *Genes* **2018**, *9*, 472. [CrossRef] [PubMed]
14. Su, H.; Liu, Y.; Liu, Y.; Birchler, J.A.; Han, F. The Behavior of the Maize B Chromosome and Centromere. *Genes* **2018**, *9*, 476. [CrossRef] [PubMed]
15. Jones, R. Transmission and Drive Involving Parasitic B Chromosomes. *Genes* **2018**, *9*, 388. [CrossRef] [PubMed]
16. Marques, A.; Klemme, S.; Houben, A. Evolution of Plant B Chromosome Enriched Sequences. *Genes* **2018**, *9*, 515. [CrossRef] [PubMed]

17. Dhar, M.K.; Kour, J.; Kaul, S. Origin, Behaviour, and Transmission of B Chromosome with Special Reference to *Plantago lagopus*. *Genes* **2019**, *10*, 152. [CrossRef]
18. Jang, T.-S.; Parker, J.S.; Weiss-Schneeweiss, H. Euchromatic Supernumerary Chromosomal Segments—Remnants of Ongoing Karyotype Restructuring in the *Prospero autumnale* Complex? *Genes* **2018**, *9*, 468. [CrossRef] [PubMed]
19. Singh, P.B.; Belyakin, S.N. L Chromosome Behaviour and Chromosomal Imprinting in *Sciara Coprophila*. *Genes* **2018**, *9*, 440. [CrossRef] [PubMed]

© 2019 by the authors. Licensee MDPI, Basel, Switzerland. This article is an open access article distributed under the terms and conditions of the Creative Commons Attribution (CC BY) license (http://creativecommons.org/licenses/by/4.0/).

G C A T
T A C G
G C A T
genes

MDPI

Review

Origin, Behaviour, and Transmission of B Chromosome with Special Reference to *Plantago lagopus*

Manoj K. Dhar *, Jasmeet Kour and Sanjana Kaul

Genome Research Laboratory, School of Biotechnology, University of Jammu, Jammu-180006, India; jassi29honey@gmail.com (J.K.); sanrozie@rediffmail.com (S.K.)
* Correspondence: manojdhar@jammuuniversity.in

Received: 11 October 2018; Accepted: 12 December 2018; Published: 18 February 2019

Abstract: B chromosomes have been reported in many eukaryotic organisms. These chromosomes occur in addition to the standard complement of a species. Bs do not pair with any of the A chromosomes and they have generally been considered to be non-essential and genetically inert. However, due to tremendous advancements in the technologies, the molecular composition of B chromosomes has been determined. The sequencing data has revealed that B chromosomes have originated from A chromosomes and they are rich in repetitive elements. In our laboratory, a novel B chromosome was discovered in *Plantago lagopus*. Using molecular cytogenetic techniques, the B chromosome was found to be composed of ribosomal DNA sequences. However, further characterization of the chromosome using next generation sequencing (NGS) etc. revealed that the B chromosome is a mosaic of sequences derived from A chromosomes, 5S ribosomal DNA (rDNA), 45S rDNA, and various types of repetitive elements. The transmission of B chromosome through the female sex track did not follow the Mendelian principles. The chromosome was found to have drive due to which it was perpetuating in populations. The present paper attempts to summarize the information on nature, transmission, and origin of B chromosomes, particularly the current status of our knowledge in *P. lagopus*.

Keywords: B chromosome; transmission; origin; drive

---

1. Introduction

B chromosomes are enigmatic elements of the genome of some organisms. Interestingly, these chromosomes do not recombine with A chromosomes and have thus followed a different pathway during evolution [1]. These are also called as non-essential, extra, or supernumerary chromosomes because of their inertness. B chromosomes have been reported in over 2000 plant, animal, and fungal species [2]. The distribution of B chromosomes within an individual or among different individuals of a population is not uniform; they are not found in all members of a population, and their copy number can also vary among individuals possessing Bs. Since the B chromosomes remain unpaired during meiosis, as a result their mode of inheritance is irregular. Due to their dispensable nature, Bs are considered non-functional or parasitic elements, since they use the cellular machinery that is required for the maintenance and inheritance of A chromosomes [2]. Although B chromosomes are not essential, some phenotypic effects have been reported, which are usually cumulative depending upon the number and not the presence or absence of Bs. When present in low numbers, Bs have little, if any, influence on the phenotype, but at higher numbers they mostly have a negative effect on the fitness and fertility of the organism [3]. It was found that Maize B chromosomes influence the A-genome transcription with stronger effect associated with an increase in the copy number of B chromosome. Huang et al., compared lines with and without the B chromosome and detected 130 differentially expressed genes [4]. Despite being deleterious to the host genome at high copy numbers, Bs survive in numerous populations across multiple eukaryotic kingdoms. There is evidence that Bs directly or indirectly influence the behavior of A chromosomes [1]. These chromosomes acquire transposons, repetitive DNA, and organellar DNA. Bs have also been shown to harbor transcribed genes [4] and noncoding loci [5]. B chromosomes have been reported to contain thousands of sequences duplicated from essentially every chromosome in the ancestral karyotype [6].

Bs are widely distributed among eukaryotes: about 15% of eukaryotes contain these chromosomes [6]. There is considerable heterogeneity between different groups and also hotspots of occurrence and their presence is correlated with genome size [7]. Though there have been a number of reports on B chromosomes but a comprehensive theory for the origin, composition, regulation, maintenance, and evolution of B chromosomes has not emerged.

Recent technological advances in sequencing and genome analysis [8,9] have significantly enhanced our insight into the biology of B chromosomes. The intent of this review is to summarize the recent findings on B chromosomes with a pre-eminent focus on the B chromosome of *Plantago lagopus*.

Plantago L. is a large genus of annual or perennial herbs and subshrubs with a wide distribution. It is the chemical properties and divergent breeding systems (gynodioecy, self-incompatibility, and male sterility), due to which *Plantago* has grabbed the attention of many researchers [10]. *Plantago lagopus* L. is a small (about 30 cm tall), annual herb. It grows as a weed in the Mediterranean region. The diploid chromosome number of the species is 2n = 2x = 12. A spontaneous trisomic plant (2n = 2x = 13, Triplo 4) of *P. lagopus* was recovered in 1982 in an experimental population [11], which was a product of a nondisjunction event. When these trisomic plants were crossed with euploids, a remarkable plasticity of the genome was revealed, and plants having 1 to 16 chromosomes in addition to the standard set of 12 chromosomes, were recovered [12]. In one of the aneuploids, a primary trisomic for chromosome 2, the extra chromosome underwent rapid changes, resulting in the formation of ring chromosomes and chromosome fragments. Finally, the extra chromosome got stabilized as the proto-supernumerary isochromosome. Dhar et al., reported a novel B chromosome in *P. lagopus*, whose main body is composed of 5S ribosomal DNA (rDNA) and has few 45S rDNA sequences at the ends [12]. Here, we would review the work that aided in the identification and characterization of B chromosome in *P. lagopus*, which in turn helped in understanding the origin and evolution of B chromosomes.

2. Distribution of B Chromosomes

B chromosomes are widely distributed among eukaryotes [13], however, due to difficulty in preparing good cytological preparations in some taxa, possibly many species possessing Bs remain unknown at present. It has been estimated that Bs occur in 15% of eukaryotic species [2] and most of these are plants [14]. As far as animals are concerned, Bs are completely absent in birds [15] and are limited in mammals, where B chromosomes have been reported in 75 species out of about 5000 known species, which have been karyotyped [16]. There are a number of factors that have been related with the distribution of Bs among various species, namely: breeding system, genome size, and also the A chromosome morphology in certain mammals [7,17]. In genus *Plantago*, B chromosomes have been reported in *P. lagopus* [12], *Plantago major*, *Plantago lanceolata*, and tetraploid cytotypes of *Plantago depressa* [18], *Plantago seraria* [19], *Plantago boissieri* [20], and *Plantago coronopus* [21].

The distribution of Bs among major groups of angiosperm lineages and among lineages within families is nonrandom [13], but widespread. While B chromosomes are present in only 3% of eudicots, 8% of studied monocots among angiosperms possess B chromosomes [17], with considerable heterogeneity between different groups in frequency at the orders, families, and generic level. There are also hot spots of occurrence in Liliales and Commelinales [13]. Within the monocots, there are apparent differences between different families in B frequency. Some monocot families, such as the Melanthiaceae, have Bs in over 40% of the studied species, while others lack Bs entirely [17].

Frequency of Bs is independent of genome ploidy, with virtually no difference in frequency between diploids and polyploids [7,22]. Bs have been found to be more common in families with larger genomes, are found more frequently in mammals with acrocentric chromosomes [7,17,22]. It was hypothesized that there is less probability of presence of B chromosome in species with small genomes as larger genomes would better tolerate additional genetic material [17]. The presence of Bs has been positively correlated with pollution and/or stressful climactic environments in vertebrates [2].

Bs are carried by all the cells within the individual. However, there are some exceptions in which the Bs show lot of instability during mitosis in somatic tissues and therefore, they are present or absent in variable numbers in specific tissues and/or organs. In plants, only one-third of the investigated species display constancy of Bs in different tissues of the plant [23]. Absence of Bs from roots has been observed in *Erianthus munja* and *Erianthus ravennae* [13]. In few species the instability of the Bs is partial and strictly defined, e.g., in *Sorghum stipoideum*, mosaicism of B chromosomes has been reported in tapetal cells and microsporocytes, while they were totally eliminated from the leaves and stem [13]. In has been observed in some animals that the presence of B chromosomes leads to irregularities. For example, in *Locusta migratoria*, inter- and intrafollicular numerical variation has been reported [24]. The number of Bs varies between the cells of the testis in *Acris crepitans* [25]. Bs are generally stable among monocots; in *Allium* Bs are mostly unstable [26]. In *Brachycome dichromosomatica*, the large B has been found to be mitotically stable and the micro B varies in number from cell to cell [27].

In *P. lagopus*, the morphology of plants with and without B chromosome was found to be highly similar, indicating that the presence or absence of the B chromosome does not affect plant growth habit or vigour [12]. The distribution of B chromosome in this genus can also be associated with genome size, as this wild species *P. lagopus* with larger genome size (1.25 Gb) possesses B chromosome as compared to the other known species with smaller genome size, like *Plantago ovata* (0.65 Gb).

3. Structure of B Chromosomes

It is expected of B chromosome to show frequent polymorphisms because of its gratuitous nature. Undeniably, apart from a number of numeric polymorphisms, considerable structural polymorphisms of B chromosomes have been reported in a number of plants, e.g., *Aegilops speltoides* [28] and *B. dichromosomatica* [27]. Nevertheless, in some species like rye, B chromosomes have shown an identical cytological and molecular structure at the level of subspecies [1].

B chromosomes show variation in size also. The morphology of B chromosomes differs from that of A and are generally heterochromatinized. In plants, these chromosomes are often smaller than the smallest A, nevertheless mammalian Bs have been found within the size range of the As [2]. There are examples where B chromosomes larger than A chromosomes have been reported e.g., in the grasshopper *Eyprepocnemis plorans* [1], the cyprinid fish *Alburnus alburnus* [29], and the neotropical fish *Astyanax paranae* [30], though they are generally smaller than A chromosomes. In some species even, different types of B chromosomes coexist, e.g., in *B. dichromosomatica* [31] and the marsupial frog *Gastrotheca espeletia* [32].

4. Molecular Composition

Though B chromosomes have been largely studied, however, information about their molecular composition is scanty. In view of the possibility of origin of Bs from As, it was expected that the B chromosome would share sequence and structural similarity with A chromosome, so that they could synapse and recombine. However, this initial similarity seems to have been lost due to the accumulation of structural changes. Though, initially, experiments like density gradient centrifugation and renaturation kinetics showed similarity between A and B chromosome DNA composition, in recent years techniques like chromosome microdissection [32] and flow sorting [33] helped in the direct isolation of B chromosome-derived DNA. Next generation sequencing (NGS) has played a major role in the characterization of B chromosomes in some species. B chromosomes in rye have been characterized by using flow sorted A and B chromosomes [34]. Next generation sequencing has also been used to study the evolution and origin of B chromosome in the Cichlid fish *Astatotilapia latifasciata* [6] and *P. lagopus* [35]. These techniques showed how the composition of B chromosomes differs from A chromosomes, which was expected as the B chromosomes follow a separate evolutionary path.

One of the many exceptional features of B chromosomes is the accumulation of repetitive DNAs. Most of the B chromosomes reported in different organisms are heterochromatic mainly due to the presence of repetitive DNA sequences made of satellite DNA, ribosomal DNA, and transposable elements [34]. The type and copy number of repeats vary in B chromosomes. In many cases, it has been observed that B chromosomes contain much larger amounts of repetitive DNA when compared to the genome from which they originated, thus suggesting the massive amplification of repeat motifs over a relatively short time-scale [12].

B chromosomes have been reported to be enriched in repetitive sequences e.g., the micro B chromosome of *B. dichromosomatica*, which is mainly composed of tandem repeats [32], mobile elements, or non-coding repetitive sequences in *Crepis capillaris*, *Nectria haematococca*, and *Drosophila* [1]. Highly repetitive sequences, including transposons, were found to be present on both A and B chromosomes, but enriched on Bs in maize [36,37]. In case of rye, B chromosomes were found to accumulate large amounts of specific repeats and insertions of organellar DNA [34]. According to Valente et al., in Cichlid fish (*A. latifasciata*), the B chromosome originated early in the evolutionary history from a small fragment of one autosome [6]. While studying the B-linked genes of *Drosophila albomicans*, it was observed that 5.5% of the genome consisted of repetitive elements as compared to 5.35% of *Drosophila melanogaster* [38]. Since B chromosomes do not take part in recombination, they acquire larger fraction of mobile elements as compared to A chromosomes, which also leads to their heterochromatization. B chromosomes because of their relaxed selective pressure provide a safe spot for mobile elements [39].

The B chromosome in *P. lagopus* is cytogenetically unique. The chromosome was found to be completely heterochromatic using C-banding analysis [12]. The structure and behavior of B chromosomes was found to be typical of many naturally occurring B chromosomes, but it was unique in having its entire body mass derived from 5S DNA sequences [12]. The authors presented the experimental evidence of de novo origin of novel B chromosome in *P. lagopus* through specific DNA sequence amplification. Southern blot analysis revealed a distinct band of about 23 kb, only in +B plants. Using fluorescence in situ hybridization (FISH) and reverse genomic in situ hybridization (GISH) techniques, the entire chromosome was found to get painted with 5S rDNA probe, while 45S rDNA sequences were localized at the two ends, just below the telomeric sequences. Using molecular cytogenetic techniques, like FISH and Fiber-FISH, Kour et al., further characterized this chromosome and reported it to be a mixture of rDNA sequences and transposable elements [10]. It was also concluded that while 45S rDNA sequences are restricted to the subtelomeric regions, the 5S rDNA sequences are interspersed with transposons in the body of the B chromosome. Recently, the DNA composition of *P. lagopus* genome with and without B chromosomes was in silico-characterized using advanced sequencing technology. With highly and moderately repeated elements, the nuclear genome (2.46 pg/2C) was found to be relatively rich in repetitive sequences, making up 68% of the genome [35]. A centromere-specific marker, a B-specific satellite and repeat enriched in polymorphic A chromosome segments were identified. The B-specific tandem repeat PLsatB was found to originate from sequence amplification including 5S rDNA fragments. The repetitive sequences found were classified into established groups of repetitive elements. Maximus/SIRE lineage of Ty1/Copia LTR-retrotransposons were reported to be the dominant repeat type, representing more than 25% of the genome (Figure 1).

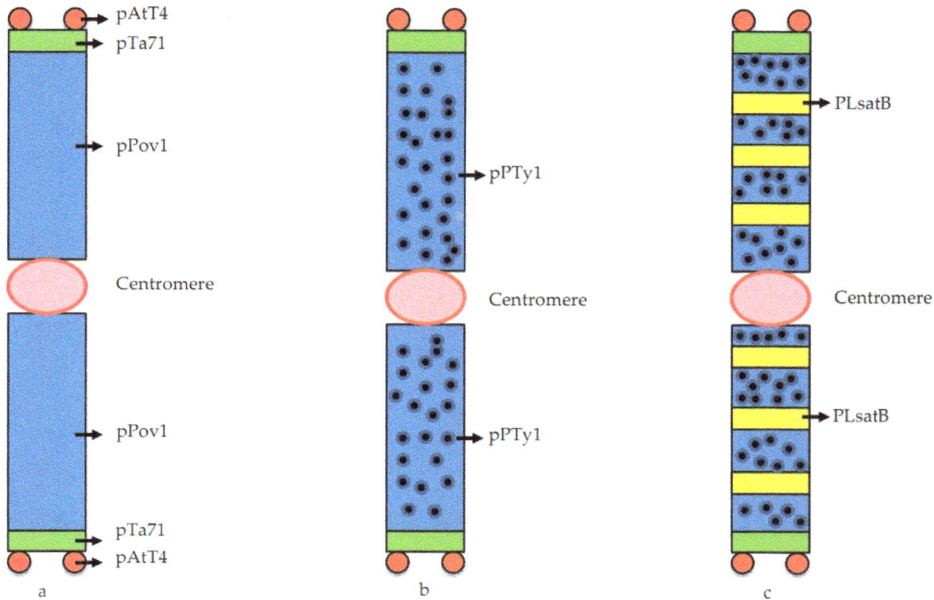

Figure 1. Deducing the composition of B chromosome in *Plantago lagopus* using FISH. (**a**) FISH signals reported on B chromosome using probes pTa71 (45S rDNA), pAtT4 (Telomere) and pPov1 (5S rDNA) [12]. Note painting of B chromosome by pPov1. (**b**) FISH signals for Ty-1 Copia elements reported over the entire length of B chromosome using probe pPTy1 [10]. (**c**) Specific localization of B specific satellite sequence, PLsatB using FISH on B chromosome [35]. FISH: Fluorescence in situ hybridization; rDNA: Ribosomal DNA.

5. Transcriptionally Active Sequences on B Chromosomes

It has been shown by various workers that B chromosomes are enriched with several classes of repetitive DNA elements [40,41]. Therefore, the identification of genes encoding proteins or the ones actively transcribing is tricky. However, the development of new techniques in recent years has considerably aided in identifying the protein encoding genes, pseudogenes, rRNA genes, and actively transcribing sequences on B chromosomes.

Based on the relationship of nucleolar phenotypes with rRNA gene expression, the first indirect evidence of active sequences on Bs was provided by Bidau [42]. However, Leach et al., provided the molecular evidence of gene activity on B chromosomes in *C. capillaris* [43]. Transcription of rRNA sequences located on B chromosomes was recently demonstrated [44,45]. Some of the important findings provided evidence that various repetitive DNAs are transcriptionally modulated by B chromosomes [37,46–50].

The proto-oncogene *c-KIT* has been mapped to B chromosomes of some animal species [51,52] and also in Siberian deer [53], where this gene is transcriptionally active. Studies on B chromosomes of Cichlid fishes have revealed many interesting facts. In *A. latifasciata*, Valente et al., have identified several genes on B chromosomes [6]. These include the genes that are involved in the chromosomal segregation and the proteins that are involved in i) microtubule organization, namely *TUBB1*, *TUBB5*; ii) kinetochore structure, such as *SKA1*, *KIF11*, and *CENP-E*, iii) recombination, including *XRCC2*, *SYCP2*, and *RTEL1*, and iv) progression through the cell cycle, such as *Separase* and *AURK* [6].

Although B chromosomes are known to affect the pairing behavior of As, recent studies have demonstrated that the presence of B chromosomes also influences the transcription of genes located on

A chromosomes [47,49,54]. Huang et al., used RNAseq technique to analyze the expression of genes in maize plants with a variable number of B chromosomes and observed that the effect of the high number of B chromosomes is more pronounced [4]. Similarly, B chromosome of *A. latifasciata* regulates the expression of noncoding RNAs, thus influencing A genome expression [50].

Interestingly, in vitro activity of a B chromosome encoded protein gene that is possibly involved in regulation has been reported in rye, thereby providing strong evidence that a B derived protein could function in silencing chromatin/DNA elements [55].

Certain phenotypic effects have been ascribed to the presence of B chromosomes. When present in low numbers, Bs have little effect on the phenotype, but at higher numbers they mostly have a negative effect on fitness and fertility of the organism [3,56]. In general, Bs negatively affect the fitness and fertility of the plants [57]. Bs directly or indirectly influence the behavior of A chromosomes. An increase in pigment production in achene's walls was observed in individuals carrying Bs in *Haplopappus gracilis* [58]. In maize, leaf stripping was correlated with the presence of Bs [59]. In *Allium schoenoprasum*, seeds carrying Bs have an advantage concerning germination over 0B seeds under drought stress conditions [60]. Bs also play a role in sex determination in cichlid fishes and in the frog *Leiopelma hochstetteri* [61]. In *P. coronopus*, the presence of Bs caused male sterility [62].

In grasshopper *Myrmeleotettix maculatus*, the presence of Bs was correlated with retardation in animal development [63] and sperm dysfunction [64]. Bs also lead to meiotic pairing in *Aegilops mutica* [65] and in hybrids between common wheat and *Aegilops variabilis* [66]. The presence of B has also been linked with crown rust resistance in *Avena sativa* [67]. In *Nasonia vitripennis*, the effect of B-like PSR chromosome constitutes changing the males into females by destroying the paternal chromosome during early embryogenesis [68]. In rice, Cheng et al., observed slight positive effect of Bs on plant height, length of panicle, length and weight of grain, and negative effect on number of tillers and width of grain [69]. In some species, individuals with Bs show better survival rate under certain stress conditions. In *A. schoenoprasum* [60], when grown under high sowing density, +B plants showed better fitness than 0B plants. A positive correlation between the mean number of B chromosomes and body mass has been observed in males of *Apodemus flavicollis* [70]. In hybrids, Bs prevent or suppress the homologous pairing of A chromosomes. This effect was observed in hybrids between *A. mutica* and *A. speltoides* with *Triticum aestivum* [65] and in a tetraploid hybrid between *Lolium temulentum* × *L. perenne* [71].

Using NGS-based approaches, it has been unfolded that Bs contain a high number of genic sequences. More than 4000 putative B-located genic sequences have been identified on rye B chromosome [34], which showed sequence polymorphism as compared to their A-located counterparts confirming pseudogenization [49]. In *D. albomicans*, one actively transcribed unit on B has been detected [38]. The evolutionary fate of these B-located genic sequences has been well addressed in a review by Houben et al. [1]. These authors have suggested that the dispensable nature of B chromosomes helps them to accumulate mutations, while they go through pseudogenization. Interestingly, all the B located genes are not inactive and also most of Bs have identical genic sequences with As, yet they still do not lead to severe phenotype. Moreover, the dosage of a chromosome is critical for normal development, but due to dosage compensation A-derived genes are likely to be downregulated [1].

6. Drive Mechanisms and Transmission of B Chromosomes

Various mechanisms by which B chromosomes are able to ensure preferential transmission to the progeny are known as accumulation or drive mechanisms. When the transmission rates of chromosomes are higher than 0.5 through male or female sex tracks, and the Mendel's law is not followed, the resulting transmission advantage is jointly referred to as drive [5]. The molecular mechanisms that are responsible for the drive are unknown [41], although these are fundamental in understanding B chromosomes. Bs successfully accumulate in populations because their transmission frequency is higher than expected, though they are mostly devoid of essential genes. There are several other genetic elements reported, which, at the expense of other components of the genome, promote their own transmission e.g., spore killer in fungi, the t haplotype in mouse and segregation distorter (SD) in *Drosophila* [72]. Drive leads to an increased number of Bs in next generation and the maximal number of Bs that can be tolerated by the host varies between species e.g., maize has 34 while rye has six B chromosomes. In this regard, *P. lagopus* is unique in having 2 B chromosomes at diploid and 5 B chromosomes at triploid levels [73]. It depends on a balance between B chromosome accumulation due to drive, and negative effects on fertility and vigor that are caused by the B chromosome [5]. However, not all B carrying species possess a drive mechanism.

The drive can occur at any stage of life cycle and has accordingly been classified as pre-meiotic, meiotic, and post-meiotic. In pre-meiotic drive, B chromosomes increase in number in the germ line cells and the mean number of B chromosomes increases, when the latter enter meiosis to form gametes. In animals, the drive occurs either before or during meiosis as the gametic nuclei are not replicating. Premeiotic drive occurs in the spermatogonial mitosis in the testes in animals [74]. It has been observed that mitotic nondisjunction occurs and the cells that possess B chromosome are preferentially included in the germ line cells of grasshopper *Locusta migratoria*. In plants, premeiotic drive has been described in *C. capillaris*, with a difference that during development the mitotic nondisjunction occurs in the meristematic cells during flower initiation with the preferential inclusion of Bs in inflorescence [75].

Meiotic drive depends on the functional symmetry of meiotic products. There are reports on the existence of meiotic drive in some grasshopper species [76]. Meiotic drive has been seen in the megaspore mother cells of *Lilium callosum* [13], a species with tetrasporic embryo sac development. Here, Bs were observed on the micropylar side of the spindle in 80% of the analyzed cells due to which they were incorporated into the resulting egg cell. Similarly, in grasshopper *M. maculatus*, Bs were located at the egg pole rather than polar bodies pole [5].

Post-meiotic drive occurs immediately after meiosis during the development of the male and female gametophyte [75]. The molecular mechanisms that are involved in drive were not known for long, until recently when Banaei-Moghaddam et al., showed meiotic drive to be due to the non-disjunction of chromatids of the B chromosome [77]. Post-meiotic drive has been frequently observed in flowering plants during the male gametophyte maturation. In rye, nondisjunction occurs during the first pollen grain mitosis, which results in the accumulation of Bs, due to which higher number of Bs are transmitted to the next generation [5]. The generative nucleus divides to produce two sperm nuclei each with unreduced number of Bs during the second pollen grain mitosis. The process of nondisjunction and frequency of transmission is controlled by B itself [5]. While analyzing the B chromosome variants, a region controlling the process of non-disjunction at the end of the long chromosome was identified, which led to the accumulation of B chromosomes. The absence of non-disjunction control region (NCR) leads to normal disjunction [78]. This factor has been shown to act in trans in rye, because when a standard B was present in the same cell that contains B lacking terminal region, the standard B mediates not only its own nondisjunction, but also of the deficient B [78]. Several B-specific satellite DNAs are present in the heterochromatic NCR. The NCR has also been labelled with H3K4me3, which is a euchromatin-specific post-translational histone mark [5].

There has been no report of a gene or a DNA sequence characterized that may be responsible for the accumulation of B chromosome. Recently, noncoding RNA has been surmised to be responsible for nondisjunction [1] because of its role in fission yeast [79] and flies [80]. It has been observed that long-non-coding RNA is produced by some NCR specific satellites predominantly in anthers [46]. There is a possibility that B chromosome encoded non-coding RNAs block necessary factors at specific genomic loci e.g., the B pericentromere or the B-derived noncoding RNAs could act as guide molecules that direct protein complexes [77]. The recent reports on identification of a notably high number of B-encoded transcripts in a number of species, furnish the basis to hypothesize the involvement of protein-coding genes, or pseudogenes for controlling non-disjunction e.g., in rye [34,54], fish [50], *Drosophila* [81], and cervids [82].

The mechanism of accumulation of the B chromosome in *P. lagopus* was recently reported in depth by Dhar et al., (2017) by performing extensive crossing experiments and calculating the transmission of B chromosome through female and male sex tracks [73]. This study not only explored the existence of drive in B chromosome of *P. lagopus*, but also helped in understanding the mechanism of perpetuation in the populations. After selfing and crossing, the progenies were raised and analyzed to determine the mode of transmission of B chromosomes. Crosses were attempted between the 0B, 1B, and 2B plants. About 531 plants were screened cytologically for the presence of the B chromosome; FISH with 5S rDNA probe was used to identify the B chromosome(s) in 1B and 2B plants.

It was reported that in *P. lagopus* when 1B plant was used as a male, transmission rate was in accordance with the expected Mendelian ratio. The differences in segregation ratio observed among various cross combinations were attributed to the heterozygous nature of *P. lagopus*—being a cross-pollinated species [73]. Alternatively, when 1B plants were used as female, there were significant deviations from the 1:1 ratio; frequency of B chromosome bearing plants was higher than the Mendelian expected value. These results indicated the preferential transmission of B chromosome through the female sex track in *P. lagopus*. Such variation in the transmission rate is a common feature of B inheritance, such that the Bs tend to distort Mendelian expectation in their favor [83]. The fact that during female meiosis in plants, only one out of the four meiotic products survives can explain this drive through the female sex tracks. The B chromosome of *P. lagopus* is perpetuated preferentially due to drive, expressed during both male and female meiosis [73]. The same phenomenon has been observed in rye where the drive happens in both male and female sex tracks [84].

In *P. lagopus*, the entire complement of the species gets duplicated in the presence of a B chromosome [73]. It was for the first time that such a phenomenon has been reported in plants, earlier, the presence of macrospermatids (> diploid chromosome number) in B containing individuals of a grasshopper *Dichroplus pratensis* were observed [85]. The formation of macrospermatids was attributed to nuclear fusion. In *P. lagopus* (2n = 12), plants with 23, 26, and 28 chromosomes were obtained from the selfed progenies of B chromosome bearing plants. The origin of these was attributed to the formation of unreduced gametes, followed by their fusion with other gametes or their endoreduplication and parthenogenetic development of the plant.

7. B Chromosome in Relation to Breeding System

The breeding system plays a fundamental role in the ecology and evolution of B chromosomes. In a comparative study performed by Burt and Trivers among different species of British flowering plants, it was demonstrated that Bs are more likely to be present in outbreeding species than in inbreeding species [86]. Besides, there are a number of other factors that play an important role in the prevalence of B chromosomes, like genome size, chromosome number, and ploidy level [22]. In several cases, experiments have verified the relationship between Bs and outbreeding. For example, when outbreeding rye (*Secale cereale*) was inbred, the B frequency was observed to decline [87]. Similarly, in inbreeding *Secale vavilovii*, when B chromosomes were experimentally introduced, their number declined rapidly [88]. Since the variation in the number of Bs among individuals gets increased in inbreeding, the latter in turn decreases B frequency by increasing the power of natural selection. Genus *Plantago* has a wide range of mating systems, from inbreeders to obligate outcrossers. Outbreeding species, such as, *P. lanceolata*, *P. lagopus*, *P. coronopus*, and *P. depressa* possess B chromosome. On the contrary, *P. ovata* and *Plantago patagonica*, representing inbreeding types are devoid of B chromosome—a fact that further substantiates the relation between the breeding system and B chromosome existence.

In order to understand the relationship between the breeding system and B chromosome, reference can be made to the data given by Levin et al., wherein phyletic hotspots for B chromosomes in Angiosperms has been discussed in detail [17]. Further, the prevalence of B chromosomes across different orders in Angiosperms is noteworthy. From there, it can be inferred that Asparagales, Liliales and Poales (Monocots), Ranunculales (Eudicot), Fabales (Rosids), and Asterales (Euasterids) had maximum number of B chromosomes. All of the species belonging to these orders having maximum Bs are outbreeders [86].

Several species of *Plantago*, including *P. lanceolata* and *P. lagopus*, which possess the B chromosome, follow outcrossing and are also gynodioecious [89]. The role of gynodioecy as an outcrossing mechanism was well demonstrated while understanding the gynodioecious breeding system in *P. lanceolata* [90], where it was observed that females produce 1.77 times the number of seeds produced by hermaphrodites over the flowering season. The superior reproductive output of females in gynodioecious population is required for the maintenance of gynodioecy [91]. However, there must be some selective advantage of the B possessing plants, which helps them in attaining and maintaining gynodioecy. It can be conjectured that organellar genes that have been reported to be present on B chromosome confer this advantage. Nuclear transfer of organelle DNA is a known process. There is a frequent transfer of mitochondrial and chloroplast DNA sequences into the nuclear genome [92]. Nuclear insertions of plastid DNA (NUPTs) and insertions of mitochondrial DNA (NUMTs) have been shown to be involved in the formation of new nuclear genes. There are a number of examples where organellar sequences have been reported on B chromosome. Large insertions of organelle DNA were found on B chromosomes of rye, *A. speltoides* [93] and wheat [34]. Recently, using techniques like next-generation sequencing and sophisticated bioinformatics tools, it was observed that in rye [34] and in the fish species *A. latifasciata* [6], B chromosomes have accumulated organellar-derived DNA [35]. There is a possibility that these captured organellar sequences have a role in affecting the sex phenotypes of gynodioecious systems, and manipulate their transmission advantage.

8. Evolution of B Chromosomes

Despite widespread occurrence of B chromosomes in all eukaryotic groups, including plants, mammals, insects, etc., the origin of B chromosome has remained an enigma. Several hypotheses have been proposed for the origin of B chromosomes [12,74,84,94]. Bs seem to have arisen in different ways in different organisms [1,74,84]. The view that is most widely accepted is that they are derived from A chromosome complement. There are some evidences that suggest that B chromosomes were generated spontaneously in response to the new genomic conditions after interspecific hybridization [74]. In some animals, the involvement of sex chromosomes has also been argued for their origin [41,74,84]. Moreover, the presence of similar B chromosome variants within related species suggests that the B chromosome arose from a single origin [34]. It has been argued that B chromosomes are selfish entities that take a distinct path of evolution and differ in sequence composition from that of the As [1]. The mobile elements and other repeats on B chromosome can easily spread and amplify, because B chromosomes are not under selective pressure. This has been reported in Bs of *B. dichromosomatica* [32], *P. lagopus* [10,12], and *Zea mays* [37].

In a historical paper from our laboratory [12], the first experimental proof of the origin of B chromosome from A chromosome in *P. lagopus* was provided. The model proposed was widely accepted and adapted [14,83]. Dhar et al., proposed that the B chromosome arose as a result of the massive amplification of sequences with similarities to 5S and 45S rDNA, as well as to other sequences [12]. The C–banding analysis clearly distinguished the B chromosome from other chromosomes in being entirely heterochromatic. Recently, NGS and Graph based clustering analysis for characterization of DNA composition of *P. lagopus* with and without B chromosomes was performed [35], that helped a great deal in ascertaining the validity of the model proposed by Dhar et al., [12]. Centromere-specific marker, a B-specific satellite, and a repeat enriched in polymorphic A chromosome segments have been identified.

In *P. lagopus*, A and B chromosomes differ significantly in composition, which is mainly due to an additional massive amplification of B-specific satellite repeats. The proportion of satellite repeats in 0B plants is 3.5% and in +B plants 6.7% [35]. The same sequences are also abundant in the A chromosomes, particularly in heterochromatic regions and it is these repetitive elements that contribute to the heterochromatic nature of the B chromosome. Cluster CL4, a satellite, with a monomer length of

325 bp, shared similarity with 5S rDNA. It had more proportion of reads from +B DNA i.e., 2.95% and very less proportion of 0B reads i.e., 0.08%. This indicates that the sequences present in this cluster were specific to B chromosome. When closely inspected, these sequences showed some similarity to 5S rDNA, but were not 5S sequences. This cluster can be fairly linked with the experimental evidence provided earlier by Dhar et al., where Southern blot analysis revealed a distinct band of about 23 kb, only in +B plants [12]. Since this cluster is specific to reads that are present on B chromosome as was the 23 kb fragment specific to +B DNA, a very important and relevant correlation can be drawn between the two.

Very early attempts to elucidate the DNA composition of Bs resulted in comparative studies of 0B versus +B genomic DNA [95–98]. However, during the last few years, microdissection [32] and flow sorting [34] have helped in the reliable isolation of B-derived DNA. B chromosomes of many species contain sequences that originated from one or more A chromosomes [32,35]. In the harvest mouse, *Reithrodontomys megalotis*, different origins for the two types of Bs found were suggested. The large B chromosome arose from centric fusion as a leftover of the centromere, whereas the small B had originated from an amplified region of an A chromosome or as an intact fragment. Karamysheva et al., proposed a two-step appearance of Bs in *Apodemus peninsulae* [99]. Destabilization of pericentromeric regions that are produced by the invasion of DNA sequences from euchromatic parts of A chromosomes leads to the formation of micro-chromosomes in high frequency, which could be considered as proto Bs. The next step is the insertion and amplification of new DNA sequences.

It is not clear how the actual process of sequence transfer from As to Bs takes place, but recent reports indicate that the transposition of mobile elements may be playing an important role [100]. B chromosomes are an ideal target for the transposition of mobile elements [48], and the structural variability that is observed in Bs can be due to the insertion of these elements [74]. In maize, when a large DNA insert clone was analyzed, Bs were observed to be composed of B-specific sequences that were interlaced with those in common with the As [37]. StarkB elements (22 kb-long B-specific repeat) had frequent insertions by LTR-type retroelement [37]. Within evolutionarily diverged species, the presence of B-enriched sequences indicated that B-specific amplification occurred after separation from the standard chromosome complement [92].

In rye that was collected from several countries, Bs have been reported in both cultivated and weedy forms [101]. In F1 hybrids of these forms, Bs had similar morphology and meiotic pairing which suggested that the rye B has a monophyletic origin [101]. The comparative sequence analysis of the A and B chromosomes in rye revealed that Bs are rich in gene fragments that are derived from multiple A chromosome fragments [102]. Many short sequences that are similar to other regions of the rye A supported the multi-chromosomal origin of B-chromosome sequences. When the composition and distribution of high-copy sequences were explored, it got affirmed that Bs contain a similar proportion of repeats as in the A chromosomes, but differ substantially in repeat composition [48]. The distribution of mobile elements, like Gypsy and Copia, in the genome of rye is similar along As and Bs [103]. Accumulation of Gypsy retrotransposons or other repeated sequences on B chromosome has also been reported in the fish *A. alburnus* [29] and the fungus *N. haematococca* [104].

In various organisms, it has been observed that Bs also contain various kinds of coding and noncoding repeats, which are very similar to those that are found in circular extrachromsomal DNA [105]. In *B. dichromosomatica*, extrachromosomal DNA with similarity to tandem repeat sequences shared by both A and B chromosomes have been identified [106]. However, whether an evolutionary link between extrachromosomal DNA and the evolution of Bs exists is still to be determined. In wild canid species, several regions of domestic dog sequences that share sequence similarity with canid B chromosomes were identified by analyzing evolutionarily conserved chromosome segments [102].

The involvement of rDNA in the evolution of B chromosomes does not appear to be accidental, as they have been detected on Bs of many plants (e.g., *C. capillaris* [107] as well as animals (e.g., *Rattus rattus* [108]). The most convincing example is of *P. lagopus* where the analysis of many generations and in situ localization with rDNA probes revealed that B originated as result of massive amplification of 5S rDNA [12,35]. It has been argued that the mobile nature of rDNA could be the reason for their presence as landing sites on the B chromosomes [1]. The reason for no or weak transcription of rDNA on B chromosomes is hazy, although differences in the histone H3 methylation between A and B chromosome may be responsible [109]. By determining the relatedness of internal transcribed spacers (ITS) between the different chromosome types, the available sequence information for B-located rRNA genes has been used to study the origin of Bs e.g., *B. dichromosomatica* [85] and *C. capillaris* [43]. According to Martis et al., Bs have been found to have accumulated large amounts of B-specific repeats and insertions of cytoplasmic organellar DNA in A-derived landscape [34]. They have termed Bs as genomic sponge, which collects and maintains sequences of diverse origin.

Some findings suggest that B chromosomes arise spontaneously in response to genomic stress following interspecific hybridization [110]. First proposed by Battaglia (1964), this mode of origin was later confirmed by other scientists [74,111]. In hybrids generated after interspecific crosses between *Coix aquaticus* (2n = 10) and *Coix gigantea* (2n = 20), a spectrum of individuals with variable number of *gigantea* and *aquaticus* chromosomes were recovered. *C. gigantea* chromosomes in *C. aquaticus* genome behaved like B chromosomes during meiosis [14]. It is presumed that an incomplete loss of one parental genome, during hybrid embryogenesis, might have played a role in the origin of Bs. There is evidence that the centromeres of the parental chromosomes undergoing elimination are the last to be lost during the uniparental chromosome elimination process [112]. A novel mechanism for the evolution of B chromosome based on the recombination of nonhomologous chromosomes, which takes place during the DNA double-strand repair process at S-phase, has been postulated for the formation of zebra chromosome, which is composed of *T. aestivum* / *Elymus trachycaulus*, structurally rearranged chromosome fragments [113]. In the gynogenetic fish *Poecilia formosa*, the sperm of male from the related bisexual species *Poecilia mexicana* or *Poecilia latipinna* are required for the initiation of embryogenesis. Eventually, paternal genome is eliminated early in the development. The progeny results in a spotted pigmentation phenotype. Supernumerary microchromosomes were observed after cytological analysis in *P. formosa* offsprings as a result of incomplete elimination of paternal genome. They are inherited and variable pigmentation results due to the different number of Bs [16].

B chromosomes may also be derived from sex chromosomes [38] during intraspecific origin. There are many features, like distribution of chromatin, the accumulation of repetitive DNA, and the loss of gene activity, which are shared by both sex and B chromosomes. The origin of Bs from sex chromosomes has also been suggested because of the similarity of morphology, heteropycnocity, and meiotic behavior between Bs and sex chromosomes in some species [74]. The similarity in the composition and localization of two DNA probes (180 bp tandem repeat and ribosomal DNA) of B chromosomes, which are mostly composed of these two sequences, with those on X chromosome suggest that the B chromosome in grasshopper *E. plorans* probably originated from X chromosomes [114]. The B chromosomes of *L. hochstetteri* showed similarity to the W sex chromosome at the DNA level [115]. Morphological similarities between Bs and the univalent W chromosome were also observed [116]. Genus *Characidium* among Neotropical fishes provides an intriguing model for cytogenetic and evolutionary studies, principally because of the presence of differentiated sex chromosome systems and B chromosomes [117]. DNA probes that were obtained from B chromosomes and sex chromosomes in three species of fish genus *Characidium* were used for chromosome painting, which showed a close resemblance in repetitive DNA content between B and sex chromosomes. Bs in Bandicoot (*Echymipera kalubu*) have been reported to be derived from sex chromosomes, based on the finding that Bs follow the same fate as sex chromosomes, which are eliminated from certain somatic tissues [15]. Frequent synapsis and recombination between Bs and the Y chromosome in *Dicrostonyx groenlandicus* advocate Y chromosome as a possible source of Bs [118].

9. Outlook

A number of efficient tools for the analysis of B chromosomes have been developed due to current breakthrough in gene sequencing and bioinformatics. Further analysis of composition and behavior of B chromosomes will help in tracing the origin of these enigmatic chromosomes. We would have a better discernment as to how selection pressure leads to genome evolution. In different organisms where the molecular studies on B chromosomes has been conducted, various theories have been propounded regarding the evolution of Bs. Having a thorough knowledge of Bs will give us exhilarating results about frequent changes in higher eukaryotes, with special reference to these supernumerary chromosomes.

Acknowledgments: The authors are grateful to Coordinator, Bioinformatics Centre (DBT Govt. of India funded), University of Jammu, Jammu for facilities.

Conflicts of Interest: The authors declare no conflicts of interest.

References

1. Houben, A.; Banaei-Moghaddam, A.M.; Klemme, S.; Timmis, J.N. Evolution and biology of supernumerary B chromosomes. *Cell. Mol. Life Sci.* **2014**, *71*, 467–478. [CrossRef]
2. Douglas, R.N.; Birchler, J.A. B Chromosomes. In *Chromosome Structure and Aberrations*; Springer: New Delhi, India, 2017; pp. 13–39.
3. Carlson, W. The B chromosome of maize. In *Handbook of Maize*; Springer: New York, NY, USA, 2009; pp. 459–480.
4. Huang, W.; Du, Y.; Zhao, X.; Jin, W. B chromosome contains active genes and impacts the transcription of A chromosomes in maize (*Zea mays* L.). *BMC Plant Biol.* **2016**, *16*, 88. [CrossRef] [PubMed]
5. Houben, A. B chromosomes—A matter of chromosome drive. *Front. Plant Sci.* **2017**, *8*, 210. [CrossRef] [PubMed]
6. Valente, G.; Conte, M.A.; Fantinatti, B.E.A.; Cabral-De-Mello, D.C.; Carvalho, R.F.; Vicari, M.; Kocher, T.; Martins, C. Origin and evolution of B chromosomes in the cichlid fish *Astatotilapia latifasciata* based on integrated genomic analyses. *Mol. Biol. Evol.* **2014**, *31*, 2061–2072. [CrossRef]
7. Palestis, B.G.; Trivers, R.; Burt, A.; Jones, R.N. The distribution of B chromosomes across species. *Cytogenet. Genome Res.* **2004**, *106*, 151–158. [CrossRef] [PubMed]
8. Novák, P.; Neumann, P.; Macas, J. Graph-based clustering and characterization of repetitive sequences in next-generation sequencing data. *BMC Bioinform.* **2010**, *11*, 378. [CrossRef] [PubMed]
9. Mayer, K.F.; Martis, M.; Hedley, P.E.; Šimková, H.; Liu, H.; Morris, J.A.; Steuernagel, B.; Taudien, S.; Roessner, S.; Gundlach, H.; et al. Unlocking the barley genome by chromosomal and comparative genomics. *Plant Cell* **2011**, *23*, 1249–1263. [CrossRef] [PubMed]
10. Kour, G.; Kaul, S.; Dhar, M.K. Molecular characterization of repetitive DNA sequences from B chromosome in *Plantago lagopus* L. *Cytogenet. Genome Res.* **2013**, *142*, 121–128. [CrossRef] [PubMed]
11. Sharma, P.K.; Koul, A.K. Genetic diversity among *Plantagos* III. Primary trisomy in *Plantago lagopus* L. *Genetica* **1984**, *64*, 135–138. [CrossRef]
12. Dhar, M.K.; Friebe, B.; Koul, A.K.; Gill, B.S. Origin of an apparent B chromosome by mutation, chromosome fragmentation and specific DNA sequence amplification. *Chromosoma* **2002**, *111*, 332–340. [CrossRef]
13. Datta, A.K.; Mandal, A.; Das, D.; Gupta, S.; Saha, A.; Paul, R.; Sengupta, S. B chromosomes in angiosperm—A review. *Cytol. Genet.* **2016**, *50*, 60–71. [CrossRef]
14. Jones, R.N.; Viegas, W.; Houben, A. A century of B chromosomes in plants: So what? *Ann. Bot.* **2007**, *101*, 767–775. [CrossRef] [PubMed]
15. Vujosevic, M.; Blagojević, J. B chromosomes in populations of mammals. *Cytogenet. Genome Res.* **2004**, *106*, 247–256. [CrossRef] [PubMed]
16. Makunin, A.I.; Dementyeva, P.V.; Graphodatsky, A.S.; Volobouev, V.T.; Kukekova, A.V.; Trifonov, V.A. Genes on B chromosomes of vertebrates. *Mol. Cytogenet.* **2014**, *7*, 99. [CrossRef] [PubMed]
17. Levin, D.A.; Palestis, B.G.; Jones, R.N.; Trivers, R. Phyletic hot spots for B chromosomes in angiosperms. *Evolution* **2005**, *59*, 962–969. [CrossRef] [PubMed]
18. Gupta, R.C.; Singh, V.; Bala, S.; Malik, R.A.; Sharma, V.; Kaur, K. Cytomorphological variations and new reports of B-chromosomes in the genus *Plantago* (Plantaginaceae) from the Northwest Himalaya. *Flora* **2017**, *234*, 69–76. [CrossRef]

19. Frost, S. The cytological behaviour and mode of transmission of accessory chromosomes in *Plantago serraria*. *Hereditas* **1959**, *45*, 191–210. [CrossRef]

20. GA, M.; Brown, G. Determination of chromosome number of Kuwaiti flora I. *Cytologia* **1999**, *64*, 181–196.

21. Brullo, S.; Pavone, P.; Terassi, M.C. Karyological considerations on the genus *Plantago* in Sicily Italy. *Candollea* **1985**, *40*, 217–230.

22. Trivers, R.; Burt, A.; Palestis, B.G. B chromosomes and genome size in flowering plants. *Genome* **2004**, *47*, 1–8. [CrossRef]

23. Jones, R.N.; Rees, H. *B Chromosomes*, 1st ed.; Academic Press: London, UK; New York, NY, USA, 1982.

24. Bernardino, A.C.; Cabral-de-Mello, D.C.; Machado, C.B.; Palacios-Gimenez, O.M.; Santos, N.; Loreto, V.B. B chromosome variants of the grasshopper *Xyleus discoideus angulatus* are potentially derived from pericentromeric DNA. *Cytogenet. Genome Res.* **2017**, *152*, 213–221. [CrossRef] [PubMed]

25. Schmid, M.; Ziegler, C.G.; Steinlein, C.; Nanda, I.; Haaf, T. Chromosome banding in Amphibia: XXIV. The B chromosomes of *Gastrotheca espeletia* (Anura, Hylidae). *Cytogenet. Genome Res.* **2002**, *97*, 205. [CrossRef] [PubMed]

26. Stevens, J.P.; Bougourd, S.M. Unstable B-chromosomes in a European population of *Allium schoenoprasum* L. (Liliaceae). *Biol. J. Linn. Soc.* **1994**, *52*, 357–363. [CrossRef]

27. Houben, A.; Thompson, N.; Ahne, R.; Leach, C.R.; Verlin, D.; Timmis, J.N. A monophyletic origin of the B chromosomes of *Brachycome dichromosomatica* (Asteraceae). *Plant Syst. Evol.* **1999**, *219*, 127–135. [CrossRef]

28. Belyayev, A.; Raskina, O. Chromosome evolution in marginal populations of *Aegilops speltoides*: Causes and consequences. *Ann. Bot.* **2013**, *111*, 531–538. [CrossRef] [PubMed]

29. Ziegler, C.G.; Lamatsch, D.K.; Steinlein, C.; Engel, W.; Schartl, M.; Schmid, M. The giant B chromosome of the cyprinid fish *Alburnus alburnus* harbours a retrotransposon-derived repetitive DNA sequence. *Chromosome Res.* **2003**, *11*, 23–35. [CrossRef] [PubMed]

30. Duílio, M.D.A.; Pansonato-Alves, J.C.; Utsunomia, R.; Araya-Jaime, C.; Ruiz-Ruano, F.J.; Daniel, S.N.; Foresti, F. Delimiting the origin of a B chromosome by FISH mapping, chromosome painting and DNA sequence analysis in *Astyanax paranae* (Teleostei, Characiformes). *PLoS ONE* **2014**, *9*, e94896.

31. Carter, C.R.; Smith-White, S. The cytology of *Brachycome lineariloba*. *Chromosoma* **1972**, *39*, 361–379. [CrossRef]

32. Houben, A.; Verlin, D.; Leach, C.R.; Timmis, J.N. The genomic complexity of micro B chromosomes of *Brachycome dichromosomatica*. *Chromosoma* **2001**, *110*, 451–459. [CrossRef] [PubMed]

33. Kubaláková, M.; Valárik, M.; Barto, J.; Vrána, J.; Cihalikova, J.; Molnár-Láng, M.; Dolezel, J. Analysis and sorting of rye (*Secale cereale* L.) chromosomes using flow cytometry. *Genome* **2003**, *46*, 893–905. [CrossRef] [PubMed]

34. Martis, M.M.; Klemme, S.; Banaei-Moghaddam, A.M.; Blattner, F.R.; Macas, J.; Schmutzer, T.; Novák, P. Selfish supernumerary chromosome reveals its origin as a mosaic of host genome and organellar sequences. *Proc. Natl. Acad. Sci. USA* **2012**, *109*, 13343–13346. [CrossRef]

35. Kumke, K.; Macas, J.; Fuchs, J.; Altschmied, L.; Kour, J.; Dhar, M.K.; Houben, A. *Plantago lagopus* B chromosome is enriched in 5S rDNA-derived satellite DNA. *Cytogenet. Genome Res.* **2016**, *148*, 68–73. [CrossRef]

36. Theuri, J.; Phelps-Durr, T.; Mathews, S.; Birchler, J. A comparative study of retrotransposons in the centromeric regions of A and B chromosomes of maize. *Cytogenet. Genome Res.* **2005**, *110*, 203–208. [CrossRef]

37. Lamb, J.C.; Riddle, N.C.; Cheng, Y.M.; Theuri, J.; Birchler, J.A. Localization and transcription of a retrotransposon-derived element on the maize B chromosome. *Chromosome Res.* **2007**, *15*, 383–398. [CrossRef]

38. Zhou, Q.; Zhu, H.M.; Huang, Q.F.; Zhao, L.; Zhang, G.J.; Roy, S.W.; Zhao, R.P. Deciphering neo-sex and B chromosome evolution by the draft genome of *Drosophila albomicans*. *BMC Genom.* **2012**, *13*, 109. [CrossRef]

39. Houben, A.; Demidov, D.; Gernand, D.; Meister, A.; Leach, C.R.; Schubert, I. Methylation of histone H3 in euchromatin of plant chromosomes depends on basic nuclear DNA content. *Plant J.* **2003**, *33*, 967–973. [CrossRef]

40. Camacho, J.P.M. B chromosomes. In *The Evolution of the Genome*; Gregory, T.R., Ed.; Elsevier: San Diego, CA, USA, 2005; pp. 223–286.

41. Burt, A.; Trivers, R. B chromosomes. In *Genes in Conflict: The Biology of Selfish Genetic Elements*; Harvard University Press: Cambridge, MA, USA, 2006; pp. 325–380.

42. Bidau, C.J. A nucleolar-organizing B chromosome showing segregation-distortion in the grasshopper *Dichroplus pratensis* (Melanoplinae, Acrididae). *Can. J. Genet. Cytol.* **1986**, *28*, 138–148. [CrossRef]

43. Leach, C.R.; Houben, A.; Field, B.; Pistrick, K.; Demidov, D.; Timmis, J.N. Molecular evidence for transcription of genes on a B chromosome in *Crepis capillaris*. *Genetics* **2005**, *171*, 269–278. [CrossRef]

44. Ruiz-Estévez, M.; López-León, M.D.; Cabrero, J.; Camacho, J.P. Ribosomal DNA is active in different B chromosome variants of the grasshopper *Eyprepocnemis plorans*. *Genetica* **2013**, *141*, 337–345. [CrossRef]

45. Ruiz-Estévez, M.; Badisco, L.; Broeck, J.V.; Perfectti, F.; Lopez-Leon, M.D.; Cabrero, J.; Camacho, J.P. B chromosomes showing active ribosomal RNA genes contribute insignificant amounts of rRNA in the grasshopper *Eyprepocnemis plorans*. *Mol. Gen. Genom.* **2014**, *289*, 1209–1216. [CrossRef]

46. Carchilan, M.; Delgado, M.; Ribeiro, T.; Costa-Nunes, P.; Caperta, A.; Morais-Cecílio, L.; Jones, R.N.; Viegas, W.; Houben, A. Transcriptionally active heterochromatin in rye B chromosomes. *Plant Cell* **2007**, *19*, 1738–1749. [CrossRef] [PubMed]

47. Carchilan, M.; Kumke, K.; Mikolajewski, S.; Houben, A. Rye B chromosomes are weakly transcribed and might alter the transcriptional activity of a chromosome sequences. *Chromosoma* **2009**, *118*, 607–616. [CrossRef] [PubMed]

48. Klemme, S.; Banaei Moghaddam, A.M.; Macas, J.; Wicker, T.; Novák, P.; Houben, A. High copy sequences reveal distinct evolution of the rye B chromosome. *New Phytol.* **2013**, *199*, 550–558. [CrossRef] [PubMed]

49. Banaei-Moghaddam, A.M.; Martis, M.M.; Macas, J.; Gundlach, H.; Himmelbach, A.; Altschmied, L.; Mayer, K.F.X.; Houben, A. Genes on B chromosomes: Old questions revisited with new tools. *Biochim. Biophys. Acta* **2015**, *1849*, 64–70. [CrossRef] [PubMed]

50. Ramos, É.; Cardoso, A.L.; Brown, J.; Marques, D.F.; Fantinatti, B.E.; Cabral-de-Mello, D.C.; Martins, C. The repetitive DNA element BncDNA, enriched in the B chromosome of the cichlid fish *Astatotilapia latifasciata*, transcribes a potentially noncoding RNA. *Chromosoma* **2016**, *126*, 313–323. [CrossRef]

51. Graphodatsky, A.S.; Kukekova, A.V.; Yudkin, D.V.; Trifonov, V.A.; Vorobieva, N.V.; Beklemisheva, V.R.; Perelman, P.L.; Graphodatskaya, D.A.; Trut, L.N.; Ferguson-Smith, M.A.; et al. The proto-oncogene c-KIT maps to canid B-chromosomes. *Chromosome Res.* **2005**, *13*, 113–122. [CrossRef] [PubMed]

52. Yudkin, D.V.; Trifonov, V.A.; Kukekova, A.V.; Vorobieva, N.V.; Rubtsova, N.V.; Yang, F.; Acland, G.M.; Ferguson-Smith, M.A.; Graphodatsky, A.S. Mapping of KIT adjacent sequences on canid autosomes and B chromosomes. *Cytogenet. Genome Res.* **2007**, *116*, 100–103. [CrossRef] [PubMed]

53. Trifonov, V.A.; Dementyeva, P.V.; Larkin, D.M.; O'Brien, P.C.; Perelman, P.L.; Yang, F.; Graphodatsky, A.S. Transcription of a protein-coding gene on B chromosomes of the Siberian roe deer (*Capreolus pygargus*). *BMC Biol.* **2013**, *11*, 90. [CrossRef] [PubMed]

54. Banaei-Moghaddam, A.M.; Meier, K.; Karimi-Ashtiyani, R.; Houben, A. Formation and expression of pseudogenes on the B chromosome of rye. *Plant Cell* **2013**, *25*, 2536–2544. [CrossRef] [PubMed]

55. Ma, W.; Gabriel, T.S.; Martis, M.M.; Gursinsky, T.; Schubert, V.; Vrána, J.; Doležel, J.; Grundlach, H.; Altschmied, L.; Scholz, U.; et al. Rye B chromosomes encode a functional Argonaute-like protein with in vitro slicer activities similar to it's a chromosome paralog. *New Phytol.* **2016**, *213*, 916–928. [CrossRef] [PubMed]

56. Bougourd, S.M.; Jones, R.N. B chromosomes: A physiological enigma. *New Phytol.* **1997**, *137*, 43–54. [CrossRef]

57. González-Sánchez, M.; González-González, E.; Molina, F.; Chiavarino, A.M.; Rosato, M.; Puertas, M.J. One gene determines maize B chromosome accumulation by preferential fertilisation; another gene(s) determines their meiotic loss. *Heredity* **2003**, *90*, 122. [CrossRef]

58. Jackson, R.C.; Newmark, P. Effects of supernumerary chromosomes on production of pigment in *Haplopappus gracilis*. *Science* **1960**, *132*, 1316–1317. [CrossRef]

59. Staub, R.W. Leaf striping correlated with the presence of B chromosomes in maize. *J. Hered.* **1987**, *78*, 71–74. [CrossRef]

60. Holmes, D.S.; Bougourd, S.M. B-chromosome selection in *Allium schoenoprasum* II. Experimental populations. *Heredity* **1991**, *67*, 117. [CrossRef]

61. Yoshida, K.; Terai, Y.; Mizoiri, S.; Aibara, M.; Nishihara, H.; Watanabe, M.; Okada, N. B chromosomes have a functional effect on female sex determination in Lake Victoria cichlid fishes. *PLoS Genet.* **2011**, *7*, e1002203. [CrossRef]

62. Paliwal, R.L.; Hyde, B.B. The association of a single B-chromosome with male sterility in *Plantago coronopus*. *Am. J. Bot.* **1959**, *46*, 460–466. [CrossRef]

63. Harvey, A.W.; Hewitt, G.M. B chromosomes slow development in a grasshopper. *Heredity* **1979**, *42*, 397. [CrossRef]

64. Hewitt, G.M.; Butlin, R.K.; East, T.M. Testicular dysfunction in hybrids between parapatric subspecies of the grasshopper *Chorthippus parallelus*. *Biol. J. Linn. Soc.* **1987**, *31*, 25–34. [CrossRef]

65. Dover, G.A.; Riley, R. Variation at two loci affecting homoeologous meiotic chromosome pairing in *Triticum aestivum* × *Aegilops mutica* hybrids. *Nat. New Biol.* **1972**, *235*, 61. [CrossRef]

66. Kousaka, R.; Endo, T.R. Effect of a rye B chromosome and its segments on homoeologous pairing in hybrids between common wheat and *Aegilops variabilis*. *Genes Genet. Syst.* **2012**, *87*, 1–7. [CrossRef] [PubMed]

67. Dherawattana, A.; Sadanaga, K. Cytogenetics of a crown rust-resistant hexaploid oat with 42 + 2 fragment chromosomes 1. *Crop Sci.* **1973**, *13*, 591–594. [CrossRef]

68. Werren, J.H. The paternal-sex-ratio chromosome of *Nasonia*. *Am. Nat.* **1991**, *137*, 392–402. [CrossRef]

69. Cheng, Z.K.; Yu, H.X.; Yan, H.H.; Gu, M.H.; Zhu, L.H. B chromosome in a rice aneuploid variation. *Theor. Appl. Genet.* **2000**, *101*, 564–568. [CrossRef]

70. Zima, J.; Piálek, J.; Macholán, M. Possible heterotic effects of B chromosomes on body mass in a population of *Apodemus flavicollis*. *Can. J. Zool.* **2003**, *81*, 1312–1317. [CrossRef]

71. Jenkins, G.; Jones, R.N. B chromosomes in hybrids of temperate cereals and grasses. *Cytogenet. Genome Res.* **2004**, *106*, 314–319. [CrossRef] [PubMed]

72. Larracuente, A.M.; Presgraves, D.C. The selfish Segregation Distorter gene complex of *Drosophila melanogaster*. *Genetics* **2012**, *192*, 33–53. [CrossRef] [PubMed]

73. Dhar, M.K.; Kour, G.; Kaul, S. B chromosome in *Plantago lagopus* Linnaeus, 1753 shows preferential transmission and accumulation through unusual processes. *Comp. Cytogenet.* **2017**, *11*, 375. [CrossRef]

74. Camacho, J.P.M.; Sharbel, T.F.; Beukeboom, L.W. B-chromosome evolution. *Philos. Trans. R. Soc. B* **2000**, *355*, 163–178. [CrossRef]

75. Jones, G.H.; Albini, S.M.; Whitehorn, J.A.F. Ultrastructure of meiotic pairing in B chromosomes of *Crepis capillaris*. *Chromosoma* **1991**, *100*, 193–202. [CrossRef]

76. Manrique-Poyato, M.I.; López-León, M.D.; Gómez, R.; Perfectti, F.; Camacho, J.P.M. Population genetic structure of the grasshopper *Eyprepocnemis plorans* in the south and east of the Iberian Peninsula. *PLoS ONE* **2013**, *8*, e59041. [CrossRef] [PubMed]

77. Banaei-Moghaddam, A.M.; Schubert, V.; Kumke, K.; Weiß, O.; Klemme, S.; Nagaki, K.; Macas, J.; Gonzalez-Sanchez, M.; Heredia, V.; Gomez-Revilla, D.; et al. Nondisjunction in favor of a chromosome: The mechanism of rye B chromosome drive during pollen mitosis. *Plant Cell* **2012**, *24*, 4124–4134. [CrossRef] [PubMed]

78. Endo, T.R.; Nasuda, S.; Jones, N.; Dou, Q.; Akahori, A.; Wakimoto, M.; Tsujimoto, H. Dissection of rye B chromosomes, and nondisjunction properties of the dissected segments in a common wheat background. *Genes Genet. Syst.* **2008**, *83*, 23–30. [CrossRef] [PubMed]

79. Volpe, T.; Schramke, V.; Hamilton, G.L.; White, S.A.; Teng, G.; Martienssen, R.A.; Allshire, R.C. RNA interference is required for normal centromere function in fission yeast. *Chromosome Res.* **2003**, *11*, 137–146. [CrossRef] [PubMed]

80. Vos, L.J.; Famulski, J.K.; Chan, G.K. How to build a centromere: From centromeric and pericentromeric chromatin to kinetochore assembly. *Biochem. Cell Biol.* **2006**, *84*, 619–639. [CrossRef] [PubMed]

81. Bauerly, E.; Hughes, S.E.; Vietti, D.R.; Miller, D.E.; McDowell, W.; Hawley, R.S. Discovery of supernumerary B chromosomes in *Drosophila melanogaster*. *Genetics* **2014**, *113*, 1007–1016. [CrossRef] [PubMed]

82. Makunin, A.I.; Kichigin, I.G.; Larkin, D.M.; O'Brien, P.C.; Ferguson-Smith, M.A.; Yang, F.; Graphodatsky, A.S. Contrasting origin of B chromosomes in two cervids (Siberian roe deer and grey brocket deer) unravelled by chromosome-specific DNA sequencing. *BMC Genom.* **2016**, *17*, 618. [CrossRef]

83. Houben, A.; Carchilan, M. Plant B chromosomes: What makes them different? In *Plant Cytogenetics*; Springer: New York, NY, USA, 2012; pp. 59–77.

84. Jones, R.N.; Houben, A. B chromosomes in plants: Escapees from the A chromosome genome? *Trends Plant Sci.* **2003**, *8*, 417–442. [CrossRef]

85. Bidau, C.J. Influence of a rare unstable B-chromosome on chiasma frequency and nonhaploid sperm production in *Dichroplus pratensis* (Melanoplinae, Acrididae). *Genetica* **1987**, *73*, 201–210. [CrossRef]

86. Burt, A.; Trivers, R. Selfish DNA and breeding system in flowering plants. *Proc. R. Soc. Lond.* **1998**, *265*, 141–146. [CrossRef]
87. Muntzing, A. Cyto-genetics of accessory chromosomes (B-chromosomes). *Caryologia* **1954**, *6*, 282–301.
88. Puertas, M.J.; Ramirez, A.; Baeza, F. The transmission of B chromosomes in *Secale cereale* and *Secale vavilovii* populations. II. Dynamics of populations. *Heredity* **1987**, *58*, 81. [CrossRef]
89. Timmis, J.N.; Ayliffe, M.A.; Huang, C.Y.; Martin, W. Endosymbiotic gene transfer: Organelle genomes forge eukaryotic chromosomes. *Nat. Rev. Genet.* **2004**, *5*, 123–135. [CrossRef] [PubMed]
90. Krohne, D.T.; Baker, I.; Baker, H.G. The maintenance of the gynodioecious breeding system in *Plantago lanceolata* L. *Am. Midl. Nat.* **1980**, *103*, 269–279. [CrossRef]
91. Charlesworth, B.; Charlesworth, D. A model for the evolution of dioecy and gynodioecy. *Am. Nat.* **1978**, *112*, 975–997. [CrossRef]
92. Marques, A.; Klemme, S.; Houben, A. Evolution of plant B chromosome enriched sequences. *Genes* **2018**, *9*, 515. [CrossRef]
93. Hosid, E.; Brodsky, L.; Kalendar, R.; Raskina, O.; Belyayev, A. Diversity of long terminal repeat retrotransposon genome distribution in natural populations of the wild diploid wheat *Aegilops speltoides*. *Genetics* **2012**, *190*, 263–274. [CrossRef]
94. Berdnikov, V.A.; Gorel, F.L.; Kosterin, O.E.; Bogdanova, V.S. Tertiary trisomics in the garden pea as a model of B chromosome evolution in plants. *Heredity* **2003**, *91*, 577. [CrossRef]
95. Rimpau, J.; Flavell, R.B. Characterisation of rye B chromosome DNA by DNA/DNA hybridisation. *Chromosoma* **1975**, *52*, 207–217. [CrossRef]
96. Timmis, K.; Cabello, F.; Cohen, S.N. Cloning, isolation, and characterization of replication regions of complex plasmid genomes. *Proc. Natl. Acad. Sci. USA* **1975**, *72*, 2242–2246. [CrossRef]
97. Sandery, M.J.; Forster, J.W.; Blunden, R.; Jones, R.N. Identification of a family of repeated sequences on the rye B chromosome. *Genome* **1990**, *33*, 908–913. [CrossRef]
98. Wilkes, T.M.; Francki, M.G.; Langridge, P.; Karp, A.; Jones, R.N.; Forster, J.W. Analysis of rye B-chromosome structure using fluorescence in situ hybridization (FISH). *Chromosome Res.* **1995**, *3*, 466–472. [CrossRef] [PubMed]
99. Karamysheva, T.V.; Andreenkova, O.V.; Bochkaerev, M.N.; Borissov, Y.M.; Bogdanchikova, N.; Borodin, P.M.; Rubtsov, N.B. B chromosomes of Korean field mouse *Apodemus peninsulae* (Rodentia, Murinae) analysed by microdissection and FISH. *Cytogenet. Genome Res.* **2002**, *96*, 154–160. [CrossRef] [PubMed]
100. Cheng, Y.M.; Lin, B.Y. Molecular organization of large fragments in the maize B chromosome: Indication of a novel repeat. *Genetics* **2004**, *166*, 1947–1961. [CrossRef] [PubMed]
101. Marques, A.; Banaei-Moghaddam, A.M.; Klemme, S.; Blattner, F.R.; Niwa, K.; Guerra, M.; Houben, A. B chromosomes of rye are highly conserved and accompanied the development of early agriculture. *Ann. Bot.* **2013**, *112*, 527–534. [CrossRef] [PubMed]
102. Becker, S.E.D.; Thomas, R.; Trifonov, V.A.; Wayne, R.K.; Graphodatsky, A.S.; Breen, M. Anchoring the dog to its relatives reveals new evolutionary breakpoints across 11 species of the Canidae and provides new clues for the role of B chromosomes. *Chromosome Res.* **2011**, *19*, 685. [CrossRef]
103. Shirasu, K.; Schulman, A.H.; Lahaye, T.; Schulze-Lefert, P. A contiguous 66-kb barley DNA sequence provides evidence for reversible genome expansion. *Genome Res.* **2000**, *10*, 908–915. [CrossRef]
104. Coleman, J.J.; Rounsley, S.D.; Rodriguez-Carres, M.; Kuo, A.; Wasmann, C.C.; Grimwood, J.; Schwartz, D.C. The genome of *Nectria haematococca*: Contribution of supernumerary chromosomes to gene expansion. *PLoS Genet.* **2009**, *5*, e1000618. [CrossRef]
105. Cohen, S.; Segal, D. Extrachromosomal circular DNA in eukaryotes: Possible involvement in the plasticity of tandem repeats. *Cytogenet. Genome Res.* **2009**, *124*, 327–338. [CrossRef]
106. Cohen, S.; Houben, A.; Segal, D. Extrachromosomal circular DNA derived from tandemly repeated genomic sequences in plants. *Plant J.* **2008**, *53*, 1027–1034. [CrossRef]
107. Maluszynska, J.; Schweizer, D. Ribosomal RNA genes in B chromosomes of *Crepis capillaris* detected by non-radioactive in situ hybridization. *Heredity* **1989**, *62*, 59. [CrossRef] [PubMed]
108. Stitou, S.; de La Guardia, R.D.; Jiménez, R.; Burgos, M. Inactive ribosomal cistrons are spread throughout the B chromosomes of *Rattus rattus* (Rodentia, Muridae). Implications for their origin and evolution. *Chromosome Res.* **2000**, *8*, 305–311. [CrossRef]

109. Marschner, S.; Kumke, K.; Houben, A. B chromosomes of *B. dichromosomatica* show a reduced level of euchromatic histone H3 methylation marks. *Chromosome Res.* **2007**, *15*, 215. [CrossRef] [PubMed]

110. Bass, H.; Birchler, J.A. *Plant Cytogenetics: Genome Structure and Chromosome Function*; Springer Science & Business Media: New York, NY, USA, 2011; Volume 4.

111. Battaglia, E. Cytogenetics of B-chromosomes. *Caryologia* **1964**, *17*, 245–299. [CrossRef]

112. Gernand, D.; Rutten, T.; Varshney, A.; Rubtsova, M.; Prodanovic, S.; Brüß, C.; Houben, A. Uniparental chromosome elimination at mitosis and interphase in wheat and pearl millet crosses involves micronucleus formation, progressive heterochromatinization, and DNA fragmentation. *Plant Cell* **2005**, *17*, 2431–2438. [CrossRef] [PubMed]

113. Zhang, P.; Li, W.; Friebe, B.; Gill, B.S. The origin of a "zebra" chromosome in wheat suggests nonhomologous recombination as a novel mechanism for new chromosome evolution and step changes in chromosome number. *Genetics* **2008**, *179*, 1169–1177. [CrossRef] [PubMed]

114. Teruel, M.; Cabrero, J.; Montiel, E.E.; Acosta, M.J.; Sánchez, A.; Camacho, J.P.M. Microdissection and chromosome painting of X and B chromosomes in *Locusta migratoria*. *Chromosome Res.* **2009**, *17*, 11. [CrossRef]

115. Rajicic, M.; Romanenko, S.A.; Karamysheva, T.V.; Blagojević, J.; Adnađević, T.; Budinski, I.; Vujošević, M. The origin of B chromosomes in yellow-necked mice (*Apodemus flavicollis*)—Break rules but keep playing the game. *PLoS ONE* **2017**, *12*, e0172704. [CrossRef]

116. Green, D.M. Structure and evolution of B chromosomes in amphibians. *Cytogenet. Genome Res.* **2004**, *106*, 235–242. [CrossRef]

117. Cioffi, M.D.B.; Bertollo, L.A.C. Chromosomal distribution and evolution of repetitive DNAs in fish. In *Repetitive DNA*; Karger Publishers: Basel, Switzerland, 2012; Volume 7, pp. 197–221.

118. Berend, S.A.; Hale, D.W.; Engstrom, M.D.; Greenbaum, I.F. Cytogenetics of collared lemmings (*Dicrostonyx groenlandicus*). II. Meiotic behavior of B chromosomes suggests a Y-chromosome origin of supernumerary chromosomes. *Cytogenet. Genome Res.* **2001**, *95*, 85–91. [CrossRef]

© 2019 by the authors. Licensee MDPI, Basel, Switzerland. This article is an open access article distributed under the terms and conditions of the Creative Commons Attribution (CC BY) license (http://creativecommons.org/licenses/by/4.0/).

GCAT
TACG
GCAT *genes*

MDPI

Article

Genomic Characterization of a B Chromosome in Lake Malawi Cichlid Fishes

Frances E. Clark, Matthew A. Conte and Thomas D. Kocher *

Department of Biology, University of Maryland, College Park, MD 20742, USA; fclark@umd.edu (F.E.C.);
mconte@umd.edu (M.A.C.)
* Correspondence: tdk@umd.edu

Received: 3 November 2018; Accepted: 30 November 2018; Published: 5 December 2018

Abstract: B chromosomes (Bs) were discovered a century ago, and since then, most studies have focused on describing their distribution and abundance using traditional cytogenetics. Only recently have attempts been made to understand their structure and evolution at the level of DNA sequence. Many questions regarding the origin, structure, function, and evolution of B chromosomes remain unanswered. Here, we identify B chromosome sequences from several species of cichlid fish from Lake Malawi by examining the ratios of DNA sequence coverage in individuals with or without B chromosomes. We examined the efficiency of this method, and compared results using both Illumina and PacBio sequence data. The B chromosome sequences detected in 13 individuals from 7 species were compared to assess the rates of sequence replacement. B-specific sequence common to at least 12 of the 13 datasets were identified as the "Core" B chromosome. The location of B sequence homologs throughout the genome provides further support for theories of B chromosome evolution. Finally, we identified genes and gene fragments located on the B chromosome, some of which may regulate the segregation and maintenance of the B chromosome.

Keywords: supernumerary chromosomes; B chromosomes; next-generation sequencing; coverage ratio analysis

1. Introduction

The genomes of eukaryotic species are typically organized into linear chromosomes, and each species has a characteristic number of chromosome pairs referred to as the A chromosomes (As). The genomes of at least 2828 eukaryotic species contain additional chromosomes commonly referred to as B chromosomes (Bs) [1]. These supernumerary B chromosomes are not essential and are found in some but not all individuals of a population [2–5]. Among species, the number of B chromosomes in each cell has been found to vary from 1 to 50 [6–8]. B chromosomes are thought to manipulate the normal mechanisms of cell division in order to increase their transmission to the next generation, a process known as drive [4,7,9].

B chromosomes often contain large amounts of highly repetitive DNA [5,10–12] and are frequently either partially or completely heterochromatic [3,5,7]. In several species, it has been shown that B chromosomes share homology with sequences from all or many of the A chromosomes [13] (the grasshopper *Podisma kanoi* [11], the fish *Astatotilapia latifasciata* [14], rye *Secale cereale* [15], and maize *Zea mays* [10]). This suggests that sequences on B chromosomes are derived from the A chromosomes through as yet uncharacterized mechanisms of gene duplication [16]. Theoretically, because they are nonessential, B chromosomes should experience relaxed selective pressures [16,17]. For this reason, they might be expected to experience high rates of sequence turnover. B chromosomes are continuously acquiring new sequences. Sequences already on the B collect mutations at a high rate, and most are eventually lost. It has been difficult to produce sequence assemblies of B chromosomes due to their repetitive nature and their high levels of homology with sequences in the A chromosomes [18–21].

Despite the fact that B chromosomes add significant amounts of genetic material to the genome, B chromosomes have rarely been associated with novel phenotypes, the most frequent exception being an effect on fertility [3,7,22–24]. With a limited list of known B-specific sequences and few or no visible phenotypes beyond drive, the prevalent view has been that B chromosomes carry few genes [16,25]. They have been thought to be composed of nonfunctional "junk" DNA together with one or two genes contributing to drive [11].

Recent advances in next-generation sequencing and bioinformatic analyses of genomic data have begun to contradict this long-standing view. These technological and analytical improvements make it possible to address many questions about B chromosome biology, including how Bs acquire sequence from the As, how these sequences evolve once on the B chromosome, whether and to what extent the B contains functional sequence, and finally, the identification of the gene(s) controlling drive. Examples of genic sequences detected on B chromosomes include the C-KIT gene in two *Canidae* species [26], ribosomal RNA (rRNA) genes and thousands of genes and gene fragments in the fish *Astatotilapia latifasciata* [14,27], protein coding genes in the grasshopper *Eyprepocnemis plorans* [28], and protein-coding genes in two mouse species from the genus *Apodemus* [29]. Furthermore, transcription has been characterized for an rRNA gene in the smooth hawksbeard *Crepis capillaris* [30], rRNA genes and a pseudogene in the grasshopper *E. plorans* [31–33], pseudogenes in rye *Secale cereale* [34], and protein coding genes in maize *Zea mays* [35,36].

Current approaches to identifying B sequences can be categorized into two types: direct and indirect [18]. Direct methods, such as the sequencing of B chromosomes isolated through flow sorting or microdissection, have a high rate of contamination [14,18] and are only possible in a few organisms. Indirect methods, such as the comparison of whole genome sequence data between samples with or without a B chromosome, can be performed on any species. For many species, the sequence reads can be aligned to a reference genome assembled from an individual lacking a B chromosome, allowing a characterization of a B sequence by its alignment to homologous portions of the A genome. While Illumina sequencing has dramatically lowered costs, there are considerable limitations to Illumina sequence data [37]. Namely, Illumina reads are very short and are not very useful for assembling the repetitive sequence of B chromosomes. However, the extent to which short reads can be used to identify B chromosome sequence has not been fully explored.

Among cichlid fishes, B chromosomes were first identified in species from South America [38–40]. More recently, they have been identified also in species from lakes Victoria and Malawi in East Africa [41,42]. B chromosomes have been found in at least seven species from Lake Malawi. In all seven species, B chromosomes are found only in females, but not all females have a B chromosome. The females that do possess a B chromosome have only a single B (haploid) per cell [42]. A karyotype of one of these species, *Metriaclima lombardoi*, shows the B chromosome is one of the largest chromosomes, representing approximately 4.5% of the genome when present. In Lake Victoria cichlids, B chromosomes are found in both males and females, and individuals carry as many as three B chromosomes per cell [41,43]. Whole-genome resequencing and the sequencing of a microdissected B chromosome were used to identify B chromosome sequences in the Lake Victoria species *A. latifaciata* [14]. Mapping of whole genome sequencing reads to a reference assembly identified thousands of gene fragments and tens of complete genes on the B chromosome. Sequence of microdissected B chromosomes detected only a small portion of the overall B chromosome in this study. Presumably, the sequences contributing to drive are among the genes and gene fragments identified in this study.

Here, we performed a sequence coverage ratio analysis of multiple individuals and species from Lake Malawi to systematically detect sequences from the B chromosome. From this, we identified genes and gene fragments on the B chromosome. We characterized the location and length of the homologous A-located sequences to understand the origin and dynamics of sequence accumulation on the B chromosome. Finally, we analyzed the proportion of B chromosome sequences that are shared among species to estimate the rate of sequence turnover on these unique chromosomes.

2. Materials and Methods

All procedures involving live animals were approved by the University of Maryland IACUC and conducted in accordance with protocol #R-13-58. The *M. lombardoi* individuals used were collected from stocks maintained at the Tropical Aquaculture Facility at the University of Maryland. These stocks were originally sourced from Lake Malawi, Africa in 2014–2016. The remaining individuals were collected directly from Lake Malawi in 2005, 2008, and 2012. Individuals were euthanized using tricaine methanesulfonate (MS-222) and inspected for testes or ovaries to confirm sex. Standard phenol chloroform methods were used in conjunction with phase-lock gel tubes (5Prime, Gaithersburg, MD, USA) for DNA extraction from fin tissue. Genotyping of B-specific single nucleotide polymorphisms (SNPs) was performed according to [42] to identify individuals carrying a B chromosome. Blood was collected from a *M. lombardoi* individual in order to prepare the high molecular weight DNA necessary for Pacific Biosciences SMRT (PacBio, Menlo Park, CA, USA) sequencing.

Illumina sequencing was performed from the fin clips of 12 female individuals with a B chromosome. To provide a comparison lacking B chromosome DNA, the sequence from pooled male individuals, previously collected and sequenced with Illumina (San Diego, CA, USA), was used. Each pooled sample contained between 10 and 20 male individuals of that species. As there was no male *M. lombardoi* sequence data available, the B female *M. lombardoi* data was compared to a pool of *Metriaclima zebra* "Boadzulu" males for the scaled coverage analysis. Sequences from two samples of pooled female individuals lacking a B chromosome (NoB), previously collected and sequenced with Illumina, were used as controls. These two NoB samples represent two types of females from the *Labeotropheus trewavasae* "Maison Reef" population, XX females, and WZ females, all lacking a B chromosome. The samples used are summarized in Table 1.

The 12 B female individual samples, the 7 NoB male pooled samples, and the 2 NoB female pooled samples were prepared for Illumina sequencing with the TruSeq DNA sample preparation kit ver.2 rev.C (Illumina Inc.). Each DNA sample was sonically sheared and selected to produce libraries of 500 bp fragments. Paired-end reads of 100 bp were obtained using an Illumina HiSeq 1500. Pacific Biosciences SMRT sequencing was performed on one *M. lombardoi* B female. DNA was extracted from nucleated blood cells using the MagAttract HMW DNA kit from Qiagen (Germantown, MD, USA). Pulse-field gel electrophoresis was performed with a Blue Pippin instrument by the University of Maryland Genomics Resource Center to select DNA fragments of the proper size. PacBio sequencing was carried out on the PacBio RS II platform with P6-C4 chemistry using nine SMRT cells and on the PacBio Sequel platform using nine additional SMRT cells. Illumina and PacBio sequencing reads were aligned to the reference assembly of a *M. zebra* "Mazinzi Reef" NoB male individual sequenced with PacBio [44], (publicly available on NCBI, Accession: GCA_000238955.4, [45]) with BWA [46] and NGM-LR [47], respectively. BWA alignments were then run through Picard (version 2.1.0) "MarkDuplicates" (http://broadinstitute.github.io/picard) to identify PCR duplicates.

Table 1. Sample Information.

Genus	Species	Locality	Sex	B?	Sample Type (#)	Sample ID	Sequencing Method	Mean Sequencing Depth
Labeotropheus	trewavasae	Thumbi	Female	B	Individual	2005-1306	Illumina	15.02
			Male	NoB	Pooled (10)	2005	Illumina	12.66
		Maison	Male	NoB	Pooled (20)	2012	Illumina	36.64
			XX Female	NoB	Pooled (20)	2012	Illumina	38.62
			WZ Female	NoB	Pooled (20)	2012	Illumina	36.63
Melanochromis	auratus		Female	B	Individual	2008-1601	Illumina	14.54
			Male	NoB	Pooled (10)	2005	Illumina	13.18
Metriaclima	greshakei		Female	B	Individual	2012-3493	Illumina	14.59
			Male	NoB	Pooled (20)	2012	Illumina	24.51
	lombardoi		Female	B	Individual	2014-1018	Illumina	16.21
			Female	B	Individual	2014-1021	Illumina	17.12
			Female	B	Individual	2014-1108	Illumina	11.75
			Female	B	Individual	2016-1012	PacBio	17.08
	mbenji		Female	B	Individual	2012-3997	Illumina	14.57
			Male	NoB	Pooled (20)	2012	Illumina	29.70
	zebra	Boadzulu	Female	B	Individual	2005-0976	Illumina	15.24
			Female	B	Individual	2005-0983	Illumina	14.78
			Female	B	Individual	2005-0986	Illumina	12.48
			Male	NoB	Pooled (20)	2012	Illumina	24.57
		Mazinzi	Male	NoB	Individual	SAMN03890374	PacBio	52.42
		Nkhata Bay	Female	B	Individual	2012-5340	Illumina	13.27
			Female	B	Individual	2012-5347	Illumina	16.20
			Male	NoB	Pooled (20)	2012	Illumina	34.39

After alignment to the reference genome, all genomic samples were analyzed with samtools (version 0.1.18) mpileup (http://samtools.sourceforge.net) to calculate read coverage depth across the genome. The raw coverage depth was scaled by dividing the raw coverage at each position by the average genome-wide coverage depth of the sample. This scaled coverage value was then used to calculate the scaled coverage ratio (SCR) between the B chromosome female and the corresponding NoB pooled male sample:

$$\text{SCR} = \frac{\text{scaled coverage of the B female}}{\text{scaled coverage of the NoB male pool}}. \tag{1}$$

For each base in the genome, a binomial test was performed to check for a statistically significant difference in coverage between the B female dataset and the NoB pooled male dataset:

$$P(X) = \frac{n!}{(n-X)!X!} \cdot (p)^X \cdot (q)^{n-X}. \tag{2}$$

In this binomial test, X represents the raw coverage depth in the B female sample and n is the sum of the raw coverage depth in the B female sample and the NoB pooled male sample. The expected frequency of B female reads, p, is calculated from the relative genome-wide sequence depth of the B female sample. The expected frequency of NoB pooled male reads, q, is calculated from the relative genome-wide sequence depth of the NoB pooled male sample. Any positions with a SCR ≥ 3 (corresponding to ≥ 4 B-located copies), a binomial test p-value ≤ 0.001, and within 300 bp of another such position were merged into a block feature with Bedtools (version 2.26.0) merge function [48]. Requiring a minimum SCR of three fails to detect any sequences with fewer than four copies on the B chromosome avoids detection of simple A chromosome duplications, which would result in a SCR of 2. These block features were filtered to remove any block feature ≤ 500 bp in length and then any block feature with $\leq 10\%$ of the positions spanned meeting the SCR ≥ 3 requirement and the p-value ≤ 0.001 requirement. The latter three parameters (merging distance of 300 bp, minimum block length of 500 bp, and minimum percent positions of 10%) were chosen after manual inspection of several preliminarily identified regions. The remaining block features are referred to as "B blocks". The B blocks of all individuals were then processed with Bedtools (version 2.26.0) intersect [48] to find B blocks common among at least 12 of the 13 B individuals (12 Illumina and 1 PacBio). These shared B blocks are referred to as the Malawi "core" blocks.

The sum of the lengths of all B blocks was calculated as an estimate of total B sequence length in the reference genome, further referred to as "A chromosome space". To account for copy number of these sequences on the B, the length of each block was multiplied by its estimated copy number, resulting in each block's contribution to the B, which was then summed to estimate the total B sequence length. Estimated copy number was calculated with one of two equations, depending on the average scaled coverage in the NoB male dataset. For NoB male scaled coverage ≥ 1, we used Equation (3):

$$(\text{SCR} * 2) - 2 \tag{3}$$

In Equation (3), SCR was multiplied by 2 to compensate for the fact that we are comparing a haploid B genome to a diploid A genome. The A chromosome copy was then accounted for by subtracting 2. To avoid overestimating the B-located copy number when the NoB male scaled coverage was less than 1, we used Equation (4):

$$\text{Female Scaled Coverage} * 2 \tag{4}$$

Here, a NoB male scaled coverage of 1 was assumed (accounting for one copy of this sequence in the A genome of the reference), allowing us to use the scaled coverage of the B female to estimate copy number, without having to account for the A chromosome copy by subtracting 2. Example scripts for our B block identification analysis are provided in Supplementary Materials (Directory S1).

3. Results

3.1. Identification of B Chromosome Sequence

3.1.1. Characterization of B Blocks

Due to the homology between A and B chromosome sequence, most sequence reads derived from the B chromosome will align to their A chromosome homologs present in the reference genome. As a result, alignments of reads from a genome with a B chromosome will have regions of increased coverage compared to an alignment from a genome lacking a B. Our analysis of coverage ratios initially identified 0.34%–1.31% of the bases in the genome as having relatively higher coverage in the B female dataset (Table 2). In comparison, the same analysis in our controls identified 0.06% and 0.44% of bases in the WZ and XX NoB females, respectively. Further analysis combined these individual bases into features referred to as B blocks, defined as consecutive sequence with increased coverage in B chromosome samples. Thousands of B blocks were identified in each B female individual. B blocks ranged in length from 500 to 100 kb, although there were multiple regions in the genome with multiple B blocks in close proximity, suggesting that a larger region was transferred to the B chromosome as a whole (Figure 1). The largest such regions were located on LG4 (~120 kb), LG9 (~250 kb), LG17 (~260 kb), and LG23 (~420 kb).

In the WZ and XX NoB females controls, we identified 343 and 2125 putative B blocks, respectively, and the longest blocks were only 3.6–5.2 kb (Table 2). As neither of these individuals carried a B chromosome, these putative B blocks represent false positives. While actual variation in A genome copy number may explain some of this error, stochastic variation in the coverage depth of Illumina data and regions of poor alignment likely also contribute to these false B block calls. Figure 2 provides representative histograms of block length, showing data for a B chromosome female (*L. trewavasae* 2005-1306), the blocks included in the core set, and the XX and WZ NoB females. Both the B female and the core set show enrichment for blocks of longer lengths when compared to the controls. The core set shows a depletion of shorter blocks. An interpretation of this is that false positive B block calls are more likely to be short in length and that a sizable portion of the shorter B blocks may be false positives (type 1 error) and do not represent actual B sequence. However, since large regions, as seen in Figure 1, are often fragmented into smaller block calls, we opted not to remove the shorter block calls at this stage of the analysis. The B block information for each dataset, including block location, coverage details, and length, is provided in Supplementary Materials (Directory S2).

The lengths of all B blocks were then summed for each sample, as well as for the set of core blocks, producing the total length of B sequence in A chromosome space (Table 3). However, since there are multiple copies of these sequences on the B, we multiplied the length of each block by the copy number of that sequence, as estimated by the difference in coverage between the B female dataset and the male dataset. These values were then summed across all blocks to produce the total estimated length of B chromosome sequence (i.e., in B chromosome space). The total length of B sequence from the core block set (not including variable blocks specific to some individuals or species) in B chromosome space was also calculated for each sample.

Table 2. B Block Sizes.

	% of A Genome Passing Both Thresholds	Number of Blocks	Mean Block Size (bp)	Standard Deviation of Block Size (bp)	Median Block Size (bp)	Maximum Block Size (bp)
L. trewavasae 2005-1306	0.59	3517	1554.8	2592.9	849	42941
M. auratus 2008-1601	0.34	2476	1415.7	1859.6	836	30172
M. greshakei 2012-3493	0.69	4392	1395.1	2618.8	805	52821
M. lombardoi 2014-1018	1.31	10918	1285.1	1845.8	824	63229
M. lombardoi 2014-1021	1.04	8251	1298.0	1954.9	822	63250
M. lombardoi 2014-1108	1.10	8684	1274.9	1902.3	809	63229
M. mbenji 2012-3997	0.68	4147	1344.7	2519.2	783	63230
M. zebra "Boadzulu" 2005-0976	0.85	5907	1264.9	2002.4	793	42941
M. zebra "Boadzulu" 2005-0983	0.79	5369	1238.5	2402.4	769	100567
M. zebra "Boadzulu" 2005-0986	0.84	5986	1228.8	2293.8	771	100079
M. zebra "Nkhata Bay" 2012-5340	0.64	4869	1419.0	2856.6	842	99928
M. zebra "Nkhata Bay" 2012-5347	0.89	7162	1420.9	2821.8	867	100026
M. lombardoi 2016-1012 (PacBio)	0.59	1904	2971.1	4569.8	1723	98620
L. trewavasae "Maison" XX females (control)	0.44	2125	714.0	243.4	642	3607
L. trewavasae "Maison" WZ females (control)	0.06	343	819.5	478.5	686	5198
Core blocks	N/A	622	2194.6	3582.8	937	32721

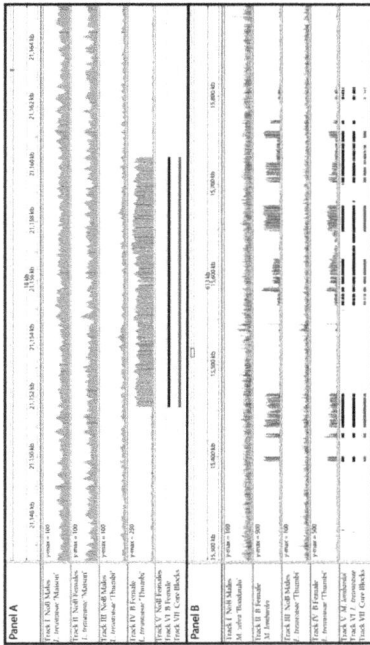

Figure 1. Read Coverage and B Blocks. B blocks from two genomic regions are shown with the corresponding read coverage. **Panel (A)** depicts an 18-kb region of LG8 with a typical B block. Tracks I and III are the male coverage (*Labeotropheus trewavasae* "Maison" and *L. trewavasae* "Thumbi", respectively), while tracks II and IV are the female coverage (NoB XX *L. trewavasae* "Maison" control and B *L. trewavasae* "Thumbi", respectively). Please note the *y*-axis maximum is 100 for tracks I, II, and III but 250 for track IV. Beneath the coverage plots are the blocks detected by our analysis; track V shows the NoB XX female blocks, track VI shows the B female *L. trewavasae* blocks, and track VII shows the core blocks. A ~8.5-kb B block can be observed in the B female *L. trewavasae* (track IV), but no such increased coverage is observable in the other coverage plots. Our B block analysis pipeline identified the B female *L. trewavasae* block (track VI) but did not identify a block in the NoB XX female control data (track V). As this B block was similarly found in at least 12 of the 13 datasets, it is included in the core block set (track VII). **Panel (B)** depicts several B blocks in close proximity to one another across a 613-kb region of LG23. Tracks I and III are again male coverage but for *M. zebra* "Boadzulu" and *L. trewavasae* "Thumbi", respectively. Tracks II and IV both depict B female coverage (B *Metriaclima lombardoi* and B *L. trewavasae* "Thumbi", respectively). Please note the *y*-axis maximum is 100 for tracks I and III but 500 for tracks II and IV. The block sets detected in the B female *M. lombardoi* (track V), B female *L. trewavasae* "Maison" (track VI), and the core blocks (track VII) are shown below. B blocks can be observed in the coverage of both B females (tracks II and IV) and correspond well with the blocks identified through our B block identification analysis (tracks V, VI, and VII). The B blocks span ~420 kb and appear to have migrated to the B as a single unit in the ancestor of *M. lombardoi* and *L. trewavasae*.

29

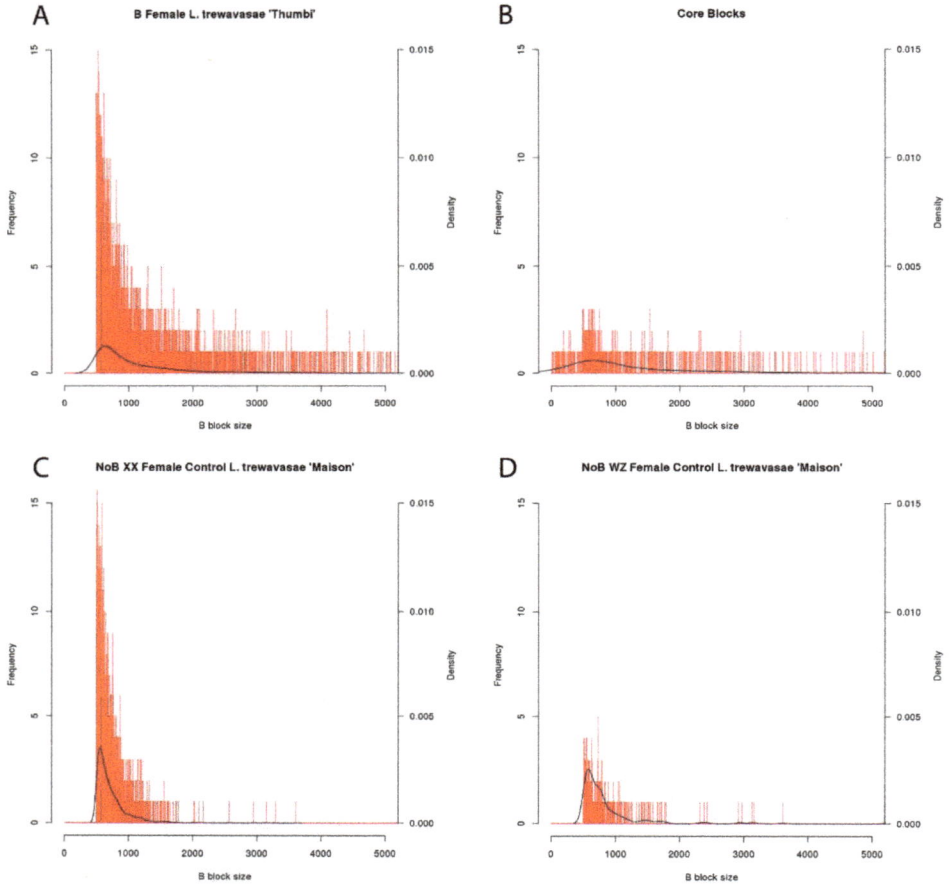

Figure 2. Block Length Histograms. Histograms of B block length for four datasets. B block size along the *x*-axis is reported in bp (bin size = 1 bp), and the *x*-axis maximum is 5000 bp to more easily view the majority of the data. The blocks not visible in this graph, larger than 5000 bp, represent only 10.3% of the core blocks and <5% of the blocks in the B female and two NoB controls. The number of blocks of each length is shown with red bars with the *y*-axis scale on the left and the density is depicted with a black line with the corresponding *y*-axis scale on the right. Because blocks shorter than 500 bp were removed during analysis, the B female (**A**) and the two NoB controls (**C**,**D**) show a lack of these smaller blocks. However, during the identification of core blocks, some larger B blocks were fragmented further, resulting in the smaller B blocks shown in the core block set histogram (**B**). All other B block length histograms are included in the Supplementary Materials as Figure S1.

The total length in A space ranges from 3.51 to 14.06 Mb among the B females and only 0.28 to 1.52 Mb in the controls. Only 1.37 Mb (in A space) is shared among at least 12 of the 13 B females. After taking copy number of these sequences into account, the total length in B space ranges from 23.19 to 99.69 Mb among B females and only 0.39 to 2.15 Mb in the controls. The 1.37 Mb of core blocks in A space translates to 12.31–44.07 Mb among B females and as little as 0.63–0.80 Mb in the controls.

Table 3. Total estimated length of B sequence identified in the coverage ratio analysis. The total length of sequence identified in the reference genome, or "in A space", is provided for all B individuals, the two NoB controls, and the core block set (the blocks shared by at least 12 of the 13 B individuals). The total length of these blocks, after accounting for their copy number on the B chromosome (i.e., "in B space") is provided for the 13 B individuals and 2 NoB controls. Similarly, the total length of just the core blocks, accounting for copy number, are provided for the 13 B individuals and 2 NoB controls. Finally, the portion of the total estimated B chromosome length that is represented by the core blocks is provided for the 13 B individuals.

	In A Space (Mb)	In B Space (Mb)	Core Blocks in B Space (Mb)	Core Blocks % of Total B in B Space
L. trewavasae 2005-1306	5.48	49.44	25.13	50.84
M. auratus 2008-1601	3.51	23.19	12.31	53.11
M. greshakei 2012-3493	6.14	58.20	30.84	52.98
M. lombardoi 2014-1018	14.06	74.53	24.09	32.31
M. lombardoi 2014-1021	10.73	62.58	23.00	36.76
M. lombardoi 2014-1108	11.09	58.48	17.67	30.22
M. mbenji 2012-3997	5.59	41.30	21.62	52.34
M. zebra "Boadzulu" 2005-0976	7.49	59.48	30.86	51.88
M. zebra "Boadzulu" 2005-0983	6.66	51.06	27.38	53.63
M. zebra "Boadzulu" 2005-0986	7.37	49.55	23.44	47.31
M. zebra "Nkhata Bay" 2012-5340	6.92	48.61	21.58	44.40
M. zebra "Nkhata Bay" 2012-5347	10.19	99.69	44.07	44.21
M. lombardoi 2016-1012 (PacBio)	5.66	35.40	15.02	42.44
L. trewavasae "Maison" XX females (control)	1.52	2.15	0.63	-
L. trewavasae "Maison" WZ females (control)	0.28	0.39	0.80	-
Core blocks	1.37	-	-	-

The consensus, or core, block set with blocks common to at least 12 of the 13 individuals successfully removed the greatest proportion of false positives (type 1 error). However, the core block set lacks any B chromosome sequence that is specific to only a few individuals or species. The B chromosome of the *M. lombardoi* individuals, sequenced with Illumina, is estimated to be 58.48–74.53 Mb in length. Considering just the most conservative B blocks (the core set), the estimated length is 17.67–24.09 Mb in these individuals. Karyotype data, available only for *M. lombardoi*, shows that the B chromosome is one of the largest chromosomes. A tentative estimate of chromosome size from karyotype data suggests a B chromosome of roughly 50 Mb. The total length of B sequence in B space in these three individuals may be inflated by false positive blocks, while the total length of core sequence is B space is slightly smaller than the length estimated from the karyotypes. The variation in estimated B chromosome length across individuals could indicate that B chromosomes vary in size among these species. This is consistent with the finding that B chromosomes vary in length within and among species of Lake Victoria cichlid. [43]. Notably, *Melanochromis auratus* consistently has the least amount of sequence detected by our analysis. The 12.31 Mb, in B space, found in *M. auratus*, compared to the 30.84 Mb found in *Metriaclima greshakei*, suggests that the B chromosome of *M. greshakei* may be twice as large as the B chromosome of *M. auratus*.

3.1.2. Comparison of Illumina and PacBio Sequence Data

To better understand the differences in B blocks called from Illumina and PacBio datasets, we compared an *M. lombardoi* B female sequenced with PacBio to the three *M. lombardoi* B females sequenced with Illumina. The Illumina reads are 100 bp in length and the PacBio reads averaged 8295 bp. The blocks identified in the individuals sequenced with Illumina ranged in total length in A space from 10.73 to 14.06 Mb, whereas the total length of blocks identified in the individual sequenced with PacBio was only 5.66 Mb in A space. As demonstrated with the block size histograms (Figure 2), we believe most falsely identified blocks are short in length. Indeed, the mean length of B blocks identified using the PacBio data was much longer than with the Illumina data (Table 2) and a depletion of shorter blocks can also be seen in the block size histogram of the PacBio data (Supplementary Materials, Figure S1). This discrepancy in length in A space could be a byproduct of the longer PacBio reads resulting in more consistent coverage and preventing the erroneous identification of shorter blocks. Additionally, longer PacBio reads will have more accurate mapping in repetitive regions than the shorter Illumina reads. These factors suggest that PacBio data would result in fewer false positives or type 1 errors. However, even when using the conservative core block set, the PacBio data identified only 15.02 Mb of core sequence in B space compared to the 17.67–24.09 Mb identified in the three Illumina datasets, suggesting the Illumina data is able to detect sequences the PacBio data does not.

While inspecting the read alignments and coverage data in detail, a few key patterns emerged. First, there were several regions of high coverage in the Illumina data, which had low coverage in the PacBio data (Figure 3, panel A). The Illumina reads in these short regions all aligned to several other locations (as indicated with white reads in Figure 3) and these regions were annotated as various repeats. Our interpretation is that these regions represent a shorter, highly repetitive sequence, with many copies found on the B chromosome. We hypothesize that the A chromosome in the *M. zebra* reference assembly experienced a recent insertion of this repeat, resulting in a lack of coverage by the *M. lombardoi* PacBio data because it does not have this insertion. Because the Illumina reads are too short to span the length of the repeat, they aligned to this insertion in the reference. This means that the Illumina data was able to detect these B-specific sequences while the PacBio data was not. However, the Illumina data wrongly places the A chromosome origin of these B sequences at the new insertion site when their existence on the B appears to predate this insertion.

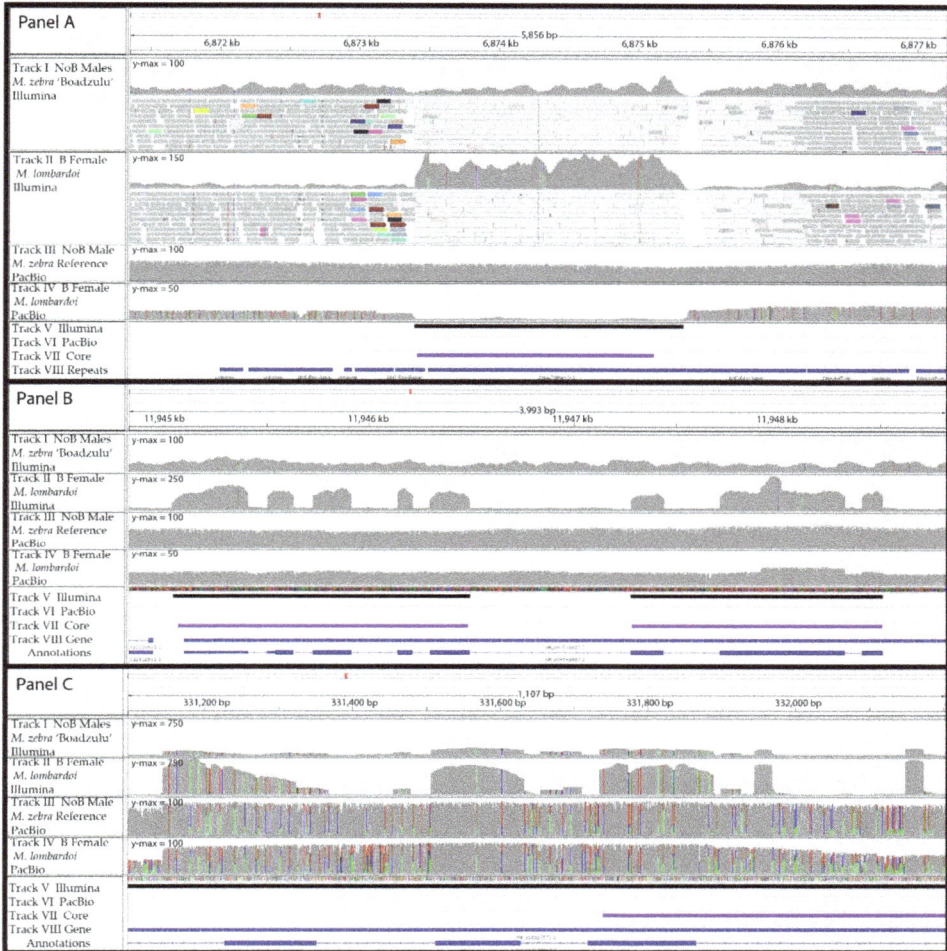

Figure 3. Comparisons of Illumina and PacBio read alignment in B blocks. Differences in the alignment of Illumina and PacBio reads affected the B block identification analysis. Panels (**A–C**) represent regions of LG20, LG22, and the unanchored scaffold 000256F_pilon_quiver, respectively. Panel (**A**) demonstrates a failure to identify a B block with PacBio data. Additionally, the localization of that block with Illumina data to a recent insertion inaccurately suggests LG20 as the A chromosome origin of this B-located sequence. Panel (**B**) demonstrates the failure of PacBio data to detect a retrogene. Panel (**C**) demonstrates a case where Illumina data suggests a retrogene, which the PacBio data reveals to be a complete gene (possessing both exons and introns). This specific example also shows increased coverage in the NoB data, suggesting the region has also experienced a duplication event within the A genome in addition to the copies present on the B chromosome. In Panels (**A–C**), tracks I and II represent the coverage of the NoB male and B female sequenced with Illumina, respectively, while tracks III and IV depict the coverage of the NoB male and B female sequenced with PacBio, respectively. In Panels (**A–C**), the B female *M. lombardoi* block sets for Illumina and PacBio are shown in tracks V and VI, respectively, while the core block set is shown in track VII. Tracks I and II in Panel (**A**) also show a portion of the reads aligning to that region. Reads shown in white have a map quality of 0 indicating multiple mapping to several regions. In Panel (**A**), track VIII displays the annotated repeat content of this region. In Panels (**B,C**), track VIII displays the gene annotations. Please note the *y*-axis maximum of the coverage plots varies to best view the variable coverage data of each plot.

A second difference between the sequence data types was in the detection of retrogene insertions (Figure 3, panel B). Again, since the PacBio reads are much longer than the retroinserted exons, they do not align well to the A reference using typical PacBio alignment software such as NGM-LR and BLASR with standard alignment parameters. In contrast, Illumina reads are usually shorter than the length of these retroinserted exons and therefore do align well to the reference. This means that standard alignment software and parameters will detect retroinserted sequences on the B chromosome with short read data but not with long read data. Proper alignment of retroinserted genes using PacBio reads requires the use of alignment tools that are splice-site aware, such as GMAP. We were able to recover this particular retroinsertion with the PacBio data by aligning with GMAP, but the majority of A genome reads did not map. Alignment software that accounted for both types of reads is needed, but to our knowledge, such tools do not yet exist.

The third difference between the two sequence data types was in the false detection (type 1 error) of retroinserted genes (Figure 3 panel C). The Illumina data showed increased coverage in the exons but not the introns of some genes, suggesting it was another retroinserted gene on the B. However, the PacBio data revealed consistently high coverage across both introns and exons, with much higher sequence polymorphism in the introns. The higher sequence polymorphism in the introns compared to the exons suggests that the B-located copy of this gene is relatively old and still experiencing purifying selection for the encoded protein. The short reads of the Illumina data failed to align to the divergent introns but did align in the less divergent exons, resulting in what appeared to be a retroinserted gene. We were only able to distinguish between 'true' and 'false' retroinserted genes on the B chromosome by comparing the Illumina data with PacBio data.

3.2. B Block Turnover

B chromosomes are thought to have a high rate of sequence turnover because they experience little purifying selection [16,17]. Because the Lake Malawi cichlid species studied here diverged less than 1 million years (MY) ago [49], we have an opportunity to study the rates and patterns of sequence turnover on the B chromosome. To gauge the amount of sequence turnover that has occurred between these species, we compared the core block set to all B blocks (core and variable) identified in each individual. The core blocks accounted for 30.22%–53.63% of total B sequence (in B space), leaving 46.37%–69.78% B sequence (in B space) variable among individuals. While some of the variable B blocks represent false positives (type 1 error), many represent sequences that are unique to a particular individual or species. These variable B blocks likely represent both sequence that was lost from a common ancestor and a new sequence acquired during the evolution of particular lineages.

3.3. B Block Origin

A comparison across these 13 B individuals has allowed us to identify sequence (the core blocks) present on the B chromosome of the most recent common ancestor to these seven cichlid species. Figure 4 depicts the position of core B blocks on the chromosome-scale assembly of the *Metriaclima zebra* A genome. Notably, each linkage group (LG), and therefore each chromosome, has at least one core B block and most have several, distantly spaced core B blocks. This is consistent with the idea that cichlid B chromosomes continue to collect A chromosome sequences over time [13–15]. No trend was observed between B block position and centromere position. There is no readily visible pattern, suggesting certain regions are more likely than others to be the source of B chromosome sequence. The longest stretch of B chromosome (along A chromosome space) corresponds to a ~420-kb region comprising several neighboring B blocks on LG23 (also shown in Figure 1). The SCR of the core blocks varies among individuals. The largest difference in SCR between these two individuals is shown on LG8.

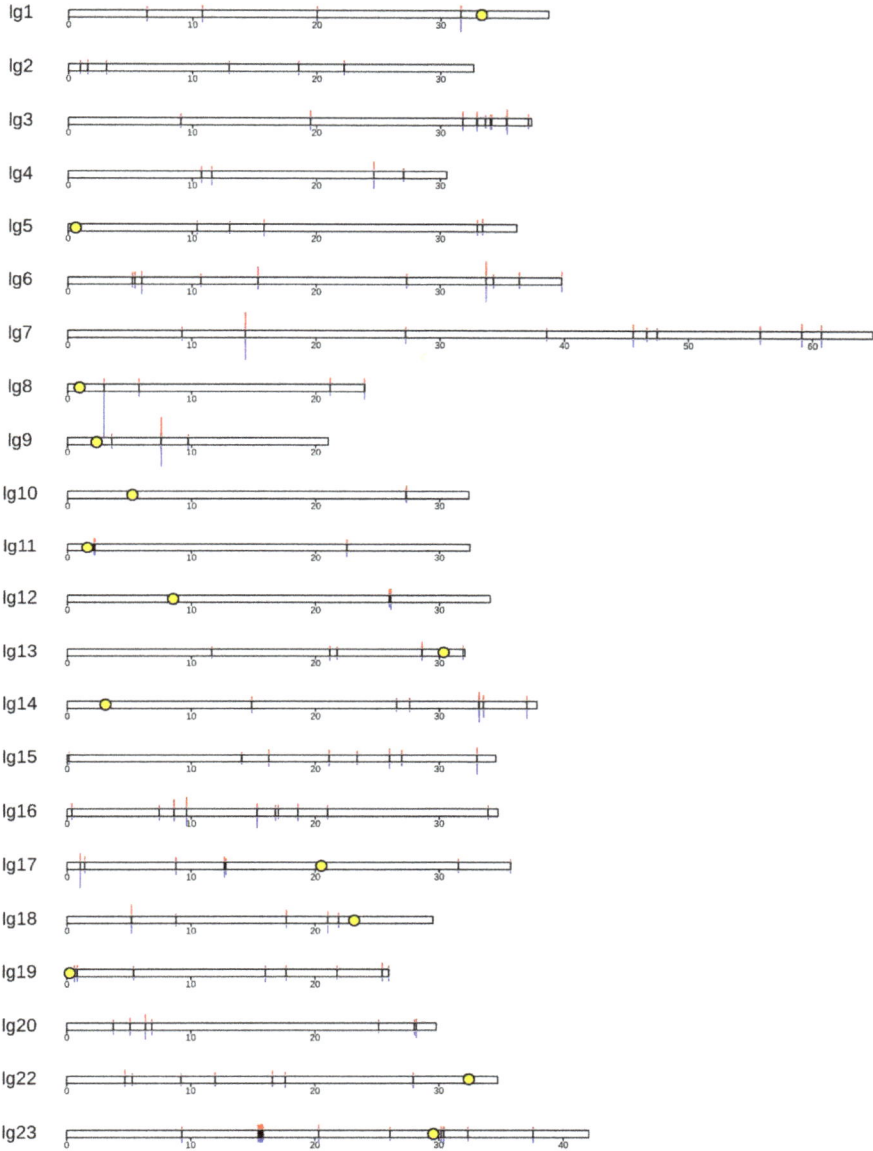

Figure 4. Karyoplot showing the A genome origins of the B chromosome. The position of B blocks (black bars) is superimposed on a karyoplot of the *Metriaclima zebra* "Mazinzi" reference genome. The A genome consists of 22 chromosomes. For simplicity, unanchored scaffolds of the genome assembly were not included. Physical distances are noted beneath each LG in Mb and the locations of centromeres (available for only some LGs) are indicated with yellow circles. Above and below each LG is a bar graph representing the scaled coverage ratio (SCR) of each core block. Above, in red, is the SCR of *M. lombardoi* 2014-1021. Below, in blue, is the SCR of *L. trewavasae* 2005-1306. The SCR of these individuals ranges from 3 to 234. The two individuals shown are arbitrarily chosen representatives of these two genera. This plot was created with the R package KaryoploteR [50].

3.4. Genes

B chromosome gene sequences were identified as overlap between RefSeq annotated genes and B blocks. Annotated genes were either partially or completely encompassed in a B block. The total number of partial or complete genes in the B chromosome blocks is listed in Table 4. The complete list of genes and gene fragments identified in each dataset is provided in Supplementary Materials (Directory S3).

Table 4. Genes and gene fragments on the B chromosome.

Sample	Number of Genes and Gene Fragments
L. trewavasae 2005-1306	702
M. auratus 2008-1601	516
M. greshakei 2012-3493	972
M. lombardoi 2014-1018	2030
M. lombardoi 2014-1021	1688
M. lombardoi 2014-1108	1664
M. mbenji 2012-3997	899
M. zebra "Boadzulu" 2005-0976	1291
M. zebra "Boadzulu" 2005-0983	1262
M. zebra "Boadzulu" 2005-0986	1260
M. zebra "Nkhata Bay" 2012-5340	1094
M. zebra "Nkhata Bay" 2012-5347	1739
M. lombardoi 2016-1012 (PacBio)	678
L. trewavasae "Maison" XX females (control)	595
L. trewavasae "Maison" WZ females (control)	132
Core blocks	132

Figure 5 includes two Venn diagrams of B-located genes shared among the three *M. zebra* "Boadzulu" individuals (0976, 0983, and 0986) and among the three *M. lombardoi* individuals (1018, 1021, and 1108). In both cases, the individuals from the same population share most of their presumed B-located genes, though there are still several hundred unique to each individual.

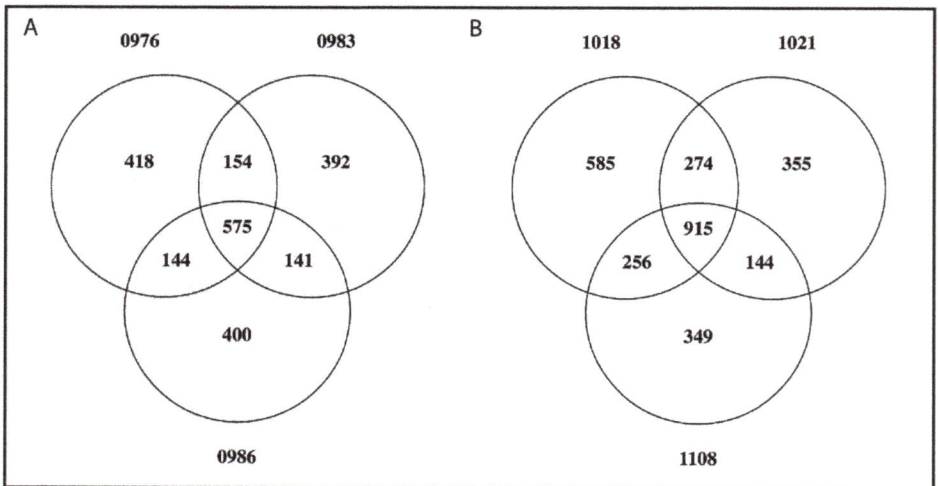

Figure 5. Shared B-located genes and gene fragments among individuals of the same population. The Venn diagrams show the number of genes and gene fragments shared by (**A**) three *M. zebra* "Nkhata Bay" individuals and (**B**) three *M. lombardoi* individuals.

4. Discussion

Using an analysis of sequence coverage, we identified 1.37 Mb of the A genome that has been copied to the B chromosome and which is now shared among several Lake Malawi cichlid species as core B chromosome sequence. In addition to this core sequence, there were many additional Mb of B chromosome sequence that were found among various subsets of individuals/species. Because the core B chromosome sequences are found in multiple copies, the total length of B-specific sequence in the three *M. lombardoi* individuals totaled 17.67–24.09 Mb. This is consistent with the size of the *M. lombardoi* B chromosome observed in karyotype data. This suggests that the coverage ratio analysis was successful in identifying an appreciable amount of sequences on the B chromosome. Using all the B blocks identified with each individual dataset (including both variable and core blocks) resulted in a size estimate of 58.48–74.53 Mb (in B space), which is slightly larger than expected from karyotype data. This suggests that some portion of the identified B blocks represent false positives, or type 1 error. Another approach to understanding the amount of type 1 error in this analysis is through the two control datasets. The percentages of individual bases in the genome passing the SCR and binomial thresholds were not markedly different for the XX NoB female control (0.44%) and the B females (0.34%–1.31%). Our downstream filtering to produce block features helped to further reduce the type 1 error, resulting in an order of magnitude fewer blocks identified in the two controls compared to the B female datasets. Further filtering of short blocks would likely continue to reduce the type 1 error but would simultaneously increase type 2 error. The length of identified B sequence, in B space, of the two controls was 0.39–2.15 Mb. Arguably, we can extrapolate from this to predict that any individual could have at least 1–5 Mb of falsely identified B sequence. Yet, the total amount of variable B sequence, in B space, ranged from 39.58–50.45 Mb for the B female datasets. From this, we conclude that B blocks identified using sequence data from a single individual likely contain some type 1 error but also correctly represent a large number of unique B blocks that are not shared among individuals or species.

In the estimation of total length in B space, proper estimation of B-located copy number is clearly crucial. For most regions, SCR can be used to estimate copy number. However, in regions of poor alignment, scaled coverage can be <1, leading to an overestimate of copy number and therefore an inflated estimate of total length in B space. To avoid this issue, the B-located copy number of any region with a scaled coverage value <1 in the NoB male dataset was instead calculated with the scaled coverage in the female B dataset. The use of multiple individuals and the identification of core sequence greatly reduces the type 1 error and we suggest that multiple individuals, if not species, be used to produce the most conservative identification of B sequence when using a coverage ratio analysis. Notably, our coverage ratio analysis ignores any sequence entirely unique to the B chromosome (not aligned to homologous A sequence) or any sequence with fewer than four B-located copies (an SCR of 3). While unique sequences are mechanistically entirely undetectable with a coverage ratio analysis, the detection of less abundant sequences on B chromosomes presents a trade off with type 1 error rates. This can be circumvented by sequencing individuals with a high number of B chromosomes per cell, when possible.

Across the 13 datasets, the core blocks represented 30.22%–53.63% of the total sequence identified in B space, leaving 46.37%–69.78% as variable among individuals. As discussed above, we believe an appreciable amount of this unshared sequence is actually type 1 error, and therefore more than 30%–54% of the B is shared among these species. Even though the individuals in this study span three genera, they are less than 1 MY diverged from one another. This suggests that the Lake Malawi B chromosome has experienced turnover of roughly half of its sequences in 1 MY. Whether the rate of sequence turnover is constant or varies over time is not yet known.

The length and position of B blocks along the A chromosomes allows us to begin unraveling the history of the B chromosome. The presence of B blocks on every chromosome supports the idea that once a proto-B forms, it somehow acquires sequence from the rest of the genome. How these sequences make their way to the B, and which types of sequences are most likely to do so, is still unknown. Most discussion of mechanisms that transfer sequence to the B involves transposable elements [9]. It is possible other mechanisms, such as nonhomologous recombination, could also be contributing to the acquisition of B sequence. B blocks range in size from a few hundred to a few hundred thousand bases. Homologous regions larger than 100 kb have been found on each of several chromosomes, suggesting that some, if not all, of these larger regions must have migrated to the B after its origin. So, the mechanisms responsible for the migration of sequences to the B must include a mechanism capable of moving and incorporating sequence blocks greater than 100 kb. While not common, transposable elements are known to move such large regions [51]. These large regions are not restricted to the distal chromosome arms, as would be expected if translocations were responsible. Furthermore, the core blocks appear to be evenly distributed across the LGs, suggesting that location along the A chromosome does not impact likelihood of migration to the B. Of course, if multiple mechanisms are involved in the acquisition of A sequence, the combination of blocks acquired via these multiple methods might obscure actual patterns in the block location data.

The most extreme divergence in SCR of core blocks between the two individuals is found on LG8 (Figure 4). The SCR of this core block is 17.5 in the *M. lombardoi* B female and 234 in the *L. trewavasae* B female. This illustrates that copy number can vary greatly and is not an indication of how long a sequence has been on the B chromosome. We suggest caution in making interpretations about the origins of the B chromosome from observations of the length and position of B blocks or their copy number on the B chromosome. In these cichlids, the longest regions of homology are dispersed over too many chromosomes to suggest they were all involved in the production of the proto-B. Similarly, regions with some of the highest SCR, therefore contributing a significant amount of sequence to the B chromosome, are regions with relatively low SCR in other species. Moreover, the rate of Malawi cichlid B sequence replacement suggests that any B chromosome more than a few million years old may have replaced the original sequence of the proto-B to the point that none remain, making assignment of origin impossible. We suggest that efforts to identify the origin of B chromosomes focus on very young B chromosomes, and then use a combination of basic sequence homology, as performed here, as well as approaches that study chromosomal rearrangements and/or centromere evolution.

The number of genes and gene fragments overlapping with B blocks ranged from 516 to 2030 among datasets. Only 132 were common to at least 12 of the 13 datasets. When comparing individuals of the same population (Figure 5), the majority of genes identified were shared. However, several hundred genes were still unique to one or two of the individuals. We believe this is the result of the higher amount of type 1 error in the unique, unshared B blocks. Again, the core blocks provide us with the most conservative estimate of gene number. Furthermore, the comparison between Illumina and PacBio datasets revealed that some blocks, while representing B sequence, are erroneously positioned in the A reference where a recent insertion of that repeat occurred. If such an insertion were to occur in the intron of a gene, our analysis would incorrectly identify that gene as being partially on the B chromosome, leading to an overestimate of gene number. Nevertheless, if the B chromosomes of these different species use the same gene(s) to achieve drive, it is reasonable to believe that gene might be found among the 132 genes common across species.

Our analysis has identified both genes and gene fragments indiscriminately. The question remains whether these genes are functional or merely pseudogenes. While it may be tempting to label the gene fragments as pseudogenes, we do not know the structure of these sequences on the B chromosome. These gene fragments may be part of a gene fusion on the B, active but with an altered function. Moreover, transcription of altered (truncated, or partially deleted) copies of these genes could function by interfering with the activity of the original gene. Further examination of the genes on B chromosomes is needed before any conclusion regarding the functionality, or lack thereof, of B-located genes. A study of B sequence function will also serve to indicate which genes among these 132 could control B chromosome behavior, namely, drive and female sex bias. A more complete understanding of the structure of the B, rather than a series of fragmented blocks, would further this goal. Future studies might benefit from using PacBio or other long read sequencing methodologies better able to assemble the repetitive sequence of the B chromosome.

5. Conclusions

Using a coverage ratio analysis, we were able to identify the sequence of a significant portion of the B chromosome in several cichlid species from Lake Malawi. An evaluation of this approach, including the comparison of sequence data types, has provided crucial insight for the future application of coverage ratio analysis to study B chromosomes in other taxa. The mapping of B blocks to their A chromosome homologs provides further support for the theory that B chromosomes collect sequences from the A genome. Both the rate of turnover and pattern of B blocks across the A genome provide important caveats to efforts to characterize the origin of B chromosomes. Finally, we identify a list of candidate genes and gene fragments located on the B chromosome that may include the gene(s) responsible for drive and female sex bias in Lake Malawi cichlids.

Supplementary Materials: The following are available online at http://www.mdpi.com/2073-4425/9/12/610/s1.

Author Contributions: DNA samples were collected, extracted, and libraries were made by all authors. PacBio alignment was performed by M.A.C. Block identification scripts were written by F.E.C., with oversight from M.A.C. and T.D.K. All authors contributed to writing of the manuscript.

Funding: This research was funded by National Science Foundation, grant number DEB-1143920.

Acknowledgments: We would like to thank the National Science Foundation for funding this study.

Conflicts of Interest: The authors declare no conflict of interest. The funders had no role in the design of the study; in the collection, analyses, or interpretation of data; in the writing of the manuscript, or in the decision to publish the results.

References

1. D'Ambrosio, U.; Alonso-Lifante, M.P.; Barros, K.; Kovarik, A.; Mas de Xaxars, G.; Garcia, S. B-chrom: A database on B-chromosomes of plants, animals and fungi. *New Phytol.* **2017**, *216*, 635–642. [CrossRef] [PubMed]
2. Wilson, E.B. The supernumerary chromosomes of Hemiptera. *Science* **1907**, *26*, 870–871.
3. Jones, R.N.; Rees, H. *B Chromosomes*; Academic Press: New York, NY, USA, 1982.
4. Jones, R.N. B-chromosome drive. *Am. Nat.* **1991**, *137*, 430–442. [CrossRef]
5. Camacho, J.P.; Sharbel, T.F.; Beukeboom, L.W. B-chromosome evolution. *Philos. Trans. R. Soc. B Biol. Sci.* **2000**, *355*, 163–178. [CrossRef]
6. Randolph, L.F. Genetic characteristics of the B chromosomes in maize. *Genetics* **1941**, *26*, 608–631. [PubMed]
7. Burt, A.; Trivers, R. *Genes in Conflict: The Biology of Selfish Genetic Elements*; Belknap Press: Cambridge, UK, 2008; pp. 325–380.
8. Uhl, C.H.; Moran, R. The chromosomes of *Pachyphytum* (Crassulaceae). *Am. J. Bot.* **1973**, *60*, 648–656. [CrossRef]
9. Houben, A. B chromosomes—A matter of chromosome drive. *Front. Plant. Sci.* **2017**, *8*. [CrossRef] [PubMed]
10. Cheng, Y.M.; Lin, B.Y. Cloning and characterization of maize B chromosome sequences derived from microdissection. *Genetics* **2003**, *164*, 299–310. [PubMed]
11. Bugrov, A.G.; Karamysheva, T.V.; Perepelov, E.A.; Elisaphenko, E.A.; Rubtsov, D.N.; Warchalowska-Sliwa, E.; Tatsuta, H.; Rubtsov, N.B. DNA content of the B chromosomes in grasshopper *Podisma kanoi* Storozh. (Orthoptera, Acrididae). *Chromosome Res.* **2007**, *15*, 315–326. [CrossRef] [PubMed]
12. Ruiz-Ruano, F.J.; Cabrero, J.; Lopez-Leon, M.D.; Sanchez, A.; Camacho, J.P.M. Quantitative sequence characterization for repetitive DNA content in the supernumerary chromosome of the migratory locust. *Chromosoma* **2018**, *127*, 45–57. [CrossRef] [PubMed]
13. Jones, N.; Houben, A. B chromosomes in plants: Escapees from the A chromosome genome? *Trends Plant Sci.* **2003**, *8*, 1360–1385. [CrossRef]

14. Valente, G.T.; Conte, M.A.; Fantinatti, B.E.A.; Cabral-de-Mello, D.C.; Carvalho, R.F.; Vicari, M.R.; Kocher, T.D.; Martins, C. Origin and evolution of B chromosomes in the cichlid fish *Astatotilapia latifasciata* based on integrated genomic analyses. *Mol. Biol. Evol.* **2014**, *31*, 2061–2072. [CrossRef] [PubMed]

15. Martis, M.M.; Klemme, S.; Banaei-Moghaddam, A.M.; Blattner, F.R.; Macas, J.; Schmutzer, T.; Scholz, U.; Gundlach, H.; Wicker, T.; Simkova, H.; et al. Selfish supernumerary chromosome reveals its origin as a mosaic of host genome and organellar sequences. *Proc. Natl. Acad. Sci. USA* **2012**, *109*, 13343–13346. [CrossRef]

16. Houben, A.; Banaei-Moghaddam, A.M.; Klemme, S.; Timmis, J.N. Evolution and biology of supernumerary B chromosomes. *Cell. Mol. Life Sci.* **2014**, *71*, 467–478. [CrossRef]

17. Klemme, S.; Banaei-Moghaddam, A.M.; Macas, J.; Wicker, T.; Novak, P.; Houben, A. High-copy sequences reveal distinct evolution of the rye B chromosome. *New Phytol.* **2013**, *199*, 550–558. [CrossRef] [PubMed]

18. Ruban, A.; Schmutzer, T.; Scholz, U.; Houben, A. How next-generation sequencing has aided our understanding of the sequence composition and origin of B chromosomes. *Genes* **2017**, *8*, 294. [CrossRef] [PubMed]

19. Makunin, A.I.; Dementyeva, P.V.; Graphodatsky, A.S.; Volobouev, V.T.; Kukekova, A.V.; Trifonov, V.A. Genes on B chromosomes of vertebrates. *Mol. Cytogenet.* **2014**, *7*. [CrossRef]

20. Banaei-Moghaddam, A.M.; Martis, M.M.; Macas, J.; Gundlach, H.; Himmelbach, A.; Altschmied, L.; Mayer, K.F.X.; Houben, A. Genes on B chromosomes: Old questions revisited with new tools. *Biochim. Biophys. Acta* **2015**, *1849*, 64–70. [CrossRef] [PubMed]

21. Makunin, A.I.; Kichigin, I.G.; Larkin, D.M.; O'Brien, P.C.M.; Ferguson-Smith, M.A.; Yang, F.; Proskuryakova, A.A.; Vorobieva, N.V.; Chernyaeva, E.N.; O'Brien, S.J.; et al. Contrasting origin of B chromosomes in two cervids (Siberian roe deer and grey brocket deer) unraveled by chromosome-specific DNA sequencing. *BMC Genom.* **2016**, *17*. [CrossRef]

22. Jones, R. New species with B chromosomes discovered since 1980. *Nucleus* **2017**, *60*, 263–281. [CrossRef]

23. Zhou, Q.; Zhu, H.; Huang, Q.; Zhao, L.; Zhang, G.; Roy, S.W.; Vicoso, B.; Xuan, Z.; Ruan, J.; Zhang, Y.; et al. Deciphering new-sex and B chromosome evolution by the draft genome of *Drosophila albomicans*. *BMC Genom.* **2012**, *13*. [CrossRef] [PubMed]

24. Gonzalez-Sanchez, M.; Chiavarino, M.; Jimenez, G.; Manzanero, S.; Rosato, M.; Puertas, M.J. The parasitic effects of rye B chromosomes might be beneficial in the long term. *Cytogenet. Genome Res.* **2004**, *106*, 386–393. [CrossRef] [PubMed]

25. Jones, R.N.; Gonzalez-Sanchez, M.; Gonzalez-Garcia, M.; Vega, J.M.; Puertas, M.J. Chromosomes with a life of their own. *Cytogenet. Genome Res.* **2008**, *120*, 265–280. [CrossRef]

26. Graphodatsky, A.S.; Kukekova, A.V.; Yudkin, D.V.; Trifonov, V.A.; Vorobieva, N.V.; Beklemisheva, V.R.; Perelman, P.L.; Graphodatskaya, D.A.; Trut, L.N.; Yang, F.; et al. The proto-oncogene C-KIT maps to canid B-chromosomes. *Chromosome Res.* **2005**, *13*, 113–122. [CrossRef] [PubMed]

27. Poletto, A.B.; Ferreira, I.A.; Martins, C. The B chromosomes of the African cichlid fish *Haplochromis obliquidens* harbor 18S rRNA gene copies. *BMC Genet.* **2010**, *11*. [CrossRef] [PubMed]

28. Navarro-Dominguez, B.; Ruiz-Ruano, F.J.; Cabrero, J.; Corral, J.M.; Lopez-Leon, M.D.; Sharbel, T.F.; Camacho, J.P.M. Protein-coding genes in B chromosomes of the grasshopper *Eyprepocnemis plorans*. *Sci. Rep.* **2017**, *7*. [CrossRef]

29. Makunin, A.I.; Rajicic, M.; Karamysheva, T.V.; Romanenko, S.A.; Druzhkova, A.S.; Blagojevic, J.; Vujosevic, M.; Rubtsov, N.B.; Graphodatsky, A.S.; Trifonov, V.A. Low-pass single-chromosome sequencing of human small supernumerary marker chromosomes (sSMCs) and *Apodemus* B chromosomes. *Chromosoma* **2018**, *127*, 301–311. [CrossRef]

30. Leach, C.R.; Houben, A.; Field, B.; Pistrick, K.; Demidov, D.; Timmis, J.N. Molecular evidence for transcription of genes on a B chromosome in *Crepis capillaris*. *Genetics* **2005**, *171*, 269–278. [CrossRef]

31. Ruiz-Estevez, M.; Lopez-Leon, M.D.; Cabrero, J.; Camach, J.P.M. B-chromosome ribosomal DNA is functional in the grasshopper *Eyprepocnemis plorans*. *PLoS ONE* **2012**, *7*, e36600. [CrossRef]

32. Ruiz-Estevez, M.; Badisco, L.; Broeck, J.V.; Perfectti, F.; Lopez-Leon, M.D.; Cabrero, J.; Camacho, J.P.M. B chromosomes showing active ribosomal RNA genes contribute insignificant amounts of rRNA in the grasshopper *Eyprepocnemis plorans*. *Mol. Genet. Genom.* **2014**. [CrossRef]

33. Navarro-Dominguez, B.; Ruiz-Ruano, F.J.; Camacho, J.P.M.; Cabrero, J.; Lopez-Leon, M.D. Transcription of a B chromosome CAP-G pseudogene does not influence normal Condensin Complex genes in a grasshopper. *Sci. Rep.* **2017**, *7*, 17650. [CrossRef] [PubMed]

34. Banaei-Moghaddam, A.M.; Meier, K.; Karimi-Ashtiyani, R.; Houben, A. Formation and expression of pseudogenes on B chromosome of rye. *Plant Cell* **2013**, *25*, 2536–2544. [CrossRef] [PubMed]

35. Huang, W.; Zhao, Y.D.X.; Jin, W. B chromosome contains active genes and impacts the transcription of A chromosomes in maize (*Zea mays* L.). *BMC Plant Biol.* **2016**, *16*, 88. [CrossRef]

36. Lin, H.Z.; Lin, W.D.; Lin, C.Y.; Peng, S.F.; Cheng, Y.M. Characterization of maize B-chromosome-related transcripts isolated via cDNA-AFLP. *Chromosoma* **2014**, *123*, 597–607. [CrossRef] [PubMed]

37. Treangen, T.J.; Salzberg, S.L. Repetitive DNA and next-generation sequencing: Computational challenges and solutions. *Nat. Rev. Genet.* **2011**, *13*, 36–46. [CrossRef] [PubMed]

38. Feldberg, E.; Bertollo, L.A.C. Discordance in chromosome number among somatic and gonadal tissue cells of *Gymnogeophagus balzanii* (Pisces: Cichlidae). *Braz. J. Genet.* **1984**, *4*, 639–645.

39. Feldberg, E.; Porto, J.I.R.; Alves-Brinn, M.N.; Mendonca, M.N.C.; Benzaquem, D.C. B chromosomes in Amazonian cichlid species. *Cytogenet. Genome Res.* **2004**, *106*, 195–198. [CrossRef] [PubMed]

40. Pires, L.B.; Sampaio, T.R.; Dias, A.L. Mitotic and meiotic behavior of B chromosomes in *Crenicichla lepidota*: New report in the family Cichlidae. *J. Hered.* **2015**. [CrossRef]

41. Poletto, A.B.; Ferreira, I.A.; Cabral-de-Mello, D.C.; Nakajima, R.T.; Mazzuchelli, J.; Ribeiro, H.B.; Venere, P.C.; Nirchio, M.; Kocher, T.D.; Martins, C. Chromosome differentiation patterns during cichlid fish evolution. *BMC Genet.* **2010**, *11*, 50. [CrossRef] [PubMed]

42. Clark, F.E.; Conte, M.A.; Ferreira-Bravo, I.A.; Poletto, A.B.; Martins, C.; Kocher, T.D. Dynamic sequence evolution of a sex-associated B chromosome in Lake Malawi cichlid fish. *J. Hered.* **2017**, *108*, 53–62. [CrossRef]

43. Yoshida, K.; Terai, Y.; Mizoiri, S.; Aibara, M.; Nishihara, H.; Watanabe, M.; Kuroiwa, A.; Hirai, H.; Hirai, Y.; Matsuda, Y.; et al. B chromosomes have a functional effect on female sex determination in Lake Victoria cichlid fishes. *PLoS Genet.* **2011**, *7*, e1002203. [CrossRef] [PubMed]

44. Conte, M.A.; Kocher, T.D. An improved genome reference for the African cichlid, *Metriaclima zebra*. *BMC Genom.* **2015**, *16*, 724. [CrossRef] [PubMed]

45. Conte, M.A.; Joshi, R.; Moore, E.C.; Nandamuri, S.P.; Gammerdinger, W.J.; Roberts, R.B.; Carleton, K.L.; Lien, S.; Kocher, T.D. Chromosome-scale assemblies reveal the structural evolution of African cichlid genomes. *bioRxiv* **2017**. preprint. [CrossRef]

46. Li, H.; Durbin, R. Fast and accurate short read alignment with Burrows-Wheeler transform. *Bioinformatics* **2009**, *25*, 1754–1760. [CrossRef] [PubMed]

47. Sedlazeck, F.J.; Rescheneder, P.; Smolka, M.; Fang, H.; Nattestad, M.; Haeseler, A.V.; Schatz, M.C. Accurate detection of complex structural variations using single-molecule sequencing. *Nat. Methods* **2018**, *15*, 461–468. [CrossRef] [PubMed]

48. Quinlan, A.R.; Hall, I.M. BEDTools: A flexible suite of utilities for comparing genomic features. *Bioinformatics* **2010**, *26*, 841–842. [CrossRef]

49. Kocher, T.D. Adaptive evolution and explosive speciation: The cichlid fish model. *Nat. Rev. Genet.* **2004**, *5*, 288–298. [CrossRef]

50. Gel, B.; Serra, E. karyoploteR: An R/Bioconductor package to plot customizable genomes displaying arbitrary data. *Bioinformatics* **2017**, *33*, 3088–3090. [CrossRef]

51. Feschotte, C.; Pritham, E.J. DNA transposons and the evolution of eukaryotic genomes. *Annu. Rev. Genet.* **2007**, *41*, 331–368. [CrossRef]

© 2018 by the authors. Licensee MDPI, Basel, Switzerland. This article is an open access article distributed under the terms and conditions of the Creative Commons Attribution (CC BY) license (http://creativecommons.org/licenses/by/4.0/).

genes

MDPI

Article

Satellite DNAs Unveil Clues about the Ancestry and Composition of B Chromosomes in Three Grasshopper Species

Diogo Milani [1], Vanessa B. Bardella [1], Ana B. S. M. Ferretti [1], Octavio M. Palacios-Gimenez [1,2], Adriana de S. Melo [3], Rita C. Moura [3], Vilma Loreto [4], Hojun Song [5] and Diogo C. Cabral-de-Mello [1,*]

[1] Instituto de Biociências/IB, Departamento de Biologia, UNESP—Universidade Estadual Paulista, Rio Claro, São Paulo 01049-010, Brazil; azafta@gmail.com (D.M.); vbbardella@gmail.com (V.B.B.); anabeatrizferretti@gmail.com (A.B.S.M.F.); octavio.palacios@ebc.uu.se (O.M.P.-G.)

[2] Department of Evolutionary Biology, Evolutionary Biology Center, Uppsala University, 75236 Uppsala, Sweden

[3] Instituto de Ciências Biológicas, Laboratório de Biodiversidade e Genética de Insetos, UPE—Universidade de Pernambuco, Recife 50100-130, Pernambuco, Brazil; adrianadesouzamelo@gmail.com (A.d.S.M.); ritamoura.upe@gmail.com (R.C.M.)

[4] Centro de Biociências/CB, Departamento de Genética, UFPE—Universidade Federal de Pernambuco, Recife 50670-901, Pernambuco, Brazil; vloreto@bol.com.br

[5] Department of Entomology, Texas A&M University, 2475 TAMU, College Station, TX 77843-2475, USA; hsong@tamu.edu

* Correspondence: mellodc@rc.unesp.br

Received: 13 September 2018; Accepted: 21 October 2018; Published: 26 October 2018

Abstract: Supernumerary (B) chromosomes are dispensable genomic elements occurring frequently among grasshoppers. Most B chromosomes are enriched with repetitive DNAs, including satellite DNAs (satDNAs) that could be implicated in their evolution. Although studied in some species, the specific ancestry of B chromosomes is difficult to ascertain and it was determined in only a few examples. Here we used bioinformatics and cytogenetics to characterize the composition and putative ancestry of B chromosomes in three grasshopper species, *Rhammatocerus brasiliensis*, *Schistocerca rubiginosa*, and *Xyleus discoideus angulatus*. Using the RepeatExplorer pipeline we searched for the most abundant satDNAs in Illumina sequenced reads, and then we generated probes used in fluorescent in situ hybridization (FISH) to determine chromosomal position. We used this information to infer ancestry and the events that likely occurred at the origin of B chromosomes. We found twelve, nine, and eighteen satDNA families in the genomes of *R. brasiliensis*, *S. rubiginosa*, and *X. d. angulatus*, respectively. Some satDNAs revealed clustered organization on A and B chromosomes varying in number of sites and position along chromosomes. We did not find specific satDNA occurring in the B chromosome. The satDNAs shared among A and B chromosomes support the idea of putative intraspecific ancestry from small autosomes in the three species, i.e., pair S11 in *R. brasiliensis*, pair S9 in *S. rubiginosa*, and pair S10 in *X. d. angulatus*. The possibility of involvement of other chromosomal pairs in B chromosome origin is also hypothesized. Finally, we discussed particular aspects in composition, origin, and evolution of the B chromosome for each species.

Keywords: fluorescent in situ hybridization; Orthoptera; satellite DNA; supernumerary chromosome; RepeatExplorer

1. Introduction

Eukaryotic genomes exhibit repetitive DNA sequences including noncoding tandemly repeated satellite DNA (satDNA). These sequences exhibit extensive variability in copy number and nucleotide sequence, even among phylogenetically related species. Arrays of satDNAs are usually located in the centromeric and telomeric heterochromatin of the chromosomes, although they have also been reported in the euchromatic region. Furthermore, satDNAs are frequently enriched on sex chromosomes and supernumerary (B) chromosomes, as they are greatly enriched in heterochromatin [1–3].

Supernumerary B chromosomes occur in approximately 15% of eukaryotes as dispensable elements (i.e., not required for normal organismal development), frequently heterochromatic and enriched repetitive DNAs, including the satDNAs, which can have implications for B chromosome evolution. Generally, B chromosomes do not recombine with A chromosomes (normal complement), and B chromosome sequences evolve at a higher evolutionary rate than A elements [4,5]. Since the first discovery of the B chromosome [6], the specific ancestry of the studied B chromosomes in eukaryotes has remained largely unknown. For decades, repetitive DNAs have been used to try to ascertain the ancestry and to describe the B chromosome composition in some species. In that way, satDNAs have helped the understanding of the evolutionary history of B chromosomes with intraspecific (from host genome) or interspecific (resultant of species hybridization) origin, for example, in grasshoppers [7,8], wasps [9], fish [10], and plants [11], among others.

Among the grasshoppers, approximately 12% of the species harbor B chromosomes. Some families seem to be hotspots for B chromosome presence, such as Acrididae, with 17.1% of the species harboring B chromosomes, unlike Romaleidae in which only 4% of the species is harboring B chromosomes [12]. As generally observed in eukaryotes, some repetitive DNAs populate the B chromosomes of grasshoppers, like the multigene families for rDNAs [13,14], histone genes [14,15], and U snDNA [16], transposable elements [8,17,18], microsatellites [19], and satDNAs [7,8,13,20]. These sequences shed light on B chromosome composition, variability, and evolutionary dynamics. Concerning satDNAs, their presence in the B chromosomes of grasshoppers is known only in a few species, including *Locusta migratoria* [8], *Abracris flavolineata* [20], *Eyprepocnemis plorans* [13], and *Eumigus monticola* [7].

The search for satDNAs in genomic data was facilitated more recently by analyzing reads from next generation sequencing (NGS) using bioinformatics approaches, like RepeatExplorer software [21]. RepeatExplorer has been a useful tool for detecting satDNAs for probe generation and chromosome mapping in species with B chromosomes, helping to unveil the composition and abundance of satDNAs in those chromosomes as well as their relationships with A elements as well [7,8,11,22].

By combining genomics and cytogenetics, we aimed to elucidate the genome content of satDNAs and used this information to track the possible ancestry of B chromosomes in three grasshopper species, *Rhammatocerus brasiliensis* (Acrididae: Gomphocerinae), *Schistocerca rubiginosa* (Acrididae: Cyrtacanthacridinae), and *Xyleus discoideus angulatus* (Romaleidae: Romaleinae) belonging to two families. The family Acrididae, which is currently most diverse lineage within the orthopteran suborder Caelifera, diverged from its sister lineage, which includes the family Romaleidae, in the late Cretaceous (~78 mya, million years ago) based on a fossil-calibrated divergence time estimate. Two acridid subfamilies included in this study, Gomphocerinae and Cyrtacanthacridinae, each belong to different clades within the family, and they are estimated to have diverged in the late Eocene [23]. For this purpose, we first made a prediction of the most abundant satDNAs in genomes by using the RepeatExplorer tool. Then we recovered the fragments by PCR and designed probes of each satDNA of the three species to use in fluorescent in situ hybridization (FISH) experiments. This allows for the investigation of spatial patterns of satDNAs that have preferentially accumulated in B chromosomes compared to autosomes. We found distinct patterns of satDNA distribution on B chromosomes that are shared with some A chromosomes or with exclusive A chromosomes. Based on these data, it is possible to hypothesize the ancestry of the B chromosome from small autosomes and to discuss aspects of the evolution of B chromosomes in the three species.

2. Material and Methods

2.1. Animal Sampling, Chromosome Preparations, and Genomic DNA Sequencing

Adult animals of *R. brasiliensis* were collected at 07°45′00″ S; 34°05′10″ W Ilha de Itamaracá/PE (Brazil) and 07°50′56″ S 35°19′14″ W Lagoa do Carro/PE (Brazil); *X. d. angulatus* at 07°12′47″ S 39°18′55″ W Juazeiro do Norte/CE (Brazil); *S. rubiginosa* at 29°25.908′ N 82°24.060′ W Levy County/Florida (USA). Testes were fixed with Carnoy's modified solution (3:1, 100% Ethanol:Glacial Acetic Acid) and stored at −20 °C until use for slides preparation. Femurs were immersed in 100% ethanol and stored at −20 °C for genomic DNA (gDNA) extraction.

The genomic DNA sequencing method for the *S. rubiginosa* (male) specimen was previously described [24]. For the *R. brasiliensis* (female) and *X. d. angulatus* (male) specimens DNA extraction was performed using the phenol/chloroform-based procedure described previously [25]. Sequencing was conducted by the Illumina company (Inc., San Diego, CA, USA) with a HiSeq 4000 to obtain paired-ends libraries (2 × 101 bp) using the service of Macrogen Inc. (Seoul, Republic of Korea).

We applied conventional staining with 5% Giemsa for chromosome observation and identification of individuals harboring B chromosomes. The C-banding for heterochromatin identification was performed according to a previously described method [26].

2.2. SatDNAs Searching by Graph-Based Clustering Method

Prior to RepeatExplorer graph-based clustering analysis, we preprocessed and checked the quality of the paired-ends reads of each species using FastQC [27]. Preprocessing of the reads was performed following default parameters using the public online platform: https://repeatexplorer-elixir.cerit-sc.cz/galaxy/. Reads were processed with a "quality trimming tool", "FASTQ interlacer" on paired end reads, "FASTQ to FASTA" converter, and "RepeatExplorer clustering" all with default recommended options [21]. First we searched and selected, by visual observation, the clusters that showed high graph density that indicated proximity with satDNAs families [28]. Contigs from the selected clusters were deeply explored and manually searched for sequences with tandem pattern confirmed by the dot plot graphics implemented using Geneious v4.8.5 software [29]. The consensus monomer of each satDNA family of each species was used as the query in two databanks, BLAST (http://www.ncbi.nlm.gov/Blast/) and Repbase (http://www.girinst.org/repbase/), to check similarity with another sequence deposited and described. Abundance of each satDNA family was calculated with the number of reads of each cluster divided by the total number of reads used in the "RepeatExplorer clustering" protocol [21]. Nucleotide divergence was calculated using the RepeatMasker package with specific parameters provided in the scripts program protocol to calculate Kimura divergence values [30]. Superfamilies (SF) were considered by comparing consensus monomer of each satDNA against all of them from each species independently using the Geneious v4.8.5 software [29] assembly tool, alternating overlap identity following the same considerations as a previous work [31]. We classified each identified satDNA family according to a previous method [31], considering the species name abbreviation and decreasing abundance, followed by the consensus monomer size; they were numbered in decreasing order of abundance. The sequences were deposited in GenBank under the accession numbers MH900339–MH900377.

2.3. Amplification of SatDNAs through PCR, Probes and Fluorescence In Situ Hybridization

We used the consensus sequences of each satDNA family of each species to design divergent primers manually or using the Primer3 tool [32] implemented in Geneious v4.8.5 software [29] (Supplementary Table S1). Polymerase chain reactions (PCRs) were performed using 10× PCR Rxn Buffer, 0.2 mM MgCl$_2$, 0.16 mM dNTPs, 2 mM of each primer, 1 U of *Taq* Platinum DNA Polymerase (Invitrogen, San Diego, CA, USA), and 50–100 ng/μL of template DNA. The PCR conditions included an initial denaturation at 94 °C for 5 min and 30 cycles at 94 °C (30 s), 55 °C (30 s), and 72 °C (80 s), plus a final extension at 72 °C for 5 min. The PCR products were visualized on a 1% electrophoresis agarose gel. The monomeric bands were isolated and purified using the Zymoclean™ Gel DNA Recovery Kit (Zymo Research Corp., The Epigenetics Company, CA, USA) according to the manufacturer's recommendations and then used as template for reamplification using the same PCR conditions. The monomers were sequenced by the Sanger method using the service of Macrogen Inc. to confirm the amplification of desired sequence.

FISH was performed in meiotic chromosomes using one or two probes according to a method described previously [33] with some adjustments as outlined previously [34]. The probes labeled with digoxigenin-11-dUTP were detected using anti-digoxigenin rhodamine (Roche, Mannheim, Germany), and probes labeled with biotin-14-dATP were detected using Streptavidin Alexa Fluor 488-conjugated (Invitrogen, San Diego, CA, USA). The preparations were counterstained using 4′,6-diamidine-2′-phenylindole (DAPI) and mounted in VECTASHIELD (Vector, Burlingame, CA, USA). FISH results were observed using an Olympus microscope BX61 (Tokyo, Japan) equipped with a fluorescent lamp and the proper filters. Images were obtained using a DP71 cooled digital camera in grayscale and then pseudo-colored in blue for chromosomes and red or green for hybridization signals, merged and optimized for brightness and contrast using Adobe Photoshop CS6. To describe the patterns of satDNA chromosomal distribution distinct cells were analyzed, including diplotene, metaphase I, metaphase II, and mitotic metaphase.

3. Results

3.1. Karyotypes, B Chromosomes, and Heterochromatin Distribution

Occurrence of a karyotype consisting of 2n = 23,X0, and presence of B chromosomes observed here for *R. brasiliensis* and *X. d. angulatus* were previously reported by different authors [14,35], including in the same population, i.e., Juazeiro do Norte/CE for *X. d. angulatus* [36]. We report for the first time the presence of B chromosomes in *R. brasiliensis* from Lagoa do Carro/PE. The karyotype of *S. rubiginosa*, described here for the first time, is also 2n = 23,X0 as observed for other species from the same genus, like *S. gregaria* [37], *S. pallens*, and *S. flavofasciata* [38]. Among the five individuals of *S. rubiginosa*, two presented B chromosomes. We classified autosomal chromosomes of the three species in three distinct groups considering size: three long chromosomes (L1–L3), five medium (M4–M8), and three small (S9–S11).

The B chromosomes of the three species are acrocentric with variable pattern of heterochromatin distribution (Figure 1). For *R. brasiliensis* pericentromeric and distal blocks were observed (Figure 1a) and for *S. rubiginona* pericentromeric and interstitial blocks, close to the centromere, were noticed (Figure 1b). In *X. d. angulatus* the B chromosome was completely heterochromatic with deeper staining in the pericentromeric region (Figure 1c). Heterochromatin blocks restricted to pericentromeric areas were noticed for A chromosomes (Figure 1).

3.2. In Silico SatDNA Analysis

By using RepeatExplorer we predicted the most abundant satDNAs as follows, twelve, nine, and eighteen satDNA families in *R. brasiliensis*, *S. rubiginosa*, and *X. d. angulatus*, respectively. Monomer lengths varied from 36 to 410 nt in *R. brasiliensis*, from 107 to 441 nt in *S. rubiginosa*, and from 8 to 289 in *X. d. angulatus*. The predominance of families with monomer length higher than 100 nt was noticeable. Only for *X. d. angulatus*, satDNA families with monomer length smaller than 50 nt was observed (Table 1). Sequence similarity analysis revealed the presence of two similar satDNAs families in the genome of *R. brasiliensis*, RbrSat01-171 and RbrSat04-168 (superfamily SF1), In *X. d. angulatus* two superfamilies were noticed each composed by two satDNA families, SF1 (XanSat05-267 and XanSat07-279) and SF2 (XanSat09-130 and XanSat14-128). No similarity between satDNAs was noticed between species.

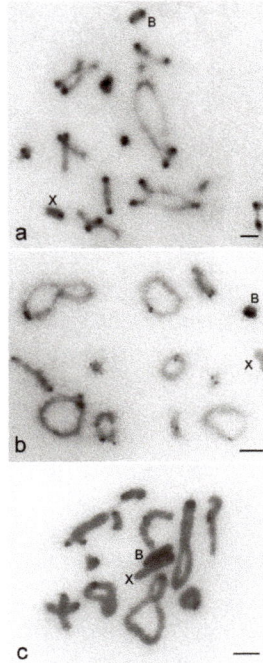

Figure 1. C-banding revealing the heterochromatin location in diplotene chromosomes of *Rhammatocerus brasiliensis* (**a**), *Schistocerca rubiginosa* (**b**), and *Xyleus discoideus angulatus* (**c**). Note the pericentromeric location of C-positive blocks on A chromosomes and the distinct patterns for the B chromosome of the three species, i.e., pericentromeric and distal in *R. brasiliensis*; pericentromeric and interstitial (close to the centromere) in *S. rubiginosa*; along the entire chromosome with darker band in pericentromeric region in *X. d. angulatus*. The X and B chromosomes are indicated. Bar = 5 μm.

A+T content was variable from 44.5% to 63.9% (mean 57.63%) in *R. brasiliensis*, from 48.9% to 61.7% (mean 55.12%) in *S. rubiginosa*, and from 28.6% to 76.2% in *X. d. angulatus* (mean 59.15%). Predominance of A+T-rich satDNA families was observed, ten in *R. brasiliensis*, eight in *S. rubiginosa*, and fifteen in *X. d. andgulatus*. Concerning total abundance, the satDNAs represented 1.499% of the genome of *R. brasiliensis*, 2.172% of *S. rubiginosa*, and 2.322% of *X. d. angulatus* genomes. In all species, even the most abundant satDNA family represented less than 1% of the genome, i.e., 0.766% in *R. brasiliensis*, 0.73% in *S. rubiginosa*, and 0.627% in *X. d. angulatus*. The lowest abundance satDNA in *R. brasiliensis* corresponded to 0.01% of the genome, in *S. rubiginosa* to 0.026%, and in *X. d. angulatus* to 0.013% (Table 1).

3.3. Chromosomal Location of SatDNAs

All satDNA families recognized by RepeatExplorer analysis were accurately amplified by PCR and sequenced; FISH mapping showed signals for most of them (Table 2; Figures 2–4). Six satDNA families revealed signals for *R. brasiliensis* (Figure 2) and *S. rubiginosa* (Figure 3), and for *X. d. angulatus* eleven satDNA families allowed identification of specific marks by FISH (Figure 4), representing clustered satDNAs. For the remaining satDNA families, six for *R. brasiliensis*, three for *S. rubiginosa*, and seven for *X. d. angulatus*, were nonclustered with no FISH signals (Table 2).

Table 1. Characteristics of satellite DNA (satDNA) families isolated from the genomes of three grasshopper species, including their monomer sizes, base pair richness, and genome abundances.

Species	SatDNA Superfamily	SatDNA Family	Monomer Size (nt)	A+T (%)	Abundance (%)	Divergence (%)
R. brasiliensis	SF1	RbrSat01-171	171	59.3	0.766	4.98
	-	RbrSat02-410	410	47.9	0.224	8.94
	-	RbrSat03-36	36	44.5	0.126	8.45
	SF1	RbrSat04-168	168	59.4	0.105	18.14
	-	RbrSat05-179	179	59.1	0.061	4.96
	-	RbrSat06-165	165	59.2	0.056	1.23
	-	RbrSat07-240	240	60.4	0.047	11.56
	-	RbrSat08-176	176	58.0	0.042	6.27
	-	RbrSat09-238	238	63.9	0.025	8.75
	-	RbrSat10-268	268	62.7	0.021	7.79
	-	RbrSat11-233	233	58.0	0.016	7.85
	-	RbrSat12-180	180	57.8	0.010	2.06
Total					1.499	
S. rubiginosa	-	SruSat01-194	194	58.3	0.730	4.30
	-	SruSat02-170	170	54.4	0.476	4.64
	-	SruSat03-170	170	59.9	0.287	7.79
	-	SruSat04-301	301	48.9	0.244	4.77
	-	SruSat05-441	441	50.6	0.135	23.18
	-	SruSat06-363	363	57.1	0.126	9.61
	-	SruSat07-232	232	61.2	0.116	9.23
	-	SruSat08-172	172	58.2	0.032	16.24
	-	SruSat09-107	107	61.7	0.026	10.85
Total					2.172	
X. d. angulatus	-	XanSat01-8	8	62.5	0.627	4.62
	-	XanSat02-21	21	28.6	0.586	4.41
	-	XanSat03-10	10	60.0	0.464	9.38
	-	XanSat04-10	10	60.0	0.228	5.22
	SF1	XanSat05-267	267	56.7	0.087	5.50
	-	XanSat06-168	168	64.3	0.069	4.53
	SF1	XanSat07-279	279	60.2	0.053	11.08
	-	XanSat08-16	16	56.2	0.033	4.38
	SF2	XanSat09-130	130	63.1	0.024	10.61
	-	XanSat10-289	289	60.2	0.022	7.27
	-	XanSat11-51	51	47.1	0.019	3.97
	-	XanSat12-246	246	59.2	0.018	11.81
	-	XanSat13-281	281	56.3	0.018	4.61
	SF2	XanSat14-128	128	62.5	0.017	14.90
	-	XanSat15-228	228	59.7	0.017	5.18
	-	XanSat16-21	21	42.9	0.014	9.79
	-	XanSat17-15	15	53.4	0.013	11.56
	-	XanSat18-21	21	76.2	0.013	6.33
Total					2.322	

The distinct clustered satDNA families were variable in number and position within A chromosomes (Figures 2–4). Only one satDNA was located exclusively on pericentromeric regions of A chromosomes in each species, RbrSat03-36 in *R. brasiliensis* (Figure 2b), SruSat02-170 in *S. rubiginosa* (Figure 3b), and XanSat03-10 in *X. d. angulatus* (Figure 4c). Heterochromatin blocks, like centromeres, were enriched in most satDNAs, but we also noticed a few satDNAs placed on the euchromatin of some chromosomes of *R. brasiliensis*: RbrSat08-176 (Figure 2e) and RbrSat09-238 (Figure 2f), and *S. rubiginosa*: SruSat03-170 (Figure 3c), SruSat06-363 (Figure 3d), SruSat07-232 (Figure 3b), and SruSat08-172 (Figure 3e). We observed that satDNA was more frequently distributed within the interstitial and distal euchromatin of *X. d. angulatus* (Figure 4a,c–g,i). The bias for pericentromeric position of satDNA was noticed by comparing the number of pericentromeric blocks with interstitial and distal ones for each species: *R. brasiliensis* twenty-eight pericentromeric and three interstitial; *S. rubiginosa* twenty-five pericentromeric, four interstitial, and five distal; *X. d. angulatus* forty-two pericentromeric, ten interstitial, and twelve distal (Table 2).

Table 2. Chromosome location of satDNAs in three grasshopper species. For each species at the bottom is indicated the number of satDNA families per chromosome and the amount of satDNAs shared with the B chromosome. p: pericentromeric, i: interstitial, d: distal, nc: nonclustered.

Species	SatDNA Family	Chromosome Location												
		1	2	3	4	5	6	7	8	9	10	11	X	B
Rhammatocerus brasiliensis	RbrSat01-171	p	p				p	p	p	p		p		p
	RbrSat02-410							nc						
	RbrSat03-36	p	p	p	p	p	p	p	p	p	p	p	p	p
	RbrSat04-168											p		p
	RbrSat05-179							p		p				
	RbrSat06-165							nc						
	RbrSat07-240							nc						
	RbrSat08-176			p	p		p,i	p			p	p	i	p
	RbrSat09-238									i				
	RbrSat10-268							nc						
	RbrSat11-233							nc						
	RbrSat12-180							nc						
Total		2	2	2	2	2	2	4	2	4	2	4	2	4
shared with B		2	2	2	2	2	2	3	2	2	2	4	2	
Schistocerca rubiginosa	SruSat01-194						d							i
	SruSat02-170	p	p	p	p	p	p	p	p	p	p	p	p	i
	SruSat03-170	p,d		p,d		p			p	p	p	p		i
	SruSat04-301							nc						
	SruSat05-441							nc						
	SruSat06-363									i,d				2i
	SruSat07-232									i,d				i
	SruSat08-172			p	i	p	p			p	p,i	p		p
	SruSat09-207							nc						
Total		2	1	3	2	3	3	1	2	5	3	3	1	6
shared with B		2	1	3	2	3	3	1	2	5	3	3	1	
Xylleus discoideus angulatus	XanSat01-8	i	i	d	i	d	i	d	d	d	d	d	2i	
	XanSat02-21	p	p	p	p	p	p	p	p	p	p,d	p	p	
	XanSat03-10	p	p	p	p	p	p	p	p	p	p	p	p	p,i,d
	XanSat04-10			p			p	p						
	XanSat05-267							p			p,i			2i
	XanSat06-168					p	i							
	XanSat07-279	i	d					p						
	XanSat08-16							d						
	XanSat09-130							nc						
	XanSat10-289							nc						
	XanSat11-51		d											
	XanSat12-246	p	p	p	p	p	p	p		p	p	p		
	XanSat13-281							p		d	i			i,d
	XanSat14-128							nc						
	XanSat15-228							nc						
	XanSat16-21							nc						
	XanSat17-15							nc						
	XanSat18-21							nc						
Total		5	6	5	4	5	9	6	3	5	6	4	3	3
shared with B		1	1	1	1	1	2	2	1	2	3	1	1	

SatDNA unique to a specific A chromosome was a rare condition in the three species. It was noticed in *R. brasiliensis* for RbrSat04-168 (Figure 2c) and RbrSat09-238 (Figure 2f) in pair S11 and S9, respectively; SruSat01-194 (Figure 3a) in pair 6, and SruSat06-363 (Figure 3d) and SruSat07-232 (Figure 3b) in pair S9 of *S. rubiginosa*; in *X. d. angulatus* the repeats XanSat08-16 (Figure 4f) and XanSat11-51 (Figure 4g) were exclusive from the pairs M6 and L2, respectively. The number of satDNAs per specific A chromosome varied from 2 to 4 (mean 2.5) in *R. brasiliensis*, from 1 to 5 (mean 2.42) in *S. rubiginosa*, and from 3 to 9 (mean 5.08) in *X. d. angulatus*.

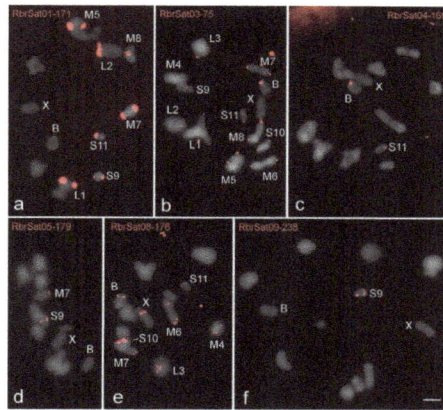

Figure 2. Fluorescent in situ hybridization (FISH) mapping on metaphase I of satDNAs identified in the genome of *R. brasiliensis*. The distinct satDNAs families are indicated. Chromosomes with signals, X and B chromosomes are identified. Note the presence on B chromosome of signals for (**a**) RbrSat01-171 (shared with some A chromosomes), (**b**) RbrSat03-36 (shared with all A chromosomes), (**c**) RbrSat04-168 (shared exclusively with pair S11), and (**e**) RbrSat08-176 (shared with some A chromosomes). For satDNAs (**d**) RbrSat05-179 and (**f**) RbrSat09-238 no signals were observed in B chromosome. Bar = 5 μm.

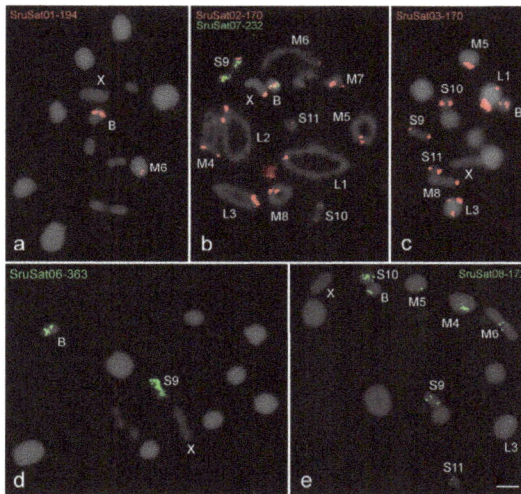

Figure 3. Chromosomal distribution of satDNAs in meiotic cells of *S. rubiginosa* (**a,c,d,e**) metaphase I and (**b**) diplotene. The distinct satDNAs families are indicated. Chromosomes with signals: X and B chromosomes are identified. Observe that the B chromosome harbors all satDNAs families, three of them shared with some A chromosomes (SruSat02-170, SruSat03-170, and SruSat08-172), and three exclusively shared with pair M6 (SruSat01-194) or pair S9 (SruSat06-363 and SruSat07-232). Bar = 5 μm.

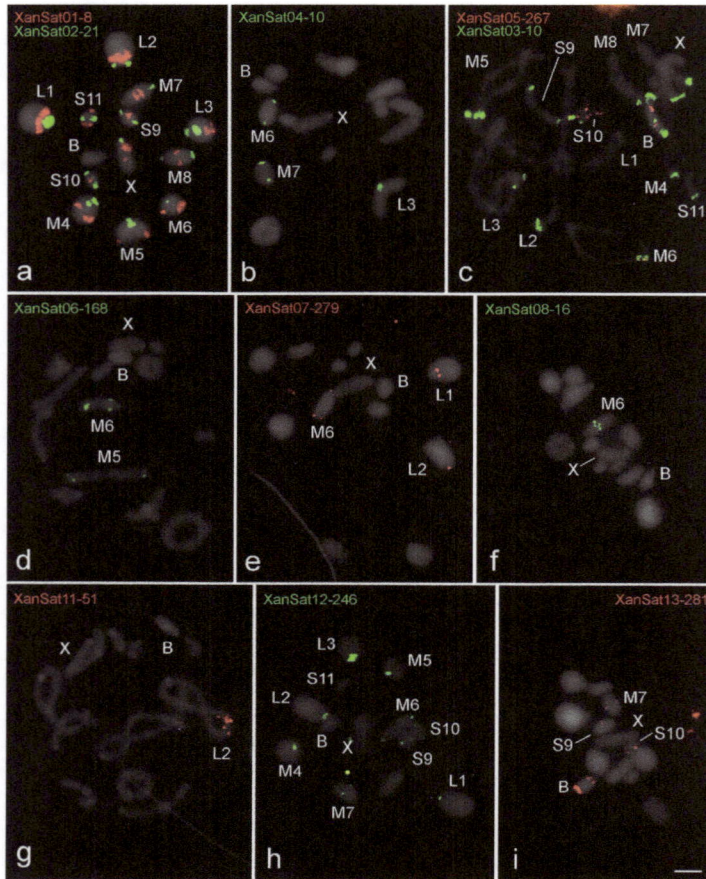

Figure 4. Patterns of chromosomal location revealed by FISH of eleven clustered satDNAs in *X. d. angulatus*. (**a,b,d–f,h,i**) metaphase I, (**c,g**) diplotene. The distinct satDNAs families are indicated. Chromosomes with signals: X and B chromosomes are identified. Note only three satDNA families on the B chromosome, XanSat03-10 (**c**), XanSat05-267 (**c**) and XanSat13-281 (**i**), all of which shared the pair S10. Multiply satDNA sites are observed for all satDNAs (**a–e,h,i**), except XanSat-08-16 (**f**) and XanSat11-51 (**g**). Bar = 5 µm.

The B chromosomes were enriched with distinct satDNA families. All of them were shared with the A chromosomes but show distinct patterns of distribution, such as occurrence in multiple chromosomes or occurrence restricted to one or few elements (Figures 2–5). Four satDNAs occupying pericentromeric regions were seen in the B chromosome of *R. brasiliensis*, RbrSat01-171 (Figure 2a), RbrSat03-36 (Figure 2b), RbrSat04-168 (Figure 2c), and RbrSat08-176 (Figure 2e). These satDNAs were shared with the chromosome S11, which accumulated them in the pericentromeric region. The satDNA RbrSat04-168 was exclusively shared between pair S11 and the B chromosome, while the others were also located in other chromosomes (Figure 2a–c,e and Figure 5a).

The B chromosome of *S. rubiginosa* harbored the six satDNAs that were found clustered in A chromosomes (Figures 3 and 5b). SruSat08-172 (Figure 3e) was located in the pericentromeric region, while the other satellites were interstitially located presenting differences in signal size (Figure 3a–d). The chromosome S9 harbored five of the six satDNAs present in the B chromosome, and among them two were exclusive of pair S9 and B chromosome, SruSat06-363 (Figure 3d) and SruSat07-232 (Figure 3b). This was the chromosome with highest number of satDNAs shared with the B chromosome. SruSat01-194 was also in the B chromosome but among the A chromosomes this repeat was only in pair 6 (Figure 3a).

Among the eleven repeats mapped by FISH in *X. d. angulatus* chromosomes only three were visualized in the B chromosome, XanSat03-10 (Figure 4c), XanSat05-267 (Figure 4c), and XanSat13-281 (Figure 4i). For these repeats more than one signal was seen in the B chromosome (Figure 4c,i and Figure 5c). Xansat03-10 was located in pericentromeric, interstitial, and distal regions (Figures 4c and 5c), XanSat05-267 presented two interstitial blocks (Figures 4c and 5c) and XanSat13-281 was placed in interstitial and distal areas (Figures 4i and 5c). We observed that none of the satDNAs located on the B chromosome were restricted to one A chromosome, XanSat03-10 (Figures 4c and 5c) was located in all pericentromeric regions, XanSat05-267 (Figures 4c and 5c) was located in pairs M6 and S10, and XanSat13-281 (Figures 4i and 5c) was located in pairs M7, S9, and S10. Chromosome S10 shared the highest amount of satDNAs with the B chromosome (Figure 4c,i and Figure 5c).

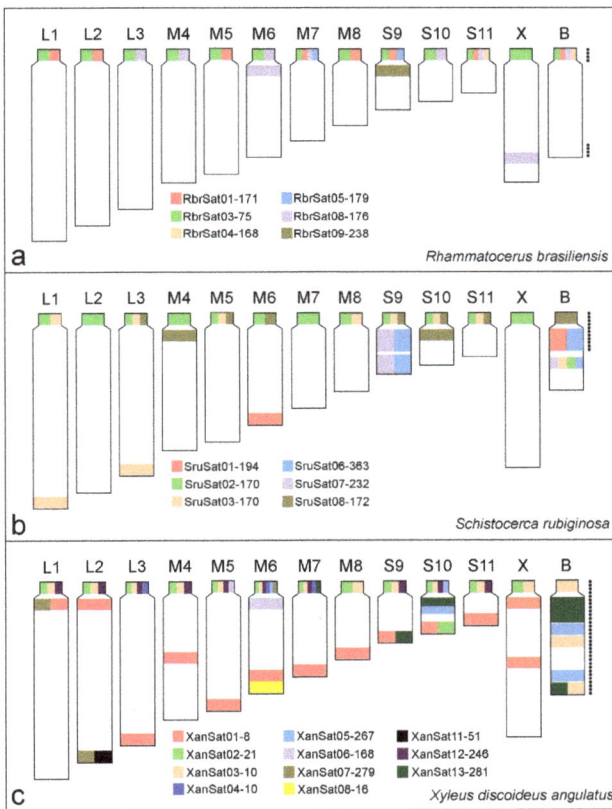

Figure 5. Ideograms summarizing the chromosomal location of clustered satDNAs in the three species of grasshoppers, (**a**) *R. brasiliensis*, (**b**) *S. rubiginosa*, and (**c**) *X. d. angulatus*. Each satDNA family is represented by one color. Black dots next to the B chromosomes indicate heterochromatin distribution in these chromosomes.

4. Discussion

Appling bioinformatic and molecular cytogenetic approaches to determine the content of satDNA allowed for a rapid increase in characterization of these kinds of elements among Orthoptera. High-throughput analysis has allowed the characterization of 234 satDNAs families in seven species [7,31,39–42]. Here we describe for first time the chromosomal organization of the most abundant satDNAs populating the genomes of three grasshopper species from the Romaleidae and Acrididae families. They correspond to a total 39 satDNAs families contributing to the knowledge of chromosomal organization of this kind of repeat on B chromosome.

SatDNAs Reveal Clues about B Chromosome Composition and Ancestry

Even though B chromosomes have been studied for a long time, we know very little about their ancestry in most species. B chromosome ancestry is known only in a few species [7], and therefore its origin still remains intriguing due in part to their high evolutionary rate. In grasshoppers, repetitive DNAs have been used to track B chromosome origin and composition in relatively few species. In *E. monticola* for instance, B chromosome ancestry is attributed to the autosomal pair S8 based on satDNAs analysis [7]. In *A. flavolineata*, the origin of B chromosome from pair 1 is attributed to the unique presence of U2 snDNA genes in these two chromosomes [16]. The ancestry of B chromosome in *L. migratoria* is related to pairs 8 and 9 due to the presence of satDNAs and histone genes in those chromosomes [8,15]. Previous works discussed the composition and putative origin of B chromosomes in two grasshoppers species studied here, *R. brasiliensis* and *X. d. angulatus*. However, it was not possible to elucidate a specific ancestral chromosome (see below). Based on satDNA content and organization we provide some clues about the ancestry of B chromosomes in these species, and additionally in *S. rubiginosa*.

The B chromosome of *R. brasiliensis* harbors four satDNAs, three of them are shared with multiple chromosomes (i.e., RbrSat01-171, RbrSat03-75, and RbrSat08-176) and, because of this, they are not good markers for ancestry determination. Most A chromosomes share two satDNAs with the B chromosome. The A chromosomes that share more satDNAs with the B chromosome are pairs M7 and S11, three and four, respectively, being good candidates to be involved in the origin of B chromosome. Three satDNAs (RbrSat01-171, RbrSat03-75, and RbrSat08-176) shared between pair M7, S11, and the B chromosome are also located on other A chromosomes. Furthermore, the pair S11 harbors the satDNA RbrSat04-168 that is an exclusive sequence shared with the B chromosome. The pericentromeric region of pair S11 fits exactly the composition of pericentromeric region of the B chromosome, supporting the hypothesis of its involvement in B chromosome origin.

Some controversial ideas were proposed for the origin of B in *R. brasiliensis*. First, the authors of a previous paper [35] proposed the origin from one or several chromosomes, including, for example, the pair S11 as supported here by satDNA mapping. Pairs L2, L3, M5, and S11 harbor 5S rDNA clusters that are shared with the B chromosome [35] and could be involved in its origin. Second, the analysis made by the authors of a previous paper [14] did not support the autosomal origin hypothesis, based on the presence of 5S rDNA and H3 gene clusters in the B chromosome that are shared with most A chromosomes (including the X chromosome), except the pair S11. However, the occurrence of these repetitive DNAs in the B chromosome could be more related to transposition events after its origin than ancestry [14]. These data, including the individuals from distinct populations, support the multiregional origin of the B chromosome in *R. brasiliensis* or dynamics for repetitive DNA organization for both A and B chromosomes, causing the emergence of new B chromosome variants. Based on a cytomolecular analysis, multiple B chromosome variants were described, for example, in rye, *Secale cereale* [43], and in the grasshoppers *X. d. angulatus* [36] and *E. plorans* [44].

Although our findings strongly support the ancestry of the B chromosome from the pair S11 in *R. brasiliensis* (at least in Lagoa do Carro/PE population), we should also point attention to the pair M7 that share three satDNAs with the B chromosome. In other populations (including Lagoa do Carro/PE) the chromosome M7 harbors H3 histone gene, which in some individuals is exclusively shared with the

B chromosome [14,45]. This suggests the involvement of M7, besides the pair S11, in B chromosome ancestry. It is similar to *L. migratoria* in which the B chromosome ancestry is putatively from two chromosomes, the pairs 8 and 9 [8], as in rye [46]. On the other hand, we should bear in mind that the pair M7 harbors one satDNA (RbrSat05-179) that is not observed in the B chromosome. Moreover, considering the high dynamism of the H3 histone gene (in number of clusters) in *R. brasiliensis*, it is possible that this gene was acquired later by the B chromosome. To shed light on this possibility, individuals from multiple populations should be studied using the distinct probes.

The satDNA mapping in *S. rubiginosa* suggests an autosomal origin for B chromosome from the pair S9. This chromosome shares five satDNAs with the B chromosome, two of them exclusive for this chromosome (SruSat06-363 and SruSat07-232), thus supporting the ancestry of the B chromosome from this bivalent. Furthermore, the pair S9 also harbors other three satDNAs present in the B chromosome, SruSat02-170, SruSat03-170, and SruSat08-172. Interestingly, those two exclusive satDNAs in S9 also are abundant in the B chromosome. This means that these repeats were massively amplified covering almost the entire length of those two chromosomes. The SruSat01-194 that is present in the pair M6 is also highly abundant in the B chromosome. We ruled out the possibility of B origin from M6 due to the absence of other three satDNAs that are present in the B chromosome, including those pericentromeric satDNA. Furthermore, if the pair M6 is involved in B chromosome origin it has a secondary contribution in comparison to the pair S9. It should be noted that SruSat01-194 corresponds to the most abundant satDNA in the *S. rubiginosa* genome, visible in the B chromosome as a large block likely due to amplification after its origin. It might be possible that the presence of this repeat in other A chromosomes (including pair S9), but arranged non-tandemly, makes it difficult to reach the FISH threshold resolution.

The satDNA content and distribution in the B chromosome of *X. d. angulatus* indicate a more complex evolution than in *R. brasiliensis* and *S. rubiginosa*, with additional chromosomal rearrangements after the B chromosome origin followed by accumulation/deletion involving repeats, as suggested by the previous analyses [35,36] (see below). Even though there are 11 satDNAs clustered on the A chromosomes of *X. d. angulatus*, only three of them are present in the B chromosome. There is no satDNA exclusively shared between the B chromosome and one chromosome of A complement. However, the pair S10 shares the most satDNAs with B chromosome (three satDNAs families), and it seems to be the ancestral pair involved in the B chromosome origin. The three satDNAs shared between B chromosome and the pair S10 are located at pericentromeric or interstitial regions (not far from the centromere), highlighting the origin of the B chromosome from about the half proximal part of the pair S10. Recently, it was suggested that pericentromeric and proximal regions enriched of repetitive DNAs were involved with the B chromosome origin in *X. d. angulatus*, followed by repetitive DNA amplification and rearrangements, like inversions [36].

Although the origin of the B chromosome in *X. d. angulatus* from the proximal part of the pair S10 is supported by current data, the presence of two other satDNAs in the pericentromeric region of pair S10 (XanSat02-21 and XanSat12-246), not shared with the B chromosome, is contrary to this hypothesis. The difference between the satDNA content in B and chromosome S10 can be explained by the changes of satDNAs amounts in the B chromosome during its evolution. In that way, the satDNA XanSat03-10 was amplified in the pericentromeric region of the B chromosome, while the other ones were completely deleted or conserved in small copy number, not detected by FISH. Interestingly, XanSat03-10 is a unique satDNA exclusively located in the pericentromeric region of all A chromosomes, likely involved in centromeric function. This could be the explanation for its amplification in the centromere of B chromosome, giving more stability through cell divisions. Besides amplification/deletion of satDNAs, the distribution of repeats in the B chromosome suggests the possibility of putative events of duplication and inversion that gave origin to the terminal region. The amplification of satDNAs after its origin and the changing satDNA repeat abundance was postulated in *E. monticola* [7]. Moreover, the putative duplication and inversion on the B chromosome of *X. d. angulatus* highlights how dynamic the repetitive DNAs are on this element, leading to the emergence of distinct morphotypes.

5. Conclusions

The present data expands the knowledge about the B chromosomes composition and their origin in grasshoppers. Our results provide support for the intraspecific origin of the B chromosome in the three species, like in the other species of grasshoppers [7,16]. Although the B chromosomes share some meiotic peculiarities with the X chromosomes that suggested origin from this chromosome, the current knowledge indicates a more common origin from autosomes in grasshoppers [7,8,16,47]. Furthermore, the species studied here and other grasshopper species with B chromosome ancestry [7,8,47] support the recurrent involvement of small chromosomes in the B chromosome origin. This could be due to the fewer number of genes and the enrichment of repetitive DNAs in small autosomes [7,31,48]. The analysis of other populations employing the repetitive DNA markers used here will shed light on the evolution of B chromosome polymorphism in the species.

Supplementary Materials: The following are available online at http://www.mdpi.com/2073-4425/9/11/523/s1, Table S1. Primers designed in this work and used for PCR amplification of satellite DNAs in the three species of grasshoppers.

Author Contributions: D.C.C.-d.-M., D.M., V.L., R.C.M., H.S. conceived and designed the experiments; D.M., A.B.S.M.F., V.B.B., O.M.P.-G., A.S.M., performed the experiments; all authors analyzed the data and wrote the paper.

Funding: Fundação de Amparo à Pesquisa do Estado de São Paulo-FAPESP (process number 2015/16661-1) and Conselho Nacional de Desenvolvimento Científico e Tecnológico-CNPq (process number 305300/2017-2).

Conflicts of Interest: The authors declare no conflicts of interest.

References

1. Charlesworth, B.; Sniegowski, P.; Stephan, W. The evolutionary dynamics of repetitive DNA in eukaryotes. *Nature* **1994**, *371*, 215–220. [CrossRef] [PubMed]
2. Garrido-Ramos, M.A. Satellite DNA: An evolving topic. *Genes* **2017**, *8*, 230. [CrossRef] [PubMed]
3. Lower, S.S.; McGurk, M.P.; Clark, A.G.; Barbash, D.A. Satellite DNA evolution: Old ideas, new approaches. *Curr. Opin. Genet. Dev.* **2018**, *49*, 70–78. [CrossRef] [PubMed]
4. Camacho, J.P.M. B chromosomes. In *The Evolution of the Genome*; Gregory, T.R., Ed.; Elsevier: San Diego, CA, USA, 2005; pp. 223–286.
5. Houben, A. B chromosomes—A matter of chromosome drive. *Front. Plant Sci.* **2017**, *8*, 210. [CrossRef] [PubMed]
6. Wilson, E.B. The supernumerary chromosomes of Hemiptera. *Science* **1907**, *26*, 870–871.
7. Ruiz-Ruano, F.J.; Cabrero, J.; López-León, M.D.; Camacho, J.P.M. Satellite DNA content illuminates the ancestry of a supernumerary (B) chromosome. *Chromosoma* **2017**, *126*, 487–500. [CrossRef] [PubMed]
8. Ruiz-Ruano, F.J.; Cabrero, J.; López-León, M.D.; Sánchez, A.; Camacho, J.P.M. Quantitative sequence characterization for repetitive DNA content in the supernumerary chromosome of the migratory locust. *Chromosoma* **2018**, *127*, 45–57. [CrossRef] [PubMed]
9. McAllister, B.F.; Werren, J.H. Hybrid origin of a B chromosome (PSR) in the parasitic wasp *Nasonia vitripennis*. *Chromosoma* **1997**, *106*, 243–253. [CrossRef] [PubMed]
10. Silva, D.M.Z.A.; Utsunomia, R.; Ruiz-Ruano, F.J.; Daniel, S.N.; Porto-Foresti, F.; Hashimoto, D.T.; Oliveira, C.; Camacho, J.P.M.; Foresti, F. High-throughput analysis unveils a highly shared satellite DNA library among three species of fish genus *Astyanax*. *Sci. Rep.* **2017**, *7*, 12726. [CrossRef] [PubMed]
11. Kumke, K.; Macas, J.; Fuchs, J.; Altschmied, L.; Kour, J.; Dhar, M.K.; Houben, A. *Plantago lagopus* B chromosome is enriched in 5S rDNA-derived satellite DNA. *Cytogenet. Genome Res.* **2016**, *148*, 68–73. [CrossRef] [PubMed]
12. Palestis, B.G.; Cabrero, J.; Trivers, R.; Camacho, J.P.M. Prevalence of B chromosomes in Orthoptera is associated with shape and number of A chromosomes. *Genetica* **2010**, *138*, 1181–1189. [CrossRef] [PubMed]

13. López-León, M.D.; Neves, N.; Schwarzacher, T.; Heslop-Harrison, J.S.; Hewitt, G.M.; Camacho, J.P.M. Possible origin of a B chromosome deduced from its DNA composition using double FISH technique. *Chromosom. Res.* **1994**, *2*, 87–92. [CrossRef]

14. Oliveira, N.L.; Cabral-de-Mello, D.C.; Rocha, M.F.; Loreto, V.; Martins, C. Chromosomal mapping of rDNAs and H3 histone sequences in the grasshopper *Rhammatocerus brasiliensis* (acrididae, gomphocerinae): Extensive chromosomal dispersion and co-localization of 5S rDNA/H3 histone clusters in the A complement and B chromosome. *Mol. Cytogenet.* **2011**, *4*, 24. [CrossRef] [PubMed]

15. Teruel, M.; Cabrero, J.; Perfectti, F.; Camacho, J.P.M. B chromosome ancestry revealed by histone genes in the migratory locust. *Chromosoma* **2010**, *119*, 217–225. [CrossRef] [PubMed]

16. Bueno, D.; Palacios-Gimenez, O.M.; Cabral-de-Mello, D.C. Chromosomal mapping of repetitive DNAs in *Abracris flavolineata* reveal possible ancestry for the B chromosome and surprisingly H3 histone spreading. *PLoS ONE* **2013**, *8*, e66532. [CrossRef] [PubMed]

17. Montiel, E.E.; Cabrero, J.; Camacho, J.P.M.; López-león, M.D. Gypsy, RTE and Mariner transposable elements populate *Eyprepocnemis plorans* genome. *Genetica* **2012**, *140*, 365–374. [CrossRef] [PubMed]

18. Palacios-Gimenez, O.M.; Bueno, D.; Cabral-de-Mello, D.C. Chromosomal mapping of two Mariner-like elements in the grasshopper *Abracris flavolineata* (orthoptera: Acrididae) reveals enrichment in euchromatin. *Eur. J. Entomol.* **2014**, *111*, 329–334. [CrossRef]

19. Milani, D.; Cabral-de-Mello, D.C. Microsatellite organization in the grasshopper *Abracris flavolineata* (Orthoptera: Acrididae) revealed by FISH mapping: Remarkable spreading in the A and B chromosomes. *PLoS ONE* **2014**, *9*, e97956. [CrossRef] [PubMed]

20. Milani, D.; Ramos, E.; Loreto, V.; Martí, D.A.; Cardoso, A.L.; Moraes, K.C.M.; Martins, C.; Cabral-de-Mello, D.C. The satellite DNA AflaSAT-1 in the A and B chromosomes of the grasshopper *Abracris flavolineata*. *BMC Genet.* **2017**, *18*, 81. [CrossRef] [PubMed]

21. Novák, P.; Neumann, P.; Pech, J.; Steinhaisl, J.; Macas, J. RepeatExplorer: A Galaxy-based web server for genome-wide characterization of eukaryotic repetitive elements from next generation sequence reads. *Bioinformatics* **2013**, *29*, 792–793. [CrossRef] [PubMed]

22. Utsunomia, R.; Silva, D.M.Z.A.; Ruiz-Ruano, F.J.; Araya-Jayme, C.; Pansonato-Alves, J.C.; Scacchetti, P.C.; Hashimoto, D.T.; Oliveira, C.; Trifonov, V.A.; Porto-Foresti, F.; et al. Uncovering the ancestry of B chromosomes in *Moenkhausia sanctaefilomenae* (Teleostei, Characidae). *PLoS ONE* **2016**, *11*, e0150573. [CrossRef] [PubMed]

23. Song, H.; Mariño-Pérez, R.; Woller, D.A.; Cigliano, M.M. Evolution, diversification, and biogeography of grasshoppers (Orthoptera: Acrididae). *Insect Sys. Div.* **2018**, *2*, 1–25. [CrossRef]

24. Song, H.; Foquet, B.; Mariño-Pérez, R.; Woller, D. Phylogeny of locusts and grasshoppers reveals complex evolution of density-dependent phenotypic plasticity. *Sci. Rep.* **2017**, *7*, 1–13. [CrossRef] [PubMed]

25. Sambrook, J.; Russell, D.W. *Molecular Cloning, a Laboratory Manual*, 3rd ed.; Cold Spring Harbor: New York, NY, USA, 2001.

26. Sumner, A.T. A simple technique for demonstrating centromeric heterochromatin. *Exp. Cell. Res.* **1972**, *75*, 304–306. [CrossRef]

27. *FastQC*, Version 0.10.1; A Quality Control Tool for High throughput Sequence Data; Babraham Bioinformatics: Cambridge, UK, 2012.

28. Novák, P.; Neumann, P.; Macas, J. Graph-based clustering and characterization of repetitive sequences in next-generation sequencing data. *BMC Bioinform.* **2010**, *11*, 378. [CrossRef] [PubMed]

29. *Geneious*, Version 4.8.5; Biomatters Ltd.: Aukland, New Zealand, 2009.

30. *RepeatMasker Open*, Version 4.0; Institute for Systems Biology: Seattle, WA, USA, 2013.

31. Ruiz-Ruano, F.J.; López-León, M.D.; Cabrero, J.; Camacho, J.P.M. High-throughput analysis of the satellitome illuminates satellite DNA evolution. *Sci. Rep.* **2016**, *6*, 28333. [CrossRef] [PubMed]

32. Rozen, S.; Skaletsky, H. Primer3 on the WWW for general users and for biologist programmers in Bioinformatics. *Methods Mol. Biol* **2000**, *132*, 365–386. [CrossRef] [PubMed]

33. Pinkel, D.; Lanlegent, J.; Collins, C.; Fuscoe, J.; Segraves, R.; Lucas, J.; Gray, J. Fluorescence in situ hybridization with human chromosome-specific libraries: Detection of trisomy 21 and translocations of chromosome 4. *Proc. Natl. Acad. Sci. USA* **1986**, *85*, 9138–9142. [CrossRef]

34. Camacho, J.P.M.; Cabrero, J.; López-León, M.D.; Cabral-de-Mello, D.C.; Ruiz-Ruano, F.J. Grasshoppers (Orthoptera). In *Protocols for Cytogenetic Mapping of Arthropod Genomes*, 1st ed.; Sharakhov, I.V., Ed.; CRC Press: Boca Raton, FL, USA, 2015; pp. 381–438.

35. Loreto, V.; Cabrero, J.; López-León, M.D.; Camacho, J.P.M. Possible autosomal origin of macro B chromosomes in two grasshopper species. *Chromosom. Res.* **2008**, *16*, 233–241. [CrossRef] [PubMed]

36. Bernardino, A.C.S.; Cabral-de-Mello, D.C.; Machado, C.B.; Palacios-Gimenez, O.M.; Santos, N.; Loreto, V. B chromosome variants of the Grasshopper *Xyleus discoideus angulatus* are potentially derived from pericentromeric DNA. *Cytogenet. Genome Res.* **2017**, *152*, 213–221. [CrossRef] [PubMed]

37. Camacho, J.P.M.; Shaw, M.W.; Cabrero, J.; Bakkali, M.; Ruíz-Estévez, M.; Ruiz-Ruano, F.J.; Martín-Blázquez, R.; López-León, M.D. Transient microgeographic clines during B chromosome invasion. *Am. Nat.* **2015**, *186*, 675–681. [CrossRef] [PubMed]

38. Souza, M.J.; Melo, N.F. Chromosome study in Schistocerca (Orthoptera-Acrididae-Cyrtacanthacridinae): Karyotypes and distribution patterns of constitutive heterochromatin and nucleolus organizer regions (NORs). *Genet. Mol. Biol.* **2007**, *30*, 54–59. [CrossRef]

39. Ruiz-Ruano, F.J.; Castillo-Martínez, J.; Cabrero, J.; Gómez, R.; Camacho, J.P.M.; López-León, M.D. High-throughput analysis of satellite DNA in the grasshopper *Pyrgomorpha conica* reveals abundance of homologous and heterologous higher-order repeats. *Chromosoma* **2018**, *127*, 3. [CrossRef] [PubMed]

40. Palacios-Gimenez, O.M.; Dias, G.B.; de Lima, L.G.; Kuhn, G.C.E.S.; Ramos, E.; Martins, C.; Cabral-de-Mello, D.C. High-throughput analysis of the satellitome revealed enormous diversity of satellite DNAs in the neo-Y chromosome of the cricket *Eneoptera surinamensis*. *Sci. Rep.* **2017**, *7*, 6422. [CrossRef] [PubMed]

41. Palacios-Gimenez, O.M.; Milani, D.; Lemos, B.; Castillo, E.R.; Martí, D.A.; Ramos, E.; Martins, C.; Cabral-de-Mello, D.C. Uncovering the evolutionary history of neo-XY sex chromosomes in the grasshopper *Ronderosia bergii* (Orthoptera, Melanoplinae) through satellite DNA analysis. *BMC Evol. Biol.* **2018**, *18*, 2. [CrossRef] [PubMed]

42. Palacios-Gimenez, O.M.; Bardella, V.B.; Lemos, B.; Cabral-de-Mello, D.C. Satellite DNAs are conserved and differentially transcribed among Gryllus cricket species. *DNA Res.* **2018**, *25*, 137–147. [CrossRef] [PubMed]

43. Marques, A.; Klemme, S.; Guerra, M.; Houben, A. Cytomolecular characterization of de novo formed rye B chromosome variants. *Mol. Cytogenet.* **2012**, *5*, 34. [CrossRef] [PubMed]

44. Cabrero, J.; López-León, M.D.; Ruíz-Estévez, M.; Gómez, R.; Petitpierre, E.; Rufas, J.S.; Massa, B.; Kamel Ben Halima, M.; Camacho, J.P.M. B1 was the ancestor B chromosome variant in the western Mediterranean area in the grasshopper *Eyprepocnemis plorans*. *Cytogenet. Genome Res.* **2014**, *142*, 54–58. [CrossRef] [PubMed]

45. Melo, A.S.; Moura, R.C. (Universidade de Pernambuco, Recife, Pernambuco, BR). Personal communication, 2018.

46. Martis, M.M.; Klemme, S.; Banaei-Moghaddam, A.M.; Blattner, F.R.; Macas, J.; Schmutzer, T.; Scholz, U.; Gundlach, H.; Wicker, T.; Simkova, H.; et al. Selfish supernumerary chromosome reveals its origin as a mosaic of host genome and organellar sequences. *Proc. Natl. Acad. Sci. USA* **2012**, *109*, 13343–13346. [CrossRef] [PubMed]

47. Teruel, M.; Ruiz-Ruano, F.J.; Marchal, J.A.; Sánchez, A.; Cabrero, J.; Camacho, J.P.M.; Perfectti, F. Disparate molecular evolution of two types of repetitive DNAs in the genome of the grasshopper *Eyprepocnemis plorans*. *Heredity* **2014**, *112*, 531–542. [CrossRef] [PubMed]

48. Hewitt, G.M. The integration of supernumerary chromosomes into the orthopteran genome. *Cold Spring Harb. Symp. Quant. Biol.* **1974**, *38*, 183–194. [CrossRef] [PubMed]

© 2018 by the authors. Licensee MDPI, Basel, Switzerland. This article is an open access article distributed under the terms and conditions of the Creative Commons Attribution (CC BY) license (http://creativecommons.org/licenses/by/4.0/).

Review

Evolution of Plant B Chromosome Enriched Sequences

André Marques [1,*], Sonja Klemme [2] and Andreas Houben [3]

[1] Laboratory of Genetic Resources, Federal University of Alagoas, Av. Manoel Severino Barbosa, 57309-005 Arapiraca—AL, Brazil
[2] Biology Centre, Czech Academy of Sciences, Institute of Plant Molecular Biology, Branišovská 31, CZ-37005 České Budějovice, Czech Republic; sonja.klemme@gmx.de
[3] Leibniz Institute of Plant Genetics and Crop Plant Research (IPK), Corrensstrasse 3, 06466 Gatersleben, Germany; houben@ipk-gatersleben.de
* Correspondence: andre.marques@arapiraca.ufal.br

Received: 24 September 2018; Accepted: 18 October 2018; Published: 22 October 2018

Abstract: B chromosomes are supernumerary chromosomes found in addition to the normal standard chromosomes (A chromosomes). B chromosomes are well known to accumulate several distinct types of repeated DNA elements. Although the evolution of B chromosomes has been the subject of numerous studies, the mechanisms of accumulation and evolution of repetitive sequences are not fully understood. Recently, new genomic approaches have shed light on the origin and accumulation of different classes of repetitive sequences in the process of B chromosome formation and evolution. Here we discuss the impact of repetitive sequences accumulation on the evolution of plant B chromosomes.

Keywords: B chromosome; satellite DNA; mobile element; organelle DNA; chromosome evolution

---

1. Introduction

Supernumerary B chromosomes (Bs) are not required for the normal development of organisms and are assumed to represent a specific type of selfish genetic elements. As a result, Bs follow their own species-specific evolutionary pathways. Several recent studies have confirmed the widely accepted view that Bs are derived from their respective A chromosome (A) complement [1–6]. Although Bs may vary in structure and chromatin properties in a species-specific way, their de novo formation is probably a rare event, because the occurrence of similar B chromosome variants within related species suggests that they arose from a single origin either from the same or from a related species [7,8].

The evolution of B chromosomes' (Bs) architecture has historically been of interest mainly at the cytogenetic level, with a recent focus on more molecular and genomic level studies (reviewed in [9–11]). Here we focus our review on the impact of repetitive DNA on the evolution of plant B chromosomes.

2. Methods and Tools to Characterize the High-Copy DNA Composition of B Chromosomes—Past and Future

First analyses of the DNA composition of B chromosomes were based on renaturation kinetics and gradient density centrifugation. These approaches showed that in rye the heterogeneity of repeats and ratio in Bs did not differ from As [12]. In maize the buoyant densities of DNA from plants with and without Bs were found to be alike [13]. Later, the use of comparative restriction endonuclease digestion of genomic DNA with and without B chromosomes of several *Glossina* species was introduced to characterize Bs. [14] Next generation sequencing (NGS), in combination with bioinformatics, led to an advance in our understanding of DNA sequence composition, functional gene content, and evolution of eukaryotic B chromosomes. In principle, NGS-based methods for the identification of B-specific sequences can be classified in two strategies [10]. In the first of these, DNA reads of microdissected or flow-sorted B chromosomes are used. Chromosome flow sorting usually produces larger amounts of DNA than microdissection. The most significant disadvantages of sequencing microdissected probes are the low amount of template DNA, contamination by surrounding material, and PCR amplification bias. Consequently, in silico purification of produced sequence reads is recommended. However, recent advances in single cell analysis demonstrated that even the DNA of a single haploid nucleus is sufficient for NGS analysis [15].

The second strategy requires sequencing of two whole genome datasets of the same species, one containing Bs (+B) and one without Bs (0B). It is an indirect approach because B-derived sequences are compared against the 0B-derived sequences as an additional step. This approach identifies B-candidate sequences where the ratio of aligned sequences is significantly increased in the +B dataset compared to the 0B dataset.

Identification of B chromosome-enriched sequences, like satellite repeats, mobile elements or organelle-derived sequences by similarity-based clustering of next generation sequence reads was achieved using the RepeatExplorer software [16,17] for rye [1] and *Plantago lagopus* [18]. With the RepeatExplorer and RepeatMasker programs the satellite DNA composition of the migratory locust (*Eyprepocnemis plorans*) B chromosomes was determined [19]. The 'satellitome' in the grasshopper *Eyprepocnemis monticola* consists of 27 satellite DNAs [6], less than half of the migratory locust, where 62 were found [19].

To identify repetitive elements or their parts the RepeatExplorer software [20] uses graph representation of read similarities to identify sequence clusters of frequently overlapping sequence reads [21]. Also, this software provides information about repeat quantities and others. The repeats are annotated based on BLASTN and BLASTX similarity searches to custom databases of repetitive elements and repeat-encoded conserved protein domains.

The in silico 'coverage ratio analysis' relies on a read alignment analysis performed for each of the 0B and +B datasets that subsequently are investigated for differences in the read coverage ratio. The strategy was used to determine the B chromosome sequence content of the cichlid *Astatotilapia latifasciata* [3]. Coverage ratio analysis revealed that the B chromosome contains thousands of sequences that have been duplicated from almost all standard chromosomes of this species, although most B-located genes are not contiguous. Subsequent sequence analysis of microdissected *A. latifasciata* Bs confirmed this conclusion.

An additional comparative approach is the *k*-mer based analysis termed '*k*-mer frequency ratio analysis'. Similar to coverage ratio analysis, it allows the analysis of 0B and +B sequence sets and investigates the differences in the *k*-mer frequency ratio. The web-based tool Kmasker [20] can be applied to run a *k*-mer frequency ratio analysis. The advantages and disadvantages of the different strategies are further discussed in Ruban, Schmutzer, Scholz and Houben [10].

Mention of specific programs in this review denotes only previous use in cited studies and does not imply endorsement.

3. Accumulation of B-Specific Repeats

Because of the high degree of evolutionary conservation and the high copy number, 5S and 45S ribosomal DNA (rDNA) satellite repeats have often been used as in situ hybridization probes for the analysis of Bs. The involvement of rDNA satellites in the evolution of plant Bs does not appear to be accidental, because rDNA loci have been detected on Bs of many species of plants (e.g., *Crepis capillaris* [22]), *Brachycome dichromosomatica* [23,24], *Aegilops* [25], and animals (e.g., *Haplochromis obliquidens* [26]) and *E. plorans* [27]. Although transcription of B chromosome-located rDNA was long believed not to occur [28], more recent studies have shown that indeed they are expressed, for instance the B-located 45S rRNA genes of the plant *C. capillaris* [29] and the grasshopper *E. plorans* [30]. In contrast, the ribosomal RNA genes specific to the B chromosomes in *B. dichromosomatica* are not transcribed [23]. Differences in posttranslational histone modifications, such as acetylation or methylation of histone, between A and B chromosomes, have been demonstrated [31–35]. Another possibility is that suppression of genes may occur due to nucleolar dominance such that the rRNA genes on the A chromosomes are active at the expense of B chromosome-located rRNA genes [24]. The inactivity of B chromosome rDNA could explain the presence of multiple ITS (internal transcribed spacer) sequences, since homogenization of rDNA spacers is thought to occur in transcribed regions only. Concerted evolution is a typical feature of the rDNA [36], but the mechanisms that control it may not include non-transcribed rDNA regions [37,38]. Since no homogenization occurs between the rDNA of A and B chromosomes, and since Bs are less active, one might expect further sequence erosion of B-located sequences. For the B chromosome-like paternal sex ratio chromosome of the wasp *Trichogramma kaykai*, it has been postulated that an increase in the number of different members of ITS sequences could be the evolutionary consequence [39].

In the herb *P. lagopus* L., the B chromosome is the product of a spontaneous amplification process of 5S ribosomal DNA derived repeats [18,40]. Interestingly, the amplification of satellite DNA has been used for the formation of engineered mammalian chromosomes ('satellite-DNA-based-artificial-chromosomes' [41]). In contrast to the situation with animals, the molecular mechanism of sequence amplification in plants is poorly understood. However, except for tobacco [42], no amplification-stimulating DNA elements from plants have been identified thus far. Alternatively, B chromosomal rDNA sites could be a consequence of the reported mobile nature of rDNA [43]. Bs may be the preferred "landing sites" because of the relative inactivity of constituent sequences and independence from selective forces on the As. Increasing evidence also indicates that rDNA can change position within the genome without corresponding changes in the surrounding sequences [44–46]. Beside rDNA, Bs accumulate chromosome specific satellite DNA (satDNA) which are listed below according to species.

3.1. Rye

The first plant B chromosome-specific satellite repeat has been identified in rye. Comparative restriction digestion of 0B and +B genomic DNA resulted in the isolation of the high-copy repeats E3900 and D1100, which seem to have de novo evolved on rye Bs [47–49]. Both repeats have classical features of satDNA such as being tandemly repeated, although atypical features such as transcriptional activity and euchromatic histone modifications have been reported for these repeats [7,35]. They also show an atypical repeat unit size with 1.1 and 3.9 kb, for D1100 and E3900, respectively [47,48]. These two repeats seem to have been assembled from fragments of a variety of sequence elements [49]. Both contain fragments of mobile elements most likely generated by chromosomal rearrangements, although no coding sequence responsible for the autonomous mobility has been found [49].

NGS studies on rye Bs have significantly increased our knowledge about B-specific repeat accumulation and evolution [1,7,50,51]. Since the sequencing of rye B, several B-specific repeats have been identified, mostly being satDNA [1,52]. Furthermore, these studies have shown that rye Bs are descended from rearrangements of the rye standard A chromosomes 3RS and 7R, with subsequent accumulation of repeats and genic fragments from other A chromosomal regions, as well as insertions of organellar DNA [1]. In silico identification of the high-copy sequence fraction revealed several B-specific repeats [52]. An accumulation of B chromosome-enriched tandem repeats was found mostly in the nondisjunction control region of the B. This unique region is late-replicating and transcriptionally active. All B-enriched repeats are not unique to the B chromosome but are also present in other species of the genus *Secale* [52]. Moreover, while it was shown that Bs contain a similar proportion of repeats to the A chromosomes in regards to their total DNA content, the two differed significantly in composition. This was due to the accumulation of B-specific satellite repeats, mostly in the nondisjunction control region at the terminal part of the long arm, as well as in the extended pericentromere [1,7,52]. The high-copy composition of the rye B seems even more conserved than that of the A chromosomes as different hybridization patterns were found for the repeats ScCl11, Sc36c82 and Sc55c1 on As of different accessions but not on Bs [7], suggesting an important role for the maintenance of the B typical structure. An overall scheme of the rye B repeat composition is shown in Figure 1A.

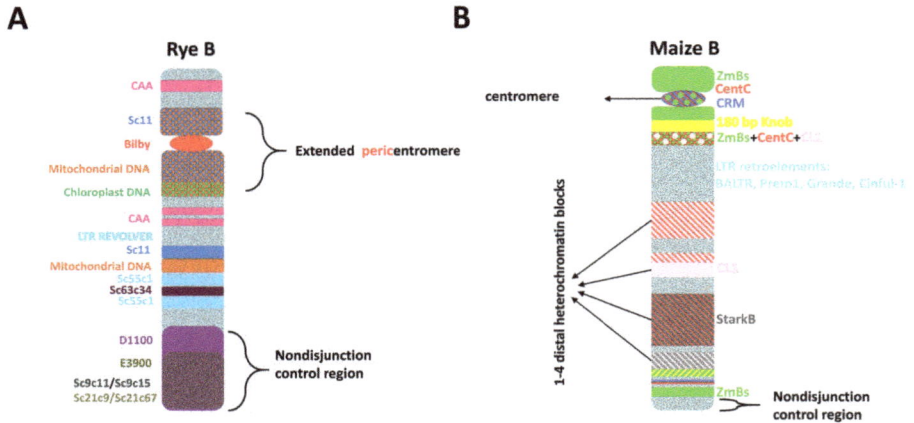

Figure 1. Model for the distribution of (**A**) rye and (**B**) maize B chromosome-enriched sequences. In both rye Bs and maize Bs only the very terminal region of B long arm is required for non-disjunction control [51,53,54].

3.2. Maize

The maize (*Zea mays*) B is also well studied and several works have been conducted aiming to understand its composition and accumulation mechanisms. Similar to rye Bs, maize Bs are also characterized by chromosome-type specific repeats [55–59]. The maize B centromere region contains a repetitive element ZmBs that is not present on the A chromosomes, which also shares homology over 90 bp with the maize knob sequence [55]. Later an approximately 700-kb domain that consists of all three previously described repeats (ZmBs repeat, CentC, and CRM) was described that is localised at the core centromere of the maize B and show enhanced association with CENH3 [60,61].

Two of the elements that are specific to the maize B are organized in long tandem arrays with repeat units of similar size. The ZmBs repeat, with approx. 1400 bp of unit length, is located in and around the B centromere as well as near the tip of the B long arm [55,56]. The CL-1 repeat, with approx. 1500 bp of unit length, is present in the first three heterochromatic blocks of the long arm [58,59]. Neither repeat has homology to any known open reading frames or sequences located on As. However, transposition of a retrotransposon and a Miniature Inverted-repeat Transposable Element (MITE) element involved in the genesis of the CL-1 repeat was detected [59] (Figure 1B).

Another B-specific repeat found on maize Bs is the Stark B element, which like D1100 and E3900 rye B elements is a retrotransposon-derived sequence. This repeat was formerly identified as B-specific sequence family with a relationship to the Prem1 family of maize retroelements, which are preferentially transcribed in pollen [57]. It is composed of repetitive sequences known from the A genome as well as novel sequences unique to the B. The StarkB element is much larger than the other B-specific elements of maize. StarkB copies vary by small insertions, deletions, and duplications as well as single-nucleotide polymorphisms. The minimum age of the StarkB repeat array was estimated to be at least 2 million years [56]. The formation of StarkB reflects a process that generates large amounts of DNA on the B chromosome. StarkB is also transcriptionally active. Therefore, the process that contributed to formation of the maize B combined pre-existing coding regions to produce novel transcripts [56].

StarkB is specifically located on heterochromatic domains, distributed throughout the third and fourth blocks of heterochromatin. Although this repeat is retrotransposon-derived it lacks the autonomous domains being characterized as non-autonomous chimeric element. StarkB is not arranged in arrays in contrast the maize B-repeats ZmBs and CL-1. Because the two blocks of heterochromatin that contain StarkB have persisted over many generations, their presence may play a role in B chromosome transmission [56].

3.3. Brachycome dichromosomatica

The daisy *B. dichromosomatica* has Bs of two different types, the larger Bs are somatically stable whereas the smaller, or micro Bs are somatically unstable. Both types of Bs contain clusters of 45S rDNA.

The large B carries a B-specific tandem repeat (Bd49) that is located mainly at the centromere [62,63]. Multiple copies of sequences related to this repeat are present on the A chromosomes of related species without Bs, whereas only a few copies exist in the As of *B. dichromosomatica* [63]. An isolated Bd49 clone was composed entirely of a tandem array of the repeat unit. However, in other clones the Bd49 repeats were linked to, or interspersed with, sequences that were repetitious and distributed elsewhere on the A and B chromosomes. One such repetitious flanking sequence had similarity to retrotransposon-like sequences and a second was similar to chloroplast DNA [64].

The micro Bs share DNA sequences with the As and the larger Bs, and they also have B-specific repeats (Bdm29 and Bdm54) [65,66]. Bdm29 is highly methylated and after in situ hybridization labelled the entire micro Bs. The Bdm29 is AT-rich with an insert of 290 bp long containing no significant subrepeats. A high number of Bdm29-like sequences was also found in the larger Bs of *B. dichromosomatica* and in other Bs within the genus, suggesting that the Bdm29 sequence is highly conserved and widespread [65]. The Bdm54 repeat is AT-rich with an insert of 477 bp long containing four copies of a subrepeat unit (TCGAAAAGTTCGAAG) as well as three perfect and four degenerate copies of a second short repeat (AGTTCGAA) that are embedded in the first unit [66]. Some micro B-located repeats have been shown to occur as clusters on the A chromosomes in a proportion of individuals within a population [67]. The observation that the genomic organization of the micro B is unlike anything found on the A chromosomes precludes their origin by simple excision from an A chromosome and also indicates that micro Bs do not integrate directly into the A complement to form polymorphic heterochromatic segments [66].

3.4. Aegilops speltoides

The *Aegilops* Bs are also known to accumulate several repetitive sequences being characterized by a number of A chromosome-localized repeats like Spelt1, pSc119.2 tandem repeats, 5S rDNA and Ty3-gypsy retroelements [26,68–70], as well as organelle-derived DNA [71]. However, no B-specific repeat has been found in this species yet.

3.5. Plantago lagopus

In *P. lagopus*, a weed of the Mediterranean region, a B-specific satellite PLsatB was identified that originated from sequence amplification including 5S rDNA fragments [41]. The satellite repeat PLsatB makes up 3.3% of the 1B genotype but only 0.09% of the 0B genotype [18]. In situ hybridization with the B-repeat revealed an almost uniform labelling of the Bs at meiotic metaphase I, while extended mitotic prometaphase chromosomes showed a more clustered distribution of the hybridization signals. In any case, no signals were detectable on the A chromosomes. Although PLsatB has evolved from 5S rDNA sequences, a 5S rDNA probe was not sufficient to label the Bs by fluorescence in situ hybridization (FISH) [18].

3.6. Cestrum

Species of *Cestrum* have shown large diversity in the accumulation and distribution of repetitive DNA families [72], and Bs have been described in six species and one interspecific hybrid: *C. strigilatum*, *C. diurnum*, *C. parqui* × *C. aurantiacum*, *C. intermedium*, *C. parqui*, *C. euanthes* and *C. nocturnum* [73–76].

Some of these repeats have already been identified and associated with B chromosomes [73]. In the hybrid *C. parqui* × *C. aurantiacum*, for instance, the B chromosome contains 35S and 5S rDNA and SSR AT-rich motifs [75]. Sequences of rDNA were also identified in Bs of *C. parqui*, *C. euanthes* and *C. nocturnum* [76]. In *C. intermedium* and *C. strigilatum*, besides C-Giemsa+/CMA+/DAPI+ bands [73], the Bs also display hybridization signals with the Gypsy-like retrotransposon probe but not with rDNA probes [73]. Some types of repetitive DNA were identified in A and B chromosomes in *C. strigilatum* and in species of this plant group, such as AT-rich SSR, 35S and 5S rDNA, C-Giemsa and C-CMA/DAPI bands and retrotransposons [77].

3.7. Crepis capillaris

In *C. capillaris* in situ hybridization of cells with labelled DNA derived from microdissected Bs confirmed that the B is composed mainly of sequences also present in the A chromosomes, but lacks the main repeats located on A chromosomes [78,79]. No B-specific repeat has been found. The highly abundant repeat B134 shows repeating units with a sequence similarity range from 69% to 90% and characterized by its richness in $(CA)_n$ repeats. Members of this family are dispersed throughout the A and B chromosomes but are more concentrated in the pericentromeric heterochromatin of the B, indicating that the molecular organization of B heterochromatin is different from that of the As. B-located B134 repeats also have diverged from those on the As [79].

4. B Chromosome-Specific Accumulation of Organelle DNA

Mitochondrial and chloroplast DNA sequences are frequently transferred into the nuclear genome. This transfer usually dependents on recombination-based insertions of organellar DNA into the nucleus. As a result nuclear insertions of plastid DNA (NUPTs) or nuclear insertions of mitochondrial DNA (NUMTs) occur [80]. Nuclear transfer of organelle DNA is a well-known process (reviewed in [81–84]. Nuclear insertions NUPTs and NUMTs have been shown to be involved in the formation of new nuclear genes [83].

Large insertions of organelle DNA were found on the rye B chromosomes [1]. While plastid DNA is not absent from the As, the B localized sequences are considerably larger. They most likely stem from several independent insertion events. However, the evolutionary forces that resulted in organelle sequences organized as clusters on Bs are not well understood. Between Bs from different geographical origin there is no difference in organelle DNA content or distribution except for a pericentric inversion detected by FISH [7]. In contrast, the B chromosomes of *Ae. speltoides* of different accessions showed differences in abundance and location of organelle DNA [70]. However, all tested accessions did show accumulation in the B chromosomes over A chromosomes. When comparing A- and B-derived organelle reads to the original organelle-genome sequence of wheat, the reads from the rye B show less similarity to the source genome [1]. The inserts on the B therefore accumulated more mutations than A located inserts. It is also noticeable that the strongest accumulation happened in the pericentromere for rye Bs, and in the distal chromosome arms for *Ae. speltoides*. Pericentric insertions of organelle DNA have also been shown in rice A chromosomes [85]. Pericentromeric regions generally contain few functional genes, and this low gene density may facilitate the repeated integration of the organelle derived DNA [85]. Alternatively, consistent with the rapid evolution of centromeres [86] after sequence integration, subsequent amplification of these sequences might have occurred within this region. The confinement to mostly one region hints at two different possibilities for enrichment in mitochondrial DNA: (i) directed repeated insertion into the surroundings of this area, or (ii) few insertions with subsequent local amplification of these sequences. Although the second explanation seems more likely, the diverse nature of the organelle sequence reads suggests many independent insertion events instead of many copies derived from one incorporation. Interestingly it has also been shown that organelle DNA often integrates several fragments into one location [83]. This would fit well with data indicating many events rather than amplification of just one insertion [1].

What mechanism could account for the accumulation of organellar DNA in B chromosomes? It has been demonstrated that environmental stresses increase the incorporation of organelle DNA into the nucleus [84]. As the Bs, especially in higher numbers, can be considered as stress factors to the cell [87], their presence might increase the basal rate of DNA transfer. Transfer of organellar DNA to the nucleus is very frequent [81,88,89], but most of the "promiscuous" DNA is also rapidly lost again via a counterbalancing removal process [90]. If this expulsion mechanism is impaired in B chromosomes, then the high turnover rates that prevent such sequences on the A chromosomes from accumulating and degrading would be absent and allow for sequence decay. Thus, the dynamic equilibrium between frequent integration and rapid elimination of organellar DNA could be imbalanced for B chromosomes. This hypothesis is supported by higher divergence of B-derived NUMT reads compared to A-derived NUMT reads in *Secale cereale* [1]. Future analyses of other B-bearing species are needed to address the question as to whether organelle-to-nucleus DNA transfer is an important mechanism that drives the evolution of B chromosomes.

Another possibility for the accumulation is the dependence on double strand breaks (DSBs). If B chromosomes are more prone to DSB, this could aid in the insertion of random available DNA [71]. Organelle DNA on Bs might be underreported in genomic studies, since organelle DNA is generally filtered out during sequence analysis, due to contamination with DNA extracted from organelles. The presence of large insertions of organelle DNA on B chromosomes might be a plant specific phenomenon, as there have not been any reports of animal B chromosomes with mitochondrial insertions.

5. B Chromosome-Specific Accumulation of Transposable Elements

Recent works have demonstrated that B chromosomes accumulate DNA from various sources existing as amalgamations of mixed repeats and single-copy regions. In such a scenario, Bs would provide a safe haven for the accumulation and spread of transposable elements (TEs). This has been suggested as a mechanism through which some of the variability in mammalian Y chromosomes has arisen, as random insertions of transposable DNA into different regions of the Y chromosome would result in elements differing with respect to DNA composition and structure [80,91].

Although TEs compose the vast majority of repetitive DNA in eukaryotic genomes, very few studies have been conducted on plants carrying Bs. Thus, little is known about TE accumulation mechanisms on Bs. Evidence that Bs can provide an ideal target for transposition of TEs comes from works on the retrotransposon NATE (*Nasonia* Transposable Element) which has been described from the PSR (paternal sex-ratio) element of *Nasonia vitripennis* [92,93]. B-specific accumulation of Ty3/gypsy retrotransposons has been also reported for the fish *Alburnus alburnus* (L.) [94]. In plants, the rye Bs show B-specific accumulation of LTR (long terminal repeat) families [1,52]. In maize B several members of LTR retroelements were found showing differential accumulation and hybridization pattern compared to the As [56,61]. A retrotransposon has also been invoked in the transposition of chloroplast DNA into the repeat element Bd49 of the B chromosomes of *B. dichromosomatica* [64]. Thus, insertion of such elements may be responsible for the generation of structural variability in Bs [95].

5.1. Rye

Although most repeats are similarly distributed along As and Bs of rye, several transposons are either amplified or depleted on the B chromosome. For instance, the ancient retroelement Sabrina, abundant in all Triticeae and transcriptionally inactive in rye [96], is highly accumulated on As but less abundant on Bs. In contrast, the active element Revolver, as well as the predicted Copia transposon Sc36c82 seem to be more amplified on the Bs [52]. The B-specific gain of active mobile elements might have its cause in the lack of selection pressure on B chromsomes. Meiotic crossing-over has been proposed to remove mobile elements [97]. The B chromosomes of rye pair frequently with each other and themselves in meiosis [98], but the bivalents of Bs are often less connected by chiasmata than the

As [99]. As proposed for plant Y chromosomes [100,101], reduced crossing-over might facilitate the accumulation of retroelements on Bs.

Based on these findings Klemme, Banaei-Moghaddam, Macas, Wicker, Novak and Houben [52] suggested a model for the selective accumulation of TEs on the B chromosomes of rye: "In the Triticeae ancestor, Sabrina was transposing and spread over the entire genome. After inactivation of Sabrina [96] before or during speciation of rye, the B was formed from the As with Sabrina still present. The newly evolving elements such as Revolver then became active and transposed throughout the rye genome. The dispensable nature of the B and the lack of selective pressure allowed for stronger accumulation of Revolver on the B, even further diluting the remnants of inactive elements which can no longer increase copy number."

Additionally, other retroelements have been also found to be accumulated on rye Bs, for instance, the Copia elements Sc11c32 and Sc11c927, similar to the centromeric sequences Bilby and Sc11 (Figure 1A), expanded more in the extended pericentromere of the B than in those of As [51]. The presumed ancestral elements are still detectable in subterminal positions on As [7,52].

5.2. Maize

Furthermore, the maize B is also known to share the same centromere-specific retrotransposons (CRM elements) with the As [60]. Another retroelement called BALTR1 (B and A LTR element 1) was also found to be hybridized throughout the A and B genome, but showed enrichment near centromeres and on the B long arm [56]. The BALTR1 element was 7892 bp in length with LTR that were 1159 and 1173 bp, and possessed the internal coding sequences and the LTRs. The LTR of BALTR1 did not show similarity to sequences in public databases. Because the coding region was most similar to the CRM family of retrotransposons, this novel retrotransposon is likely a member of the Ty3/gypsy class [56]. Additionally, several other members of retroelements of LTR Ty3/Gypsy class were found on the maize Bs. For instance while both Prem1 and Cinful-1 elements showed hybridization signals more intense than the As with a similar distribution pattern as the centromere diffuse elements, the Huck element only hybridized weakly to the euchromatic region on the B [61]. The maize B is also uniformly labelled with the Grande element at the same level as the As [102]. An overview of maize B-enriched repeats is shown in Figure 1B.

6. Rapid Evolution of Repeats on Bs

The accumulation of repeats accompanies the evolution of Bs in several plant species, suggesting that the Bs most likely represents a suitable chromosomal context for satellite expansion. But why do Bs often accumulate repeats?

There are at least two possible pathways for Bs to undergo repeat accumulation. One alternative source for the accumulation of repeats on Bs comes from studies conducted on rye where a restriction of meiotic recombination in Bs was observed, which showed variation in the frequency of bivalents formation among different genotypes [103,104]. This restriction of recombination can be considered as starting point for the independent evolution of Bs. The presence of fast-evolving repetitive sequences, could predispose a nascent B to undergo further rapid structural modifications required to establish and amplify new B-specific repeats.

A second alternative is that Bs are under reduced selective pressure due to their non-essential nature, as far as it does not affect its accumulation mechanism. Such chromosomal environment may be considered as a safe haven for non-coding fast-evolving sequences, such as promiscuous DNA as satDNA, organelle DNA and TEs, which are frequent genome hitchhikers able to settle in non-recombining regions. Thus, taking in account the features of Bs, these two pathways may explain the great diversification of repeats found on Bs.

Shrinkage and expansion of the Y chromosome is influenced by the sex-specific regulation of repetitive DNA. It was suggested that the dynamics of Y chromosome evolution is an interplay of genetic and epigenetic processes [80]. Many eukaryotic genomes contain a large proportion of repetitive

DNA sequences. These sequences often colonize specialist chromosomes (Y, W or B chromosomes). In particular, the non-recombining regions of the Y chromosome, are subject to different evolutionary forces compared with autosomes. Repetitive DNA sequences often accumulate in the non-recombining regions of the Y chromosome [105]. A similar mechanism may take place in Bs, since it is known that, at least for rye, the frequency of meiotic recombination and proper bivalent formation may vary greatly among different genotypes as discussed above.

Because Bs are assumed to be non-essential they are also presumed to be evolutionarily neutral in host genomes and, thus, it is expected that high B sequence variability will be observed among different samples. The origin of B structural variants is most likely from a monophyletic origin from a unique type of ancestral B chromosome which afterwards diverged in different types through generations [7,8,106]. Indeed, there are several cases of B polymorphisms whether numerical or structural [107,108]. For the B chromosome of the grasshopper *E. plorans,* a large variety of structural variants has been demonstrated among many populations [107]. In plants, B polymorphisms have been mainly attributed to numerical polymorphisms [108,109]. Although, in a few cases B structural variants in natural populations have been identified e.g., *B. dichromosomatica* [106], *Ae. speltoides* [71] and *S. autumnalis* [110].

7. Evolutionary Aspects of Repeat Accumulation

The B-specific repeats E3900 and D1100 which are located in the non-disjunction controlling region of the rye B are characterized by unusual properties including a predisposition to instability [49]. There are similarities between StarkB element of maize and E3900. Both elements share sequences with the respective A genome and are composed of a complex mixture of unique sequences. Also, both are found in clusters intermixed with other B-specific sequences near the end of the chromosome [49,56]. These common features may display similarities in how these elements evolved. It has been suggested that alterations to the E3900/D1100 region of rye could influence the degree of chromatin packaging and affect the efficiency of B chromosome drive [49]. When the number of Bs in a population is too high, so that the fitness of the host is adversely affected, B variants with lower transmission would be selected [56]. Indeed, rye genotypes with different B chromosome transmission rates exist [103,104].

As a general model for the accumulation and evolution of repeats on plant Bs, we propose the following: (1) A proto-B chromosome was derived as a result of multiple translocations and duplications of A chromosome fragments and, therefore, shows a similar sequence composition as the original A chromosome fragments. The presence of a chromosome fragment processing a functional centromere to assure mitotic and meiotic segregation is essential. (2) Gene silencing/erosion followed by restriction of meiotic recombination and reduced selective pressure triggers a B-specific repeat accumulation due to the non-essential nature of the B chromosome. At this point the formation of a B chromosome-specific drive/accumulation mechanism is essential for the survival of the B chromosome. (3) Mature Bs are characterized by B-specific repeats and a tolerable impact of the Bs on the fitness of the host organism. Unless the B is eliminated, it could constantly accumulate sequences and change its structure (neutral mutations) and enhance its drive mechanism (positive selection) along its evolution. Furthermore, it is important to notice that the B centromeres at least in rye and maize show specific features to assure chromosome drive via nondisjunction at first or second pollen grain mitosis, respectively (reviewed in Houben [111]). Thus, the evolution of B-specific (peri)centromeric properties seems to be a key step in the evolution of Bs. Figure 2 shows a diagrammatic summary for the proposed model of repeat accumulation and evolution of Bs.

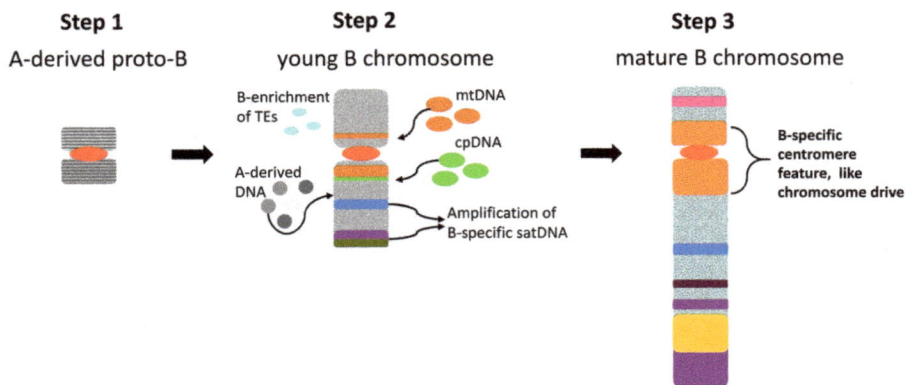

Figure 2. Model for the repeat accumulation and evolution of plant B chromosomes. (**1**) Proto-B chromosome derived as a result of multiple translocations and duplications of A chromosome fragments (harboring a functional centromere). (**2**) Gene erosion/silencing followed by restriction of meiotic recombination triggers B-specific repeat accumulation. (**3**) Mature Bs could achieve a high degree of B-specific repeats, an efficient drive mechanism and a tolerable impact on the fitness of the host organism. mtDNA: mitochondrial DNA; cpDNA: chloroplast DNA; satDNA: satellite DNA.

8. Concluding Remarks and Future Perspectives

As discussed above, Bs are expected to undergo reduced selective pressure due to their non-essential nature and, thus, exhibit larger variation than A chromosomes. However, it is remarkable to see that in fact some Bs have a relatively conserved structure with rare occurrence of structural variants across plant species [7,106]. In some cases, it is remarkable to see that the variation in the As might be even higher than that observed in the Bs, as for instance in *B. dichromosomatica* [106] and for the B-specific repeat ScCl11 in rye [7]. This suggests the existence of a control mechanism for the maintenance of a standard B structure. Although Bs seem always to have an A-derived architecture, they are frequently found to have accumulated several B-specific repeats and, at least in some cases, insertions of organellar DNA. Furthermore, they may show notable variation in their quantitative repeat composition. These unique features observed on Bs highlight on the one hand the origin of Bs from As, and on the other a different evolutionary pathway of As and Bs. It seems that the B acts like a "genomic sponge" which collects and maintains sequences of diverse origins.

Author Contributions: A.M. and A.H. designed and wrote the manuscript. S.K. wrote the manuscript.

Funding: This research was funded by CNPq/FAPEAL by providing the DCR fellowship for A.M. and by the Deutsche Forschungsgemeinschaft DFG (grant number HO1779/26-1).

Acknowledgments: We thank the reviewers for fruitful comments on the manuscript.

Conflicts of Interest: The authors declare no conflict of interest.

References

1. Martis, M.M.; Klemme, S.; Banaei-Moghaddam, A.M.; Blattner, F.R.; Macas, J.; Schmutzer, T.; Scholz, U.; Gundlach, H.; Wicker, T.; Simkova, H.; et al. Selfish supernumerary chromosome reveals its origin as a mosaic of host genome and organellar sequences. *Proc. Natl. Acad. Sci. USA* **2012**, *109*, 13343–13346. [CrossRef] [PubMed]

2. Silva, D.M.; Pansonato-Alves, J.C.; Utsunomia, R.; Araya-Jaime, C.; Ruiz-Ruano, F.J.; Daniel, S.N.; Hashimoto, D.T.; Oliveira, C.; Camacho, J.P.; Porto-Foresti, F.; et al. Delimiting the origin of a B chromosome by FISH mapping, chromosome painting and DNA sequence analysis in *Astyanax paranae* (Teleostei, Characiformes). *PLoS ONE* **2014**, *9*, e94896. [CrossRef] [PubMed]

3. Valente, G.T.; Conte, M.A.; Fantinatti, B.E.; Cabral-de-Mello, D.C.; Carvalho, R.F.; Vicari, M.R.; Kocher, T.D.; Martins, C. Origin and evolution of B chromosomes in the Cichlid fish *Astatotilapia latifasciata* based on integrated genomic analyses. *Mol. Biol. Evol.* **2014**, *31*, 2061–2072. [CrossRef] [PubMed]

4. Croll, D.; McDonald, B.A. The accessory genome as a cradle for adaptive evolution in pathogens. *PLoS Pathog.* **2012**, *8*, e1002608. [CrossRef] [PubMed]

5. Navarro-Domínguez, B.; Ruiz-Ruano, F.J.; Cabrero, J.; Corral, J.M.; López-León, M.D.; Sharbel, T.F.; Camacho, J.P.M. Protein-coding genes in B chromosomes of the grasshopper *Eyprepocnemis plorans*. *Sci. Rep.* **2017**, *7*, 45200. [CrossRef] [PubMed]

6. Ruiz-Ruano, F.J.; Cabrero, J.; Lopez-Leon, M.D.; Camacho, J.P.M. Satellite DNA content illuminates the ancestry of a supernumerary (B) chromosome. *Chromosoma* **2017**, *126*, 487–500. [CrossRef] [PubMed]

7. Marques, A.; Banaei-Moghaddam, A.M.; Klemme, S.; Blattner, F.R.; Niwa, K.; Guerra, M.; Houben, A. B chromosomes of rye are highly conserved and accompanied the development of early agriculture. *Ann. Bot.* **2013**, *112*, 527–534. [CrossRef] [PubMed]

8. Muñoz-Pajares, A.; Martinez-Rodriguez, L.; Teruel, M.; Cabrero, J.; Camacho, J.P.M.; Perfectti, F. A single, recent origin of the accessory B chromosome of the grasshopper *Eyprepocnemis plorans*. *Genetics* **2011**, *187*, 853–863. [CrossRef] [PubMed]

9. Valente, G.T.; Nakajima, R.T.; Fantinatti, B.E.; Marques, D.F.; Almeida, R.O.; Simoes, R.P.; Martins, C. B chromosomes: From cytogenetics to systems biology. *Chromosoma* **2017**, *126*, 73–81. [CrossRef] [PubMed]

10. Ruban, A.; Schmutzer, T.; Scholz, U.; Houben, A. How next-generation sequencing has aided our understanding of the sequence composition and origin of B chromosomes. *Genes* **2017**, *8*, 294. [CrossRef] [PubMed]

11. Makunin, A.I.; Dementyeva, P.V.; Graphodatsky, A.S.; Volobouev, V.T.; Kukekova, A.V.; Trifonov, V.A. Genes on B chromosomes of vertebrates. *Mol. Cytogenet.* **2014**, *7*, 99. [CrossRef] [PubMed]

12. Timmis, J.N.; Ingle, J.; Sinclair, J.; Jones, R.N. Genomic quality of rye B chromosomes. *J. Exp. Bot.* **1975**, *26*, 367–378. [CrossRef]

13. Chilton, M.D.; Mccarthy, B.J. DNA from Maize with and without B-Chromosomes—Comparative Study. *Genetics* **1973**, *74*, 605–614. [PubMed]

14. Amos, A.; Dover, G. The distribution of repetitive DNAs between regular and supernumerary chromosomes in species of *Glossina* (Tsetse)—A 2-step process in the origin of supernumeraries. *Chromosoma* **1981**, *81*, 673–690. [CrossRef] [PubMed]

15. Dreissig, S.; Fuchs, J.; Himmelbach, A.; Mascher, M.; Houben, A. Sequencing of single pollen nuclei reveals meiotic recombination events at megabase resolution and circumvents segregation distortion caused by postmeiotic processes. *Front. Plant Sci.* **2017**, *8*, 1620. [CrossRef] [PubMed]

16. Novak, P.; Avila Robledillo, L.; Koblizkova, A.; Vrbova, I.; Neumann, P.; Macas, J. TAREAN: A computational tool for identification and characterization of satellite DNA from unassembled short reads. *Nucleic Acids Res.* **2017**, *45*, e111. [CrossRef] [PubMed]

17. Novak, P.; Neumann, P.; Pech, J.; Steinhaisl, J.; Macas, J. RepeatExplorer: A Galaxy-based web server for genome-wide characterization of eukaryotic repetitive elements from next-generation sequence reads. *Bioinformatics* **2013**, *29*, 792–793. [CrossRef] [PubMed]

18. Kumke, K.; Macas, J.; Fuchs, J.; Altschmied, L.; Kour, J.; Dhar, M.K.; Houben, A. *Plantago lagopus* B Chromosome Is Enriched in 5S rDNA-Derived Satellite DNA. *Cytogenet. Genome Res.* **2016**, *148*, 68–73. [CrossRef] [PubMed]

19. Ruiz-Ruano, F.J.; Lopez-Leon, M.D.; Cabrero, J.; Camacho, J.P. High-throughput analysis of the satellitome illuminates satellite DNA evolution. *Sci. Rep.* **2016**, *6*, 28333. [CrossRef] [PubMed]

20. Schmutzer, T.; Ma, L.; Pousarebani, N.; Bull, F.; Stein, N.; Houben, A.; Scholz, U. Kmasker—A tool for in silico prediction of single-copy FISH probes for the large-genome species *Hordeum vulgare*. *Cytogenet. Genome Res.* **2014**, *142*, 66–78. [CrossRef] [PubMed]

21. Novak, P.; Neumann, P.; Macas, J. Graph-based clustering and characterization of repetitive sequences in next-generation sequencing data. *BMC Bioinform.* **2010**, *11*, 378. [CrossRef] [PubMed]

22. Maluszynska, J.; Schweizer, D. Ribosomal RNA genes in B chromosomes of *Crepis capillaris* detected by non-radioactive in situ hybridization. *Heredity* **1989**, *62*, 59–65. [CrossRef] [PubMed]

23. Donald, T.M.; Houben, A.; Leach, C.R.; Timmis, J.N. Ribosomal RNA genes specific to the B chromosomes in *Brachycome dichromosomatica* are not transcribed in leaf tissue. *Genome* **1997**, *40*, 674–681. [CrossRef] [PubMed]

24. Donald, T.M.; Leach, C.R.; Clough, A.; Timmis, J.N. Ribosomal RNA genes and the B chromosome of *Brachycome dichromosomatica*. *Heredity* **1995**, *74*, 556–561. [CrossRef] [PubMed]

25. Friebe, B.; Jiang, J.; Gill, B. Detection of 5S-rDNA and other repeated DNA on supernumerary B-chromosomes of *Triticum* species (*Poaceae*). *Plant Syst. Evol.* **1995**, *196*, 131–139. [CrossRef]

26. Poletto, A.B.; Ferreira, I.A.; Martins, C. The B chromosomes of the African cichlid fish *Haplochromis obliquidens* harbour 18S rRNA gene copies. *BMC Genet.* **2010**, *11*, 1. [CrossRef] [PubMed]

27. Lopez-Leon, M.D.; Neves, N.; Schwarzacher, T.; Heslop-Harrison, J.S.; Hewitt, G.M.; Camacho, J.P. Possible origin of a B chromosome deduced from its DNA composition using double FISH technique. *Chromosome Res.* **1994**, *2*, 87–92. [CrossRef] [PubMed]

28. Ishak, B.; Jaafar, H.; Maetz, J.L.; Rumpler, Y. Absence of transcriptional activity of the B chromosomes of Apodemus peninsulae during pachytene. *Chromosoma* **1991**, *100*, 278–281. [CrossRef]

29. Leach, C.R.; Houben, A.; Field, B.; Pistrick, K.; Demidov, D.; Timmis, J.N. Molecular evidence for transcription of genes on a B chromosome in *Crepis capillaris*. *Genetics* **2005**, *171*, 269–278. [CrossRef] [PubMed]

30. Ruiz-Estevez, M.; Lopez-Leon, M.D.; Cabrero, J.; Camacho, J.P. B-chromosome ribosomal DNA is functional in the grasshopper *Eyprepocnemis plorans*. *PLoS ONE* **2012**, *7*, e36600. [CrossRef] [PubMed]

31. Carchilan, M.; Kumke, K.; Mikolajewski, S.; Houben, A. Rye B chromosomes are weakly transcribed and might alter the transcriptional activity of A chromosome sequences. *Chromosoma* **2009**, *118*, 607. [CrossRef] [PubMed]

32. Kumke, K.; Jones, R.N.; Houben, A. B chromosomes of *Puschkinia libanotica* are characterized by a reduced level of euchromatic histone H3 methylation marks. *Cytogenet. Genome Res.* **2008**, *121*, 266–270. [CrossRef] [PubMed]

33. Marschner, S.; Kumke, K.; Houben, A. B chromosomes of *B. dichromosomatica* show a reduced level of euchromatic histone H3 methylation marks. *Chromosome Res.* **2007**, *15*, 215–222. [CrossRef] [PubMed]

34. Jin, W.W.; Lamb, J.C.; Zhang, W.L.; Kolano, B.; Birchler, J.A.; Jiang, J.M. Histone modifications associated with both A and B chromosomes of maize. *Chromosome Res.* **2008**, *16*, 1203–1214. [CrossRef] [PubMed]

35. Carchilan, M.; Delgado, M.; Ribeiro, T.; Costa-Nunes, P.; Caperta, A.; Morais-Cecilio, L.; Jones, R.N.; Viegas, W.; Houben, A. Transcriptionally active heterochromatin in rye B chromosomes. *Plant Cell* **2007**, *19*, 1738–1749. [CrossRef] [PubMed]

36. Dover, G. Concerted evolution, molecular drive and natural selection. *Curr. Biol.* **1994**, *4*, 1165–1166. [CrossRef]

37. Lim, K.Y.; Kovarik, A.; Matyasek, R.; Bezdek, M.; Lichtenstein, C.P.; Leitch, A.R. Gene conversion of ribosomal DNA in *Nicotiana tabacum* is associated with undermethylated, decondensed and probably active gene units. *Chromosoma* **2000**, *109*, 161–172. [PubMed]

38. Dadejova, M.; Lim, K.Y.; Souckova-Skalicka, K.; Matyasek, R.; Grandbastien, M.A.; Leitch, A.; Kovarik, A. Transcription activity of rRNA genes correlates with a tendency towards intergenomic homogenization in *Nicotiana* allotetraploids. *New Phytol.* **2007**, *174*, 658–668. [CrossRef] [PubMed]

39. Van Vugt, J.; de Nooijer, S.; Stouthamer, R.; de Jong, H. NOR activity and repeat sequences of the paternal sex ratio chromosome of the parasitoid wasp *Trichogramma kaykai*. *Chromosoma* **2005**, *114*, 410–419. [CrossRef] [PubMed]

40. Dhar, M.K.; Friebe, B.; Koul, A.K.; Gill, B.S. Origin of an apparent B chromosome by mutation, chromosome fragmentation and specific DNA sequence amplification. *Chromosoma* **2002**, *111*, 332–340. [CrossRef] [PubMed]

41. Csonka, E.; Cserpan, I.; Fodor, K.; Hollo, G.; Katona, R.; Kereso, J.; Praznovszky, T.; Szakal, B.; Telenius, A.; de Jong, G.; et al. Novel generation of human satellite DNA-based artificial chromosomes in mammalian cells. *J. Cell Sci.* **2000**, *113*, 3207–3216. [PubMed]

42. Borisjuk, N.; Borisjuk, L.; Komarnytsky, S.; Timeva, S.; Hemleben, V.; Gleba, Y.; Raskin, I. Tobacco ribosomal DNA spacer element stimulates amplification and expression of heterologous genes. *Nat. Biotechnol.* **2000**, *18*, 1303–1306. [CrossRef] [PubMed]

43. Schubert, I.; Wobus, U. In situ hybridisation confirms jumping nucleolus organizing regions in *Allium*. *Chromosoma* **1985**, *92*, 143–148. [CrossRef]

44. Abirached-Darmency, M.; Prado-Vivant, E.; Chelysheva, L.; Pouthier, T. Variation in rDNA locus number and position among legume species and detection of 2 linked rDNA loci in the model *Medicago truncatula* by FISH. *Genome* **2005**, *48*, 556–561. [CrossRef] [PubMed]

45. Dubcovsky, J.; Dvorak, J. Ribosomal RNA multigene loci: Nomads of the Triticeae genomes. *Genetics* **1995**, *140*, 1367–1377. [PubMed]

46. Datson, P.M.; Murray, B.G. Ribosomal DNA locus evolution in *Nemesia*: Transposition rather than structural rearrangement as the key mechanism? *Chromosome Res.* **2006**, *14*, 845–857. [CrossRef] [PubMed]

47. Sandery, M.J.; Forster, J.W.; Blunden, R.; Jones, R.N. Identification of a family of repeated sequences on the rye B chromosome. *Genome* **1990**, *33*, 908–913. [CrossRef]

48. Blunden, R.; Wilkes, T.J.; Forster, J.W.; Jimenez, M.M.; Sandery, M.J.; Karp, A.; Jones, R.N. Identification of the E3900 family, a second family of rye B chromosome specific repeated sequences. *Genome* **1993**, *36*, 706–711. [CrossRef] [PubMed]

49. Langdon, T.; Jenkins, G.; Seago, C.; Jones, R.N.; Ougham, H.; Thomas, H.; Forster, J.W. De novo evolution of satellite DNA on the rye B chromosome. *Genetics* **2000**, *154*, 869–884. [PubMed]

50. Marques, A.; Klemme, S.; Guerra, M.; Houben, A. Cytomolecular characterization of de novo formed rye B chromosome variants. *Mol. Cytogenet.* **2012**, *5*, 34–37. [CrossRef] [PubMed]

51. Banaei-Moghaddam, A.M.; Schubert, V.; Kumke, K.; Weibeta, O.; Klemme, S.; Nagaki, K.; Macas, J.; Gonzalez-Sanchez, M.; Heredia, V.; Gomez-Revilla, D.; et al. Nondisjunction in favor of a chromosome: The mechanism of rye B chromosome drive during pollen mitosis. *Plant Cell* **2012**, *24*, 4124–4134. [CrossRef] [PubMed]

52. Klemme, S.; Banaei-Moghaddam, A.M.; Macas, J.; Wicker, T.; Novak, P.; Houben, A. High-copy sequences reveal distinct evolution of the rye B chromosome. *New Phytol.* **2013**, *199*, 550–558. [CrossRef] [PubMed]

53. Lamb, J.C.; Han, F.; Auger, D.L.; Birchler, J. A trans-acting factor required for non-disjunction of the B chromosome is located distal to the TB-4Lb breakpoint on the B chromosome. *Maize Genet. Coop. Newsl.* **2006**, *80*, 51–54.

54. Ward, E.J. Nondisjunction: Localization of the controlling site in the maize B chromosome. *Genetics* **1973**, *73*, 387–391. [PubMed]

55. Alfenito, M.R.; Birchler, J.A. Molecular characterization of a maize B chromosome centric sequence. *Genetics* **1993**, *135*, 589–597. [PubMed]

56. Lamb, J.C.; Riddle, N.C.; Cheng, Y.M.; Theuri, J.; Birchler, J.A. Localization and transcription of a retrotransposon-derived element on the maize B chromosome. *Chromosome Res.* **2007**, *15*, 383–398. [CrossRef] [PubMed]

57. Stark, E.A.; Connerton, I.; Bennett, S.T.; Barnes, S.R.; Parker, J.S.; Forster, J.W. Molecular analysis of the structure of the maize B-chromosome. *Chromosome Res.* **1996**, *4*, 15–23. [CrossRef] [PubMed]

58. Cheng, Y.M.; Lin, B.Y. Cloning and characterization of maize B chromosome sequences derived from microdissection. *Genetics* **2003**, *164*, 299–310. [PubMed]

59. Cheng, Y.M.; Lin, B.Y. Molecular organization of large fragments in the maize B chromosome: Indication of a novel repeat. *Genetics* **2004**, *166*, 1947–1961. [CrossRef] [PubMed]

60. Jin, W.; Lamb, J.C.; Vega, J.M.; Dawe, R.K.; Birchler, J.A.; Jiang, J. Molecular and functional dissection of the maize B chromosome centromere. *Plant Cell* **2005**, *17*, 1412–1423. [CrossRef] [PubMed]

61. Lamb, J.C.; Kato, A.; Birchler, J.A. Sequences associated with A chromosome centromeres are present throughout the maize B chromosome. *Chromosoma* **2005**, *113*, 337–349. [CrossRef] [PubMed]

62. John, U.P.; Leach, C.R.; Timmis, J.N. A sequence specific to B chromosomes of Brachycome dichromosomatica. *Genome* **1991**, *34*, 739–744. [CrossRef] [PubMed]

63. Leach, C.R.; Donald, T.M.; Franks, T.K.; Spiniello, S.S.; Hanrahan, C.F.; Timmis, J.N. Organisation and origin of a B chromosome centromeric sequence from *Brachycome dichromosomatica*. *Chromosoma* **1995**, *103*, 708–714. [CrossRef] [PubMed]

64. Franks, T.K.; Houben, A.; Leach, C.R.; Timmis, J.N. The molecular organisation of a B chromosome tandem repeat sequence from *Brachycome dichromosomatica*. *Chromosoma* **1996**, *105*, 223–230. [CrossRef] [PubMed]

65. Houben, A.; Leach, C.R.; Verlin, D.; Rofe, R.; Timmis, J.N. A repetitive DNA sequence common to the different B chromosomes of the genus *Brachycome*. *Chromosoma* **1997**, *106*, 513–519. [CrossRef] [PubMed]

66. Houben, A.; Verlin, D.; Leach, C.R.; Timmis, J.N. The genomic complexity of micro B chromosomes of *Brachycome dichromosomatica*. *Chromosoma* **2001**, *110*, 451–459. [CrossRef] [PubMed]

67. Houben, A.; Wanner, G.; Hanson, L.; Verlin, D.; Leach, C.R.; Timmis, J.N. Cloning and characterisation of polymorphic heterochromatic segments of *Brachycome dichromosomatica*. *Chromosoma* **2000**, *109*, 206–213. [CrossRef] [PubMed]

68. Belyayev, A.; Raskina, O. Chromosome evolution in marginal populations of *Aegilops speltoides*: Causes and consequences. *Ann. Bot.* **2013**, *111*, 531–538. [CrossRef] [PubMed]

69. Raskina, O.; Brodsky, L.; Belyayev, A. Tandem repeats on an eco-geographical scale: Outcomes from the genome of *Aegilops speltoides*. *Chromosome Res.* **2011**, *19*, 607–623. [CrossRef] [PubMed]

70. Hosid, E.; Brodsky, L.; Kalendar, R.; Raskina, O.; Belyayev, A. Diversity of long terminal repeat retrotransposon genome distribution in natural populations of the wild diploid wheat *Aegilops speltoides*. *Genetics* **2012**, *190*, 263–274. [CrossRef] [PubMed]

71. Ruban, A.; Fuchs, J.; Marques, A.; Schubert, V.; Soloviev, A.; Raskina, O.; Badaeva, E.; Houben, A. B Chromosomes of *Aegilops speltoides* are enriched in organelle genome-derived sequences. *PLoS ONE* **2014**, *9*, e90214. [CrossRef] [PubMed]

72. Paula, A.A.; Fernandes, T.; Vignoli-Silva, M.; Vanzela, A.L.L. Comparative cytogenetic analysis of *Cestrum* (Solanaceae) reveals different trends in heterochromatin and rDNA sites distribution. *Plant Biosyst.* **2015**, *149*, 976–983. [CrossRef]

73. Fregonezi, J.N.; Rocha, C.; Torezan, J.M.; Vanzela, A.L. The occurrence of different Bs in Cestrum intermedium and *C. strigilatum* (Solanaceae) evidenced by chromosome banding. *Cytogenet. Genome Res.* **2004**, *106*, 184–188. [CrossRef] [PubMed]

74. Sobti, S.N.; Verma, V.; Rao, B.L.; Pushpangadan, P. In IOPB chromosome number reports LXV. *Taxon* **1979**, *28*, 627.

75. Sykorova, E.; Lim, K.Y.; Fajkus, J.; Leitch, A.R. The signature of the *Cestrum* genome suggests an evolutionary response to the loss of (TTTAGGG)n telomeres. *Chromosoma* **2003**, *112*, 164–172. [CrossRef] [PubMed]

76. Urdampilleta, J.D.; Chiarini, F.; Stiefkens, L.; Bernardello, G. Chromosomal differentiation of Tribe Cestreae (Solanaceae) by analyses of 18-5.8-26S and 5S rDNA distribution. *Plant Syst. Evol.* **2015**, *301*, 1325–1334. [CrossRef]

77. Vanzela, A.L.L.; de Paula, A.A.; Quintas, C.C.; Fernandes, T.; Baldissera, J.; de Souza, T.B. *Cestrum strigilatum* (Ruiz & Pavon, 1799) B chromosome shares repetitive DNA sequences with A chromosomes of different *Cestrum* (Linnaeus, 1753) species. *Comp. Cytogenet.* **2017**, *11*, 511–524. [PubMed]

78. Jamilena, M.; Ruiz Rejon, C.; Ruiz Rejon, M. A molecular analysis of the origin of the *Crepis capillaris* B chromosome. *J. Cell Sci.* **1994**, *107*, 703–708. [PubMed]

79. Jamilena, M.; Garrido-Ramos, M.; Ruiz Rejon, M.; Ruiz Rejon, C.; Parker, J.S. Characterisation of repeated sequences from microdissected B chromosomes of *Crepis capillaris*. *Chromosoma* **1995**, *104*, 113–120. [CrossRef] [PubMed]

80. Hobza, R.; Cegan, R.; Jesionek, W.; Kejnovsky, E.; Vyskot, B.; Kubat, Z. Impact of repetitive elements on the Y chromosome formation in plants. *Genes* **2017**, *8*, 302. [CrossRef] [PubMed]

81. Timmis, J.N.; Ayliffe, M.A.; Huang, C.Y.; Martin, W. Endosymbiotic gene transfer: Organelle genomes forge eukaryotic chromosomes. *Nat. Rev. Genet.* **2004**, *5*, 123–135. [CrossRef] [PubMed]

82. Kleine, T.; Maier, U.G.; Leister, D. DNA transfer from organelles to the nucleus: The idiosyncratic genetics of endosymbiosis. *Annu. Rev. Plant Biol.* **2009**, *60*, 115–138. [CrossRef] [PubMed]

83. Lloyd, A.H.; Timmis, J.N. The origin and characterization of new nuclear genes originating from a cytoplasmic organellar genome. *Mol. Boil. Evol.* **2011**, *28*, 2019–2028. [CrossRef] [PubMed]

84. Wang, D.; Lloyd, A.H.; Timmis, J.N. Environmental stress increases the entry of cytoplasmic organellar DNA into the nucleus in plants. *Proc. Natl. Acad. Sci. USA* **2012**, *109*, 2444–2448. [CrossRef] [PubMed]

85. Matsuo, M.; Ito, Y.; Yamauchi, R.; Obokata, J. The rice nuclear genome continuously integrates, shuffles, and eliminates the chloroplast genome to cause chloroplast-nuclear DNA flux. *Plant Cell* **2005**, *17*, 665–675. [CrossRef] [PubMed]

86. Hall, A.E.; Keith, K.C.; Hall, S.E.; Copenhaver, G.P.; Preuss, D. The rapidly evolving field of plant centromeres. *Curr. Opin. Plant Biol.* **2004**, *7*, 108–114. [CrossRef] [PubMed]

87. Jones, R.N.; Viegas, W.; Houben, A. A century of B chromosomes in plants: So what? *Ann. Bot.* **2008**, *101*, 767–775. [CrossRef] [PubMed]

88. Huang, C.Y.; Ayliffe, M.A.; Timmis, J.N. Direct measurement of the transfer rate of chloroplast DNA into the nucleus. *Nature* **2003**, *422*, 72–76. [CrossRef] [PubMed]

89. Sheppard, A.E.; Ayliffe, M.A.; Blatch, L.; Day, A.; Delaney, S.K.; Khairul-Fahmy, N.; Li, Y.; Madesis, P.; Pryor, A.J.; Timmis, J.N. Transfer of plastid DNA to the nucleus is elevated during male gametogenesis in tobacco. *Plant Physiol.* **2008**, *148*, 328–336. [CrossRef] [PubMed]

90. Sheppard, A.E.; Timmis, J.N. Instability of plastid DNA in the nuclear genome. *PLoS Genet.* **2009**, *5*, e1000323. [CrossRef] [PubMed]

91. Graves, J.A. The origin and function of the mammalian Y chromosome and Y-borne genes—An evolving understanding. *BioEssays* **1995**, *17*, 311–320. [CrossRef] [PubMed]

92. McAllister, B.F. Isolation and characterization of a retroelement from B chromosome (PSR) in the parasitic wasp *Nasonia vitripennis*. *Insect Mol. Boil.* **1995**, *4*, 253–262. [CrossRef]

93. McAllister, B.F.; Werren, J.H. Hybrid origin of a B chromosome (PSR) in the parasitic wasp Nasonia vitripennis. *Chromosoma* **1997**, *106*, 243–253. [CrossRef] [PubMed]

94. Ziegler, C.G.; Lamatsch, D.K.; Steinlein, C.; Engel, W.; Schartl, M.; Schmid, M. The giant B chromosome of the cyprinid fish *Alburnus alburnus* harbours a retrotransposon-derived repetitive DNA sequence. *Chromosome Res.* **2003**, *11*, 23–35. [CrossRef] [PubMed]

95. Camacho, J.P.; Sharbel, T.F.; Beukeboom, L.W. B-chromosome evolution. *Philos. Trans. R. Soc. B Boil. Sci.* **2000**, *355*, 163–178. [CrossRef] [PubMed]

96. Shirasu, K.; Schulman, A.H.; Lahaye, T.; Schulze-Lefert, P. A contiguous 66-kb barley DNA sequence provides evidence for reversible genome expansion. *Genome Res.* **2000**, *10*, 908–915. [CrossRef] [PubMed]

97. Charlesworth, B.; Sniegowski, P.; Stephan, W. The evolutionary dynamics of repetitive DNA in eukaryotes. *Nature* **1994**, *371*, 215–220. [CrossRef] [PubMed]

98. Diez, M.; Jimenez, M.M.; Santos, J.L. Synaptic patterns of rye B chromosomes. II. The effect of the standard B chromosomes on the pairing of the A set. *Theor. Appl. Genet.* **1993**, *87*, 17–21. [CrossRef] [PubMed]

99. Jimenez, G.; Manzanero, S.; Puertas, M.J. Relationship between pachytene synapsis, metaphase I associations, and transmission of 2B and 4B chromosomes in rye. *Genome* **2000**, *43*, 232–239. [CrossRef] [PubMed]

100. Charlesworth, D. Plant sex chromosomes. *Genome Dyn.* **2008**, *4*, 83–94. [PubMed]

101. Charlesworth, D. Plant Sex Chromosomes. *Annu. Rev. Plant Biol.* **2016**, *67*, 397–420. [CrossRef] [PubMed]

102. Lamb, J.C.; Meyer, J.M.; Corcoran, B.; Kato, A.; Han, F.; Birchler, J.A. Distinct chromosomal distributions of highly repetitive sequences in maize. *Chromosome Res.* **2007**, *15*, 33–49. [CrossRef] [PubMed]

103. Jimenez, M.M.; Romera, F.; Gonzalez-Sanchez, M.; Puertas, M.J. Genetic control of the rate of transmission of rye B chromosomes. III. Male meiosis and gametogenesis. *Heredity* **1997**, *78*, 636–644. [CrossRef]

104. Jimenez, M.M.; Romera, F.; Gallego, A.; Puertas, M.J. Genetic control of the rate of transmission of rye B chromosomes. II. 0Bx2B crosses. *Heredity* **1995**, *74*, 518–523. [CrossRef]

105. Kejnovsky, E.; Hobza, R.; Cermak, T.; Kubat, Z.; Vyskot, B. The role of repetitive DNA in structure and evolution of sex chromosomes in plants. *Heredity* **2009**, *102*, 533–541. [CrossRef] [PubMed]

106. Houben, A.; Thompson, N.; Ahne, R.; Leach, C.R.; Verlin, D.; Timmis, J.N. A monophyletic origin of the B chromosomes of *Brachycome dichromosomatica* (Asteraceae). *Plant Syst. Evol.* **1999**, *219*, 127–135. [CrossRef]

107. Bakkali, M.; Camacho, J.P.M. The B chromosome polymorphism of the grasshopper *Eyprepocnemis plorans* in North Africa: III. Mutation rate of B chromosomes. *Heredity* **2004**, *92*, 428–433. [CrossRef] [PubMed]

108. Lia, V.V.; Confalonieri, V.A.; Poggio, L. B chromosome polymorphism in maize landraces: Adaptive vs. demographic hypothesis of clinal variation. *Genetics* **2007**, *177*, 895–904. [CrossRef] [PubMed]

109. Sieber, V.K.; Murray, B.G. Structural and numerical chromosomal polymorphism in natural populations of *Alopecurus* (Poaceae). *Plant Syst. Evol.* **1981**, *139*, 121–136. [CrossRef]

110. Parker, J.S.; Lozano, R.; Taylor, S.; Rejon, M.R. Chromosomal structure of populations of *Scilla autumnalis* in the Iberian Peninsula. *Heredity* **1991**, *67*, 287–297. [CrossRef]

111. Houben, A. B Chromosomes—A Matter of Chromosome Drive. *Front. Plant Sci.* **2017**, *8*, 210. [CrossRef] [PubMed]

© 2018 by the authors. Licensee MDPI, Basel, Switzerland. This article is an open access article distributed under the terms and conditions of the Creative Commons Attribution (CC BY) license (http://creativecommons.org/licenses/by/4.0/).

GCAT
TACG
GCAT

genes

MDPI

Article

B Chromosomes in Grasshoppers: Different Origins and Pathways to the Modern B_s

Ilyas Yerkinovich Jetybayev [1,2,*], Alexander Gennadievich Bugrov [2,3],
Victoria Vladimirovna Dzyubenko [3] and Nikolay Borisovich Rubtsov [1,3]

[1]   The Federal Research Center Institute of Cytology and Genetics, Russian Academy of Sciences,
     Siberian Branch, Lavrentjev Ave., 10, 630090 Novosibirsk, Russia; rubt@bionet.nsc.ru
[2]   Institute of Systematics and Ecology of Animals, Russian Academy of Sciences, Siberian Branch,
     Frunze str. 11, 630091 Novosibirsk, Russia; bugrov@fen.nsu.ru
[3]   Novosibirsk State University, Pirogov str., 2, 630090 Novosibirsk, Russia; victoriad@mail.ru
*   Correspondence: jetybayev@mail.ru; Tel.: +7-383-363-49-63 (ext. 1027)

Received: 28 August 2018; Accepted: 11 October 2018; Published: 18 October 2018

Abstract: B chromosomes (B_s) were described in most taxa of eukaryotes and in around 11.9% of studied Orthopteran species. In some grasshopper species, their evolution has led to many B chromosome morphotypes. We studied the B_s in nine species (*Nocaracris tardus*, *Nocaracris cyanipes*, *Aeropus sibiricus*, *Chorthippus jacobsoni*, *Chorthippus apricarius*, *Bryodema gebleri*, *Asiotmethis heptapotamicus songoricus*, *Podisma sapporensis*, and *Eyprepocnemis plorans*), analyzing their possible origin and further development. The studied B_s consisted of C-positive or C-positive and C-negative regions. Analyzing new data and considering current hypotheses, we suggest that B_s in grasshoppers could arise through different mechanisms and from different chromosomes of the main set. We gave our special attention to the B_s with C-negative regions and suggest a new hypothesis of B chromosome formation from large or medium autosomes. This hypothesis includes dissemination of repetitive sequences and development of intercalary heterochromatic blocks in euchromatic chromosome arm followed by deletion of euchromatic regions located between them. The hypothesis is based on the findings of the *Eyprepocnemis plorans* specimens with autosome containing numerous intercalary repeat clusters, analysis of C-positive B_s in *Eyprepocnemis plorans* and *Podisma sapporensis* containing intercalary and terminal C-negative regions, and development of heterochromatic neo-Y chromosome in some Pamphagidae grasshoppers.

Keywords: B chromosomes; grasshoppers; DNA composition; repeat clusters; euchromatin degradation; microdissected DNA probes

---

1. Introduction

The B chromosomes (B_s) were initially described in 1907 as additional elements to the standard karyotype in species of the *Metapodius* genus (Hemiptera) [1]. The number of eukaryotic species with B_s were estimated from 1685 [2] in 1980 to more than 2828 [3] in 2017. Among Orthopteran species, B_s were described in 191 species, which constitutes 11,9% of studied species [4]. Often B_s are enriched with repetitive sequences and are highly heterochromatic. The presence of B_s is non-essential for organism development and provides little or no effect [5–8].

Despite intensive studies and wide range of used methods, the origin and further evolution of B_s remains an intriguing question. None of the approaches can provide compelling evidence on this topic. Modern views on these questions are still full of hypotheses and assumptions. However, all researchers agree, that the B_s can arise either from rearranged chromosomes of a basic set or from material resulted from interspecific hybridization [5].

In the former case, the B_s often derive from pericentric regions of A chromosomes (A_s) which resulted from deletion of the whole or almost of the whole euchromatic part of a chromosome.

The B$_s$ originated from this event are small and mostly C-positive. However, there are many other morphotypes of B$_s$ in grasshoppers, where the mechanism of their origination and further evolution is not so clear.

Frequently the X chromosome and smallest autosomes are suggested as the possible ancestors of the B$_s$ in grasshoppers: the X shows special meiotic behavior and dosage compensation mechanisms that can reduce problems caused by partial aneuploidy [5,6,9,10]. Small autosomes traditionally were also considered as a possible ancestor of the B$_s$ because they contain fewer genes and thus additional small autosome could be better tolerated than other large A$_s$ [9]. Recently it was shown that some small autosomes usually contain many repetitive DNA [11–13] strands and often they contain large additional C-positive blocks on their distal part [14,15].

Because of heteropicnosity and univalency of many B$_s$, they resemble the X in meiosis. This similarity is a circumstantial evidence in favor of the X chromosome origin of B$_s$ but it cannot be acknowledged to be the direct proof [10]. Molecular markers also do not provide compelling evidence of the B chromosome origin. For instance, mutual location of 180 bp satellite DNA (satDNA) and ribosomal DNA (rDNA) cluster on the B$_s$ and on the X in *Eyprepocnemis plorans* supported the hypothesis that the X was the ancestor of these B$_s$ [16,17]. However, sequencing of internal transcribed spacer (ITS) of 45S rRNA gene revealed unique variants of the ITS present in each chromosome. The ITS from the autosome S11 appeared to be the closest one to the ITS from the B chromosome [18]. We should also note that sequencing of B$_s$ in six mammalian species showed that B$_s$ can contain DNA fragments from different A$_s$ [19–21], which significantly puzzles the question of the B$_s$ origin.

The large autosomes are rarely considered as possible ancestors of the B$_s$ [22]. Furthermore, most of the B chromosome studies are focused on C-positive B$_s$ and rarely discuss the origin C-negative B chromosome regions. Here, we described B$_s$ in 9 grasshopper species, examined their various origins including the large or medium autosomes, and proposed that they also might be considered as an ancestor of such C-negative B$_s$.

2. Material and Methods

Specimens of nine grasshopper species were collected during field season of 1987 and 2000–2017 in different locations indicated in the Table 1.

Table 1. Studied species and locations of their collection.

Taxa	Species Name	Location	Year	N
Acrididae				
Gomphocerinae	*Chorthippus apricarius* (L.)	Foothills of Trans-Ili Alatau mountains near Almaty, Kazakhstan	2005	3 ☉ from 2♀
	Chorthippus jacobsoni (Ikonn.)	Foothills of Trans-Ili Alatau mountains near Almaty, Kazakhstan	2007	7 ☉ from 3♀
	Aeropus sibiricus (L.)	Kurai steppe, Altay mountains, Russian Federation	2007	6♂
Oedipodinae	*Bryodema gebleri* (F-W.)	Kurai steppe, Altay mountains, Russian Federation	2017	12♂
Melanoplinae	*Podisma sapporensis sapporensis* Shir.	Hokkaido, Japan	2000, 2005	157♂, 37 ☉

Table 1. *Cont.*

Taxa	Species Name	Location	Year	N
Eyprepocnemidinae	*Eyprepocnemis plorans plorans* (Charp.)	Megri and Yerevan, Armenia, Antalia, Turkey, Malaga, Spain (kindly provided by JPM Camacho)	2003, 2007	381♂ 551♂
	Eyprepocnemis plorans meridionalis Uv.	Springbok, South Africa	2004	33♂
Pamphagidae				
Pamphaginae	*Nocaracris tardus* Ünal, Bugrov & Jetybayev	Sultandag mountains, Turkey	2014	7♂
	Nocaracris cyanipes (F-W.)	Prielbrusie National park Kabardino-Balkaria, Russia	1987	5♂
Trinchinae	*Asiotmethis heptapotamicus songoricus* Shum.	Ayagoz, East Kazakhstan	2015	18♂

N—Number of studied samples; ⊙—eggpods.

2.1. Chromosome Preparation

Testes of males were fixed in the field according to standard methods [23]. Females were kept alive and reared in our laboratory to obtain eggpods. The meiotic chromosomes and metaphase chromosomes of somatic cells were prepared from male testes and embryos according to standard procedures [24].

2.2. DNA Probes Generation

Chromosome microdissection was performed according to standard procedure as described earlier [24]. Microdissected DNA libraries (Table 2) were generated according to standard protocol from 8–28 copies of the chromosome or chromosome region collected by extended glass needle using a micromanipulator MR (Zeiss, Oberkochen, Germany). The WCPNtaB (Whole Chromosome Paint of *Nocaracris tardus* B) DNA probe was generated from 20 copies of B_s of the *Nocaracris tardus* specimen with the 4 B_s per cell. The PCPCapBq (Partial Chromosome Paint of *Chorthippus apricarius* B q arm) DNA probe was generated from the 8 copies of long arm of the B chromosome and PCPCapAc (Partial Chromosome Paint of *Chorthippus apricarius* A centromeric region)) DNA probe was generated from 8 copies of pericentromeric C-positive region of one of autosome of *Chorthippus apricarius*. A microdissected WCPEplBa4 (Whole Chromosome Paint of *Eyprepocnemis plorans* Ba4) DNA probe was generated from eight copies of the Ba4 of *Eyprepocnemis plorans*. The amplification of DNA isolated from the microdissected material was carried out using polymerase chain reaction (PCR) with partly degenerated primer MW6 (DOP-PCR) [25] as described earlier [26]. The PCPPsaB1-B2dist (Partial Chromosome Paint of *Podisma sapporensis* distal parts of B1 and B2) DNA probe was generated from 28 copies of distal C-negative regions of the B1 and B2 chromosomes of *Podisma sappoernsis*. Dissection was performed from central part of B1-B2 bivalent that formed chiasma in the distal parts of the B_s. Dissected material was amplified using a GenomePlex Single Cell Whole Genome Amplification Kit (WGA4) (Sigma-Aldrich, St. Louis, MO, USA) according to the manufacturer's protocol. DNA labelling was carried out in additional 15 cycles of PCR in presence of Flu- or TAMRA-dUTP (Genetyx, Novosibirsk, Russia) in additional 20 PCR cycles using WGA3 kit (Sigma-Aldrich, St. Louis, MO, USA) or with 20 high temperature cycles of DOP-PCR. DNA probe for rDNA and telomeric repeats was prepared as previously described [27].

Table 2. Microdissected DNA probes.

DNA probe	Species	Dissected Chromosome/Region
WCPNtaB	*Nocaracris tardus*	Whole B
PCPCapBq	*Chorthippus apricarius*	Long arm of the B
PCPCapAc	*Chorthippus apricarius*	Pericentric region of autosome
WCPEplBa4	*Eyprepocnemis plorans*	Whole Ba4
PCPPsaB1-B2dist	*Podisma sapporensis*	Distal parts of B1 and B2

2.3. C-Banding and Fluorescence In Situ Hybridization (FISH) Technique

C-banding and fluorescence in situ hybridization (FISH) were performed as described earlier [28,29].

2.4. Microscopy

Microscopy was performed at the Centre for Microscopy of Biological Objects (Institute of Cytology and Genetics SB RAS, Novosibirsk, Russia) with an AxioImager.M1 (Zeiss) fluorescence microscope equipped with #49, #46HE, #43HE filter sets (Zeiss), ProgRes MF CCD camera (JenaOptik, Jena, Germany), using Software package ISIS5 (MetaSystems GmbH, Altlussheim, Germany).

3. Results

B Chromosome Morphology in Studied Grasshopper Species

We analyzed B_s in nine species of grasshoppers: *Nocaracris tardus, Nocaracris cyanipes, Aeropus sibiricus, Chorthippus jacobsoni, Chorthippus apricarius, Bryodema gebleri, Asiotmethis heptapotamicus songoricus, Podisma sapporensis,* and *Eyprepocnemis plorans.*

In *Nocaracris tardus* (2n = 18 + neo-XX♀/neo-XY♂) B_s were found in 6 out of 7 karyotyped specimens. Number of the B_s varied in different specimens from 0 to 4 per cell. All revealed B_s were C-positive and very small. Their size was approximately equal to C-positive blocks detected in the most of autosomes (Figure 1a). On early stages of meiosis, these B_s conjugated with the terminal region of the short arm of the neo-X chromosome (Figure 1b). Two-color FISH with labelled rDNA and telomeric DNA probe showed only telomeric repeats on the B chromosome termini (Figure 1b). FISH with WCPNtaB revealed strong hybridization signal only on the B_s. We observed weaker hybridization signal in centromeric regions of L1 and S8 autosomes (Figure 1c).

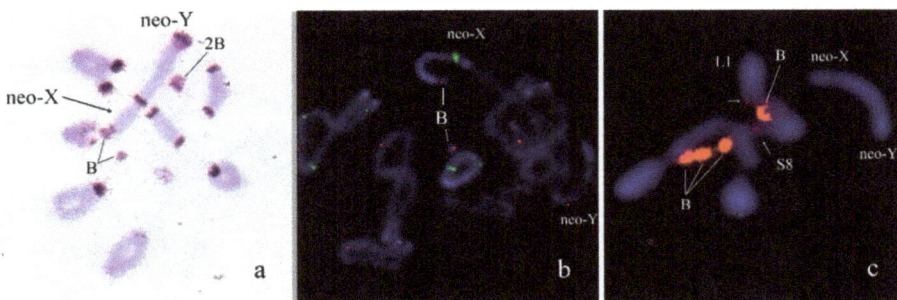

Figure 1. B_s in meiotic cells of *Nocaracris tardus*: (**a**) C-banding; (**b**) fluorescence in situ hybridization (FISH) with ribosomal DNA (rDNA) (green) and telomeric DNA (red) probes; (**c**) FISH with WCPPtaB (red) DNA probe. Arrows indicate localization of hybridization signals in pericentric regions of L1 and S8 autosomes. Chromosomes counterstained with 4′,6-Diamidino-2-Phenylindole, Dihydrochloride (DAPI) (blue).

In *Nocaracris cyanipes* (2n = 18 + neo-XX♀/neo-XY♂), the only B chromosome was found in the male from Elbrus region of Russia. It was the same as the B_s of *N. tardus* in size and C-banding pattern. Two-color FISH with labelled rDNA and telomeric DNA probe also showed only telomeric repeats clusters on the B chromosome termini. In meiosis, this B chromosome was often located close to a terminal region of the short arm of the neo-X chromosome.

In *Aeropus sibiricus* (2n = 16 + XX♀/X0♂), four B_s were found in one male out of six analyzed specimens. All four B_s were present in all studied cells. All of them were medium-sized acrocentrics with pericentromeric C-positive region similar in size to pericentromeric C-positive regions of A_s. They contain also large C-negative regions. In meiosis, they can be univalents or formed bi-, tri-, and tetravalent (Figure 2a). Two-color FISH with labelled rDNA and telomeric DNA probe revealed on the B_s rDNA in C-positive pericentric region and telomeric repeat clusters on the termini of B_s (Figure 2b). We observed the same distribution of these repeats in all chromosomes of the set.

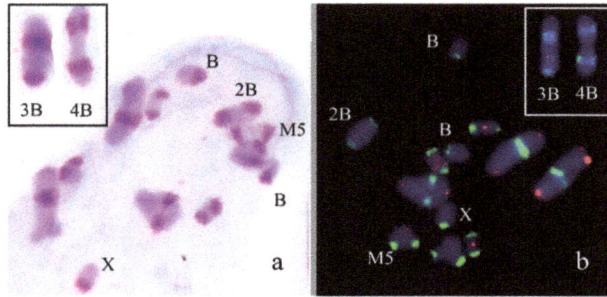

Figure 2. B_s in meiotic cells of *Aeropus sibiricus*: (**a**) C-banding, inset shows trivalent and tetravalent of B_s; (**b**) FISH with rDNA (green) and telomeric DNA (red) probes, chromosomes counterstained with DAPI (blue), inset shows trivalent and tetravalent of B_s.

In *Chorthippus jacobsoni* (2n = 16 + XX♀/X0♂), the B_s were found in 15 out of 108 analyzed embryos, with one B chromosome per cell. They were metacentrics equal to the X in size with arms similar to arms of M6 autosome. They have large pericentromeric C-positive block and C-negative arms (Figure 3a). They looked like typical isochromosomes. Two-color FISH with rDNA and telomeric DNA probes showed telomeric clusters on the ends of the B_s, the interstitial site of rDNA in two large metacentrics (Figure 3b). We did not observe any interstitial telomeric signals in pericentric C-positive block of B chromosome, in contrast to pericentric C-positive regions of large autosomes that contained the large cluster of these repeats.

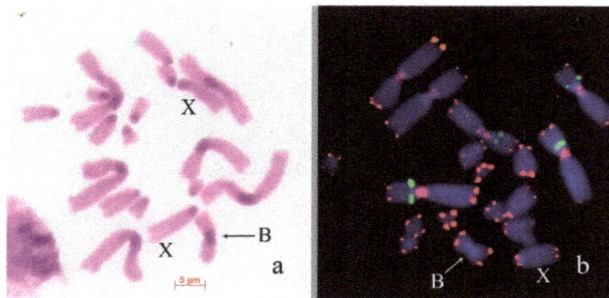

Figure 3. B_s in mitotic cells of *Chorthippus jacobsoni*: (**a**) C-banding; (**b**) FISH with rDNA (green) and telomeric DNA (red) probes, chromosomes counterstained with DAPI (blue). Arrows indicate B_s.

In *Chorthippus apricarius* (2n = 16 + XX♀/X0♂), B$_s$ were found in seven out of 52 analyzed embryos, with one B chromosome per cell. The B$_s$ were submetacentrics equal to the X in size. C-banding revealed medium-sized pericentric C-positive block and large interstitial C-positive block in the long arm of the B$_s$ (Figure 4a). FISH with labeled rDNA did not reveal clusters of rDNA in the B$_s$ (Figure 4b). PCPCapBq intensively painted interstitial C-band in the B chromosome arm and distal parts of pericentric regions of most of A$_s$. PCPCapAc painted all pericentric C-blocks (Figure 4c). In biarmed L1–L3 chromosomes, hybridization signals were significantly weaker and were observed only in distal part of the pericentric C-bands in their short arms. In L3 autosome, hybridization signals were weaker than in the L1 and L2 autosomes (Figure 4c).

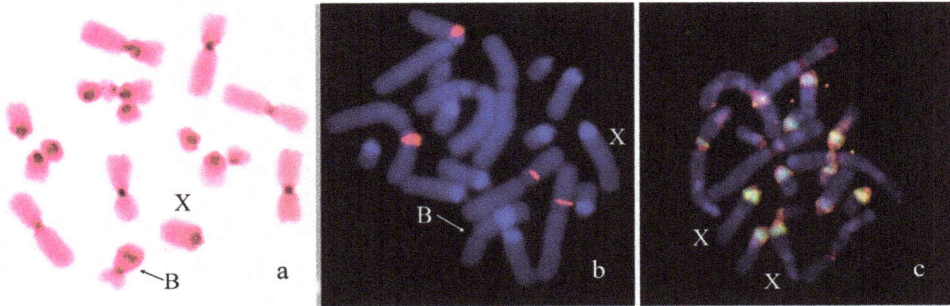

Figure 4. B$_s$ in mitotic cells of *Chorthippus apricarius*: (**a**) C-banding; (**b**) FISH with rDNA (red) probe; (**c**) FISH of PCPCapBq (green) and PCPCapAc (red) with main chromosome set. Chromosomes counterstained with DAPI (blue). Arrows indicate B$_s$.

In *Bryodema gebleri* (2n = 22 + XX♀/X0♂), the only B was found in one out of 12 males from Kurai steppe in Altay region of Russia. It was large acrocentric chromosome with very small C-negative short arm, equal to L3 in size. The B chromosome contained small pericentromeric and six intercalary C-positive blocks in the long arm. Terminal large C-positive block was approximately 1/3rd of the B chromosome size (Figure 5a). Two-color FISH with labelled rDNA and telomeric DNA probe showed small telomeric repeat cluster on the end of long arm and a large cluster of telomeric repeats in short arm of the B. (Figure 5b).

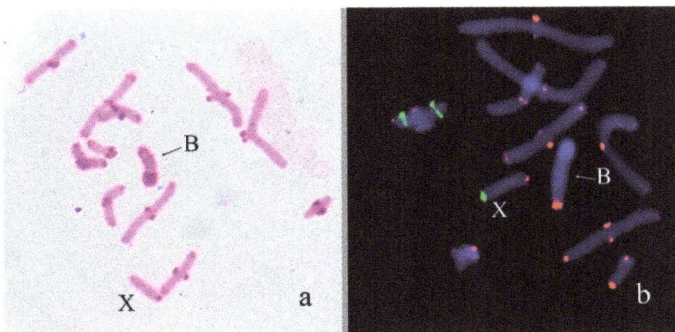

Figure 5. B$_s$ in meiotic cells of *Bryodema gebleri*: (**a**) C-banding; (**b**) FISH with rDNA (green) and telomeric DNA (red) probes, chromosomes counterstained with DAPI (blue). Arrows indicate B$_s$.

In *Asiotmethis heptapotamicus songoricus* (2n = 18 + neo-XX♀/neo-XY♂), the B_s were found in three out of 20 males from a population near the Ayaguz river, eastern Kazakhstan. In one specimen, two B chromosomes per cell were observed. The B_s were approximately equal to M6-S7 autosomes in size. They were partly C-negative and a proximal half of the B_s contained two interstitial C-positive blocks in addition to pericentromeric C-positive region. In meiosis of one specimen, two B_s formed bivalents (Figure 6a). Two-color FISH with labelled rDNA and a telomeric DNA probe showed only telomeric repeat clusters on the termini of the B_s (Figure 6b).

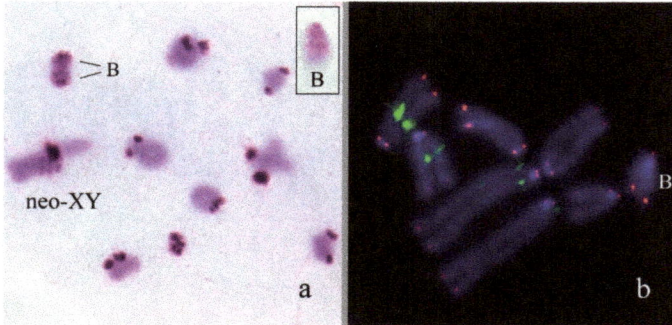

Figure 6. B_s in meiotic and mitotic cells of *Asiotmethis heptapotamicus songoricus*: (**a**) C-banding, inset shows mitotic B chromosome; (**b**) FISH with rDNA (green) and telomeric DNA (red) probes, chromosomes counterstained with DAPI (blue).

We previously analyzed the populations of *Podisma sapporensis* (2n = 22 + XX♀/X0♂ and 2n = 20 + neo-XX♀/neo-XY♂) and described seven morphotypes of the B_s [24,30,31]. The B1, B2, B3, B4 and B6iso morphotypes correspond to heavily heterochromatic B_s with C-negative region(s) in the distal part of the chromosomes. In B1 and B2 morphotypes, interstitial C-negative regions were interlaced with C-positive ones. In the B3 morphotypes C-negative regions were small. The B6iso morphotype was characterized with terminal C-negative regions in both chromosome arms. The B5iso exhibited multiple small C-positive regions dispersed along the whole chromosome. Finally, the B7 chromosome was almost fully C-negative (Figure 7). In meiosis, the B1 and B2 conjugated with their distal C-negative regions (Figure 8c).

Podisma sapporensis

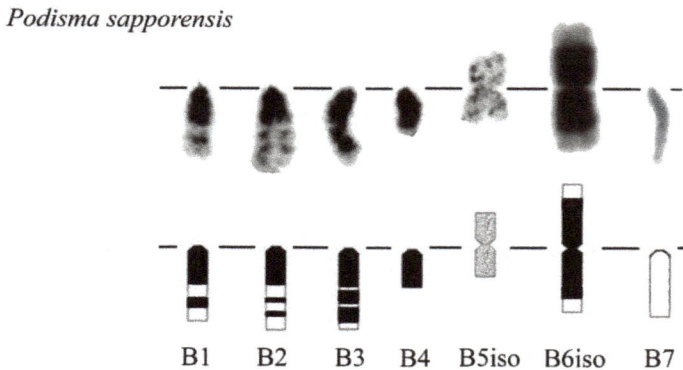

| B1 | B2 | B3 | B4 | B5iso | B6iso | B7 |

Figure 7. Morphotypes of B chromosome in *Podisma sapporensis* according to [30].

Fluorescence in situ hybridization (FISH) with rDNA and telomeric DNA probes showed that C-positive regions of B1 and B2 morphotypes were enriched with rDNA. We observed telomeric repeats on the ends of all chromosomes. In addition to these clusters, an interstitial telomeric cluster was revealed, on the border of their distal interstitial C-positive block and interstitial C-negative region in the B1 (Figure 8a,b). The PCPPsaB1-B2dist DNA probe was probably contaminated with material of C-positive regions. Consequently, FISH with PCPPsaB1-B2dist gave a very strong signal on C-positive regions of the B$_s$, masking the signal in original C-negative region (Figure 8c). Patterns of FISH signal with this microdissected DNA probe and labeled rDNA on the B$_s$ were very similar, suggesting that FISH with this microdissected DNA probe showed mainly the regions highly enriched with rDNA (Figure 8b).

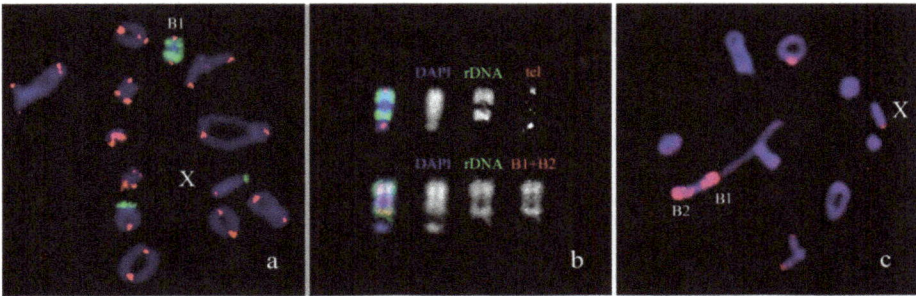

Figure 8. Fluorescence in situ hybridization (FISH) of DNA probes with meiotic cells of *Podisma sapporensis*: (**a**) rDNA (green) and telomeric DNA (red) probes; (**b**) localization of DNA probes on B1 chromosome; (**c**) PCPPsaB1-B2dist (red). Chromosomes counterstained with DAPI (blue).

In *Eyprepocnemis plorans* (2n = 22 + XX♀/X0♂), five morphotypes of the B$_s$ were revealed in eastern populations from Armenia and Turkey (Figure 9) [32]. The Ba1 and Ba2 were large acrocentrics, equal or larger to the X and mostly C-positive. Small C-negative arm was revealed in Ba2 chromosome. The Ba3 chromosome was fully C-negative, as well as Ba3iso. The Ba4 was equal in size to the smallest autosome and was mostly C-negative with small pericentric C-positive region [32]. The Ba5 was medium-sized with multiple interstitial C-positive blocks along the whole chromosome (Figure 9). Two additional morphotypes of the B$_s$ were found in populations from Turkey [32]. The Bt chromosome was small C-positive acrocentric. The Btmini chromosome was smaller than Bt and C-negative (Figure 9). In addition to eastern morphotypes of B$_s$, we studied B24 morphotype from Spain. It is a C-positive medium sized B chromosome with interstitial and distal C-negative regions. Regions of these B$_s$ painted with labeled rDNA and WCPEplBa4 DNA probe are shown in Figures 9 and 10.

Eyprepocnemis plorans

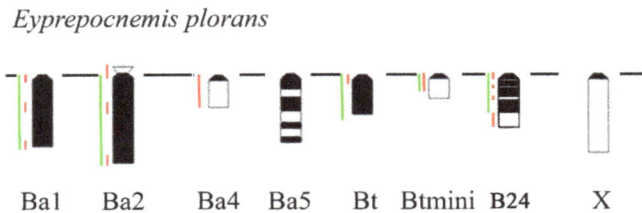

Figure 9. Schematic image of C-banding patterns and localization of rDNA (**green line**) and WCPEplBa4 DNA probes (**red line**) of different morphotypes of B$_s$ in *Eyprepocnemis plorans*. The X is shown for size comparison.

Fluorescence in situ hybridization with WCPEplBa4 also intensively painted the whole S10 autosome on metaphase plates of specimens from Armenian (Figure 10a,c,d), Spanish (Figure 10b), Turkish (Figure 10e,f) and South African (Figure 10g) populations. In the Spanish population the S10 carried a large C-positive pericentric region that was not painted with the WCPEplBa4 while its C-negative part was strongly painted with the WCPEplBa4 (Figure 10b).

FISH with WCPEplBa4 also gave a dot-like signal in centromeric regions of all other A chromosomes of Armenian (Figure 10c) specimens. C-negative material of other chromosomes exhibited a weaker hybridization signal after FISH with WCPEplBa4.

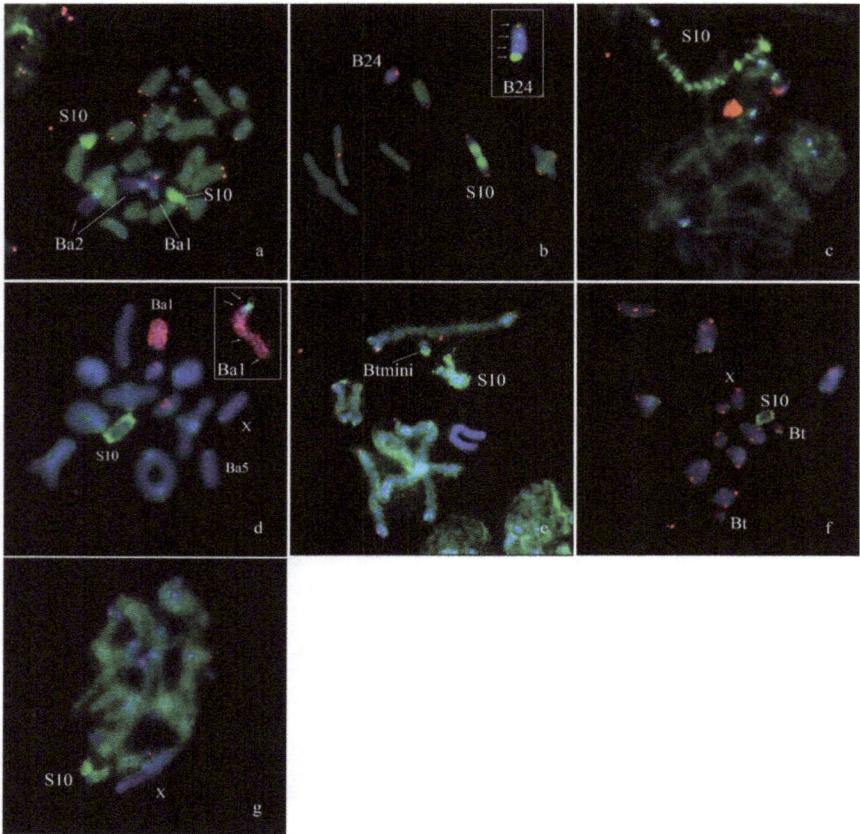

Figure 10. Fluorescence in situ hybridization of DNA probes with chromosomes of *Eyprepocnemis plorans*: (**a**) WCPEplBa4 (green) and telomeric DNA probe (red) with individual from Armenian population containing Ba1 and Ba2; (**b**) WCPEplBa4 (green) and telomeric DNA probe (red) with individual from Spanish population containing B24 (inset shows mitotic B24, arrows indicate signals from WCPEplBa4); (**c**) WCPEplBa4 (green) and rDNA probe (red) with pachytene chromosomes of individual from Armenian population; (**d**) WCPEplBa4 (green) and rDNA probe (red) with individual from Armenian population containing Ba1 and Ba5, inset shows mitotic Ba1, arrows indicate signals from WCPEplBa4; (**e**) WCPEplBa4 (green) and telomeric DNA probe (red) with individual from Turkish population containing Btmini; (**f**) WCPEplBa4 (green) and telomeric DNA probe (red) with individual from Turkish population containing BT chromosome; (**g**) WCPEplBa4 (green) and telomeric DNA probe (red) with individual of *Eyprepocnemis plorans meridionalis* from South Africa.

We revealed unexpected painting pattern after FISH with the WCPEplBa4 and labeled rDNA on the Ba1. FISH with WCPEplBa4 gave a signal on the pericentromeric region, a weaker signal on the distal region of the long arm and a dot-like signal in the proximal region and in the middle of the long arm. At the same time, labeled rDNA painted the whole Ba1 but a proximal part of the pericentromeric region (Figure 10d). On the extended meiotic chromosomes, a hybridization signal of the WCPEplBa4 on S10 autosome was split into multiple intensive signals interlaced with weaker labeled regions (Figure 10c). The same FISH pattern was observed on the S10 of all studied individuals from Armenian, Turkish, Spanish and South African populations. A FISH pattern of the Ba2 obtained with WCPEplBa4 and labeled rDNA was similar to the pattern of the Ba1 (Figure 10a). The only distinct difference of the Ba2 was associated with its short arm that was painted with the WCPEplBa4. The FISH pattern of the Ba5 after FISH with WCPEplBa4 was similar to the FISH patterns of the As (Figure 10d). FISH with the WCPEplBa4 on the Bs from Turkish populations gave hybridization signal in distal part of the Btmini and in pericentric region of the Bt (Figure 10e,f). The FISH of WCPEplBa4 with the B24 from Spanish populations exhibited two interstitial hybridization signals in C-negative regions in 1/3rd and 2/3rd of the length and one in distal terminal region (Figure 10b).

4. Discussion

4.1. Origin of Dot-Like B_s

The B_s in grasshoppers vary in size from minute dot-like elements to a large chromosome equal to the largest autosome (Figures 1–10). We observed the dot-like B_s in *N. tardus* and *N. cyanipes* (Figure 1). The mechanism of chromosome rearrangement proving these dot-like B_s remains unknown; however, these chromosomes resemble human small supernumerary marker chromosomes (SMCs). Many small SMCs are dot-like chromosomes and usually they exhibit mitotic and meiotic instability leading to mosaicism of small SMCs in tissues [33]. The B_s of *N. tardus* also showed mosaicism. Two hypotheses were previously suggested for the mechanism of human small SMC origin: SMCs arise as a result of deletion of almost the whole arm(s) or by forming the inverted duplication of pericentromeric region of acrocentrics followed by inactivation of one of the centromeres [33]. In our study, the results of the FISH of WCPNtaB probe with metaphase chromosomes of *N. tardus* supports the hypothesis of their origination from a pericentromeric region of the acrocentrics. The DNA of the B chromosome showed homology to the pericentromeric regions of the L1 and S8 autosomes (Figure 1c) indicating them as the possible B chromosome ancestor. However, taking in account rapid evolution of repetitive DNA in the pericentromeric regions, we cannot regard these repeats as reliable phylogenetic markers.

The special features of *N. tardus* and *N. cyanipes* karyotype made us look for another hypothesis of the B chromosome origin in these species. In contrast to many grasshopper species, *N. tardus* and *N. cyanipes* have neo-XX/neo-XY sex chromosomes. They derived from the fusion of the ancestral X with one of autosomes. The small fragment containing centromere could arise as the by-product of the fusion. We suppose that this new small chromosome could evolve into modern B_s. The association of these B_s with the terminal region of a short arm of neo-X chromosome (arm derived from the ancestral X) in meiosis (Figure 1b) is an additional argument for this hypothesis.

4.2. From Dot-Like B_s to the Large Ones

Traditional view on B chromosome evolution suggests that insertions into dot-like B_s followed by DNA amplification lead to large B_s [6]. Data on the B_s in *Podisma* species were involved earlier in the discussion of mechanisms of B chromosome evolution [34]. The DNA amplification leads to enlargement of their C-positive regions significantly increasing the size of the B_s. The DNA amplified in the B_s includes DNA fragments from ancestor chromosome and fragments inserted into B_s on various stages of their evolution. As a result, large C-positive B_s contain different types of repeats, such as satDNA, mobile elements, and rDNA [34–36].

The high mobility of rDNA was shown for many species and in many grasshopper species the regions of B_s appeared to be enriched with rDNA [5]. In *E. plorans* and *P. sapporensis* many C-positive regions of B_s are enriched with rDNA [34,37]. Most part of rDNA in B_s is non-functional [38]. Probably, the amplification of rDNA is frequently involved into formation of large C-positive regions of A_s and B_s. These repeats can contribute to molecular composition of pericentric C-positive regions [28,39] and in formation of numerous additional C-positive arms of A_s, for instance, in *P. sapporensis* [31] and *Eremopeza festiva* [40]. However, the amplification of rDNA does not always occur in C-positive B_s of grasshoppers. In other species involved in this study, the B_s did not contain rDNA. It is possible that the amplification of rDNA takes place in the B_s that initially contained these repeats or in species with highly mobile genetic elements containing DNA homologous rDNA.

SatDNA is another type of repetitive sequences mainly present in pericentomeric C-positive regions of A_s in many species including grasshoppers [11,13,37,41]. In grasshoppers, containing no B chromosome satDNA represents small portion of their genomes. In specimens of *Pyrgomorpha conica* without B_s in their karyotypes, high-throughput analysis of satDNA revealed 87 satDNA variants, representing 9.4% of its genome [42]. In the main genome of *Locusta migratoria*, satDNA comprise only 2.4% of genomic DNA [42]. In contrast to the main genome, C-positive B_s are enriched with satDNA. In studied B chromosome of *Locusta migratoria*, 65.2% of its DNA is consisted of this type of repeats [42]. Abundance of a certain type of satellite repeats can be altered by the DNA amplification. In the B chromosome of *L. migratoria* more than a half of DNA refer to one type of satellite repeat [42]. In *E. plorans*, B_s are significantly enriched with 180 bp satDNA repeat [43]. However, satellitome studies on grasshopper B_s are in their beginning stage and additional intensive studies are required for understanding of the role of satDNA in B chromosome evolution.

The third type of repeats that can be involved in B chromosome development are mobile elements. It was thought that genetically inert B_s could serve as the targets for mobile element insertions and their accumulation [6]. However, earlier it was shown that mariner-like elements are located mostly in euchromatic regions of A_s and C-negative regions of B_s. They were undetected in the regions enriched with rDNA and tandem repeats and in heterochromatic parts of the B_s [35,36]. Furthermore, study of the B chromosome in *L. migratoria* showed that transposable elements (TE) comprise only 18% of its content. It is significantly lower than TE content in A_s [42]. Probably, the role of TE in growth of C-positive regions of B chromosomes is nonessential.

Besides the insertion of rDNA, satDNA, and mobile elements, other repeats could also contribute to the B chromosome development. We had expected that in some B_s, amplification of telomeric repeats could also take place like in development of large C-positive regions of chromosomes in stick insects [44]. However, to date large C-positive regions of B_s enriched with telomeric repeats have not been observed. The only exception was a region in the B chromosome of *B. gebleri* (Figure 5). Small unique DNA fragments of the main genome could probably also be inserted into the B_s without forming C-negative region. This phenomenon was described in studies of some B_s in mammals [19–21].

The factors determining the frequency of the dot-like B chromosome arising and their further evolution are associated with the basic genome particularities: location of hotspots of chromosome rearrangements close to centromere; high mobility of some genetic elements; predisposition to DNA amplification. All these genome characteristics and processes might play a similar role in different species. For instance, hotspots of chromosome rearrangements close to pericentromeric regions are associated with the small SMC arising in humans [33,45], the mobility of DNA fragments leads to formation of duplicon clusters [46–48], insertions of DNA fragments and their amplification could lead to the development of additional C-positive regions [49].

4.3. The B_s with C-Negative Regions

Karyotyping of some grasshopper species revealed mostly C-negative B_s and B_s with C-negative regions. These B_s can be divided into at least three groups according to their morphology and hypothetical evolutionary stage: (*i*) mostly C-negative B_s (Figures 2–4); (*ii*) the B_s containing large

C-positive and C-negative region(s) (Figures 5–10); (*iii*) mostly C-positive B_s and C-positive B_s with small C-negative region(s) (Figures 7–10). We consider the transposition of large C-negative regions from A_s into the B_s on an advanced stage of their evolution being at least rare events and for explanation of the existence of B_s with large C-negative region(s) we looked for different mechanisms of such B chromosome development.

Cell functioning is regulated by complex interactions of multiple genes and gene dosage balance is crucial in this process. Usually aneuploidy has detrimental effects on the ontogenesis [50]. The example of a high frequency of aneuploidy of the large chromosome in free-living flatworm *Macrostomum lignano* turned out to be hidden polyploidy [51–53]. At least partial inactivation of the B_s containing C-negative regions is required to avoid genetic imbalance [9] otherwise they should be prone to negative selection. In mostly heterochromatic B_s, genes in their small C-negative regions are probably inactivated due to their close proximity to heterochromatic blocks. Human small SMCs with small euchromatic regions also show no visible effect on phenotype of their carriers [54]. This suggests that such B_s can appear to be neutral or almost neutral elements for natural selection. But for the mostly C-negative B_s the problem of dosage balance should still exist. Earlier, the X and small autosomes were considered the most probable sources for B_s. The extra X chromosome could be tolerated due to dosage compensation mechanisms while small autosomes could be tolerated due to the low number of genes [6,9]. However, we should take into account that in some grasshoppers a high level of polysomy of some autosomes was observed and these polysomic elements do not cause problems in meiosis exhibiting signs of heteropicnocity and chromosome inactivation [55]. This suggests that even extra copies of autosomes in grasshoppers can escape from negative selection. These data inspired us to come back to consideration of large and medium autosomes as ancestors of B_s in more detail.

We revealed and described B_s in *Aeropus sibiricus* that are probably on their early evolutionary stage. The morphology and C-banding patterns of these B_s were very similar to the chromosomes of the basic set (Figure 2). Similar B_s were observed in other population of *A. sibiricus* and even initially misinterpreted as aneuploidy [56]; however, later they were referred to as B_s [57,58]. In meiosis, they formed multivalents with each other and did not interfere with bivalents of A_s. Other early evolutionary stages of B_s were observed in *Chorthipps jacobsoni* and *Chorthippus apricarius* (Figures 3 and 4), however, these B_s are probably on a more advanced stage of their evolution. In *C. jacobsoni*, the B_s underwent rearrangements resulted in iso-B_s. More complex reorganization (such as inversions, new C-positive region development, and other rearrangements) has occurred in evolution of the B_s in *C. apricarius*. Occurrence of the whole C-negative B_s are rare; however, C-negative morphotypes of B_s were also described in Greek, Turkish and Armenian populations of *E. plorans* [16,32] and in *Abracris flavolineatus* [59].

The B_s containing large C-positive and C-negative regions are probably on a more evolutionary advanced stage. The B_s of this type are more frequent. They were revealed among of B_s in *E. plorans* [32,60–62] (Figure 9), *P. sapporensis* [30] (Figure 7), *Bryodema geblery* (Figure 5), *Myrmeleotettix maculatus* [63], *Xyleus discoideus angulatus* [22], and in *L. migratoria* [64] showing various morphotypes. Often these chromosomes contain a large C-positive region and smaller alternating C-positive and C-negative regions. The morphology of these chromosomes can be acrocentric or metacentric (i.e., isochromosomes) (Figures 7 and 9).

Finally, the massive part of the advanced B_s are mostly C-positive B_s or C-positive B_s with small C-negative region(s). Many described B_s of this type look like completely or almost completely heterochromatic. This stage of B chromosome evolution is indistinguishable from large B_s evolved from dot-like B_s after amplification of their DNA. However, some of these entirely C-positive B_s could contain small euchromatic regions visible only in very stretched chromosomes. For example, the C-positive Ba1 in *E. plorans* contained small regions painted with euchromatic DNA probe (Figure 10a,d). The sequencing of microdissected libraries of mammalian B_s revealed DNA fragments homologous to a unique DNA located in different chromosomes of the basic set [19–21]. The question "are the C-negative regions of the B_s the remnants of large euchromatic regions earlier present in the B or are they the result of multiple insertions/duplications of a small euchromatic chromosome region?"

remains open. Probably both types of B_s are present in natural populations. Furthermore, in some B_s small insertions of DNA fragments from euchromatic regions of A_s seem to be too small to form visible C-negative regions in these B_s [19,20].

4.4. DNA Content of C-Negative Regions in B_s

Grasshopper genomes are the largest genomes among insects and are significantly enriched with repeats [65–68]. These repeats are differentially distributed among C-positive and C-negative regions. For instance, C-negative regions contain more mobile elements than C-positive ones [35,36]. This allowed generating a microdissected DNA probe from C-negative regions that paints only C-negative chromatin [31] even in different closely related species [29,34,69]. However, in B_s C-negative regions can be enriched with repeats that differ from dispersed repeats of C-negative regions of A_s. The C-negative B7 in *P. sapporensis* is a perfect example of such B_s [31]. FISH with DNA probe generated from a C-negative region of A chromosome gave signal only in a distal region of the B7. The rest of its C-negative regions remained unpainted. DNA content of the proximal part of the B7 remains unknown. If it was formed with DNA amplification of C-negative region of the A chromosome, it should have usual interspersed repeats and should be painted with a DNA probe derived from the euchromatic region of A chromosome [70]. The loss of all interspersed repeats from these regions is unlikely. In addition, we cannot accept the B chromosome origin from interspecific hybridization of closely related species. Interspersed repeats of euchromatic regions in *Podisma* species show high homology [34]. As concerns the C-negative B_s in the Greek population of *E. plorans*, they showed enrichment for rDNA and 180 bp satDNA [16]. Maybe these B_s are on the early stages of repeat expansion and they are not able to convert C-negative regions to C-positive ones. In *E. plorans*, we revealed the mostly C-negative autosome containing numerous dispersed clusters of repeats homologous to DNA of the Ba4 (Figure 10c), which supports this hypothesis. These intercalary repeat clusters are too small to form visible C-positive blocks and their distribution was limited with the only autosome, S10.

The C-negative regions of different sizes in the B_s pose additional questions about the DNA content of B_s and transcriptional activity of these genes. There is only a little information about their transcriptional activity. Recently, ten genes were discovered in the B_s of *E. plorans*. The *CIP2A*, *GTPB6*, *KIF20A*, and *MTG1* were complete while the *CKAP2*, *CAP-G*, *HYI*, *MYCB2*, *SLIT*, and *TOP2A* were truncated. At least a half of these genes are transcriptionally active [71]. It should be noted that these genes are involved in cell division processes as well as the genes of B_s of vertebrates [19–21,72–74]. However, the exact location and flanking regions of these genes are mostly unknown. Their location in C-positive regions could not be excluded. We should note that DNA transcription in heterochromatic regions was shown in many species on different stages of individual development and in some other special situations [75,76]. The role and significance of transcription taking place in the B_s should be thoroughly studied in the future.

4.5. Mechanisms of C-Negative B Chromosome Evolution

Two forces driving B chromosome evolution can be generally described as the DNA amplification from one side and the degradation of some chromosome regions from another side. We suppose that in further development of C-negative B_s, their degradation will be probably the main process, while in dot-like B_s we should expect the amplification of their DNA. The mechanisms of degradation may include generation and distribution of repeat clusters along the euchromatic arm of the precursor chromosome like the distribution of repeat clusters along the euchromatic arm of the S10 in *E. plorans* (Figure 10c). The expansion of the repeats cluster along the chromosome probably occurred through inversions, transferring the part of cluster along the chromosome arm. After transposition, amplification of its DNA resulted in growing the repeat cluster in size. We suggest at least two consequences of this process. Due to inversions, crossing over between original and rearranged autosomes should be suppressed and the frequency of meiotic abnormalities involving them should be increased leading to a higher frequency of aneuploidy on this chromosome. Additionally, transcription activity of genes located between heterochromatic blocks could be decreased, saving the specimens with trisomy from negative natural selection. Subsequent deletions of C-negative regions located between repeat clusters could decrease genetic imbalance on this part of the genome and lead to the B chromosome arising. Smaller size of the Ba4 can be explained by the advanced stages of elimination of the euchromatic regions located between these repeat clusters. The elimination of euchromatic regions may eventually lead to a fusion of repeat clusters into a continuous region composed of repetitive DNA, showing pattern of C-positive staining. Among the studied B_s there are chromosomes containing proximal or distal enlarged C-positive regions. Probably, the distribution of repeat clusters along the chromosome arm followed by the elimination of euchromatic regions can start from the proximal or distal part of the chromosome, leading to the formation of a visible C-positive block in a corresponded chromosome region. A strong support of this hypothesis was provided by finding of the neo-Y degradation in Pamphagidae grasshoppers leading morphotypes of the neo-Ys that are similar to morphotypes of some studied B_s [27–29].

4.6. Lesson from neo-Y Chromosomes in Pamphagidae Grasshoppers

Degradation of euchromatic part of ancestor chromosome played a crucial role in neo-Y chromosome evolution in Pamphagidae grasshoppers. In contrast to standard XX♀/X0♂ sex determination chromosome system the neo-sex chromosomes are rare for grasshoppers. However, the frequency of the fusion of the autosome with the X chromosome in Pamphagidae grasshoppers appeared to be unusually high. The neo-Y chromosome in these grasshoppers undergoes a process of intensive degradation [27,29,40,77–79]. Different stages of the neo-Y evolution can be observed in species of this family. In early stages, the neo-Ys are fully homologous to the arm of the neo-X derived from the autosome. In the next stage of its development, the neo-Ys are characterized with small interstitial C-positive regions accumulated in its proximal part. More advanced neo-Ys appear to be smaller due to the deletion of C-negative regions located between C-positive blocks. The fusion of the small interstitial C-blocks forms the large ones [27,29]. Dissemination of repeats followed by the development of repeat clusters and the deletion of the euchromatic regions located between them on further stages may be a universal mechanism of the chromosome degradation in grasshoppers that could be also involved in process of the B chromosome development.

4.7. Many Ways to the B_s

We cannot discard any hypothetical mechanisms suggested earlier [5,6,10,13,71,80] and mentioned above. On the contrary, we suppose that many of them or even a combination of them can be involved in the appearance and further evolution of the B_s. Involvement of different mechanisms depends on particularities of the basic genome. Their diversity results in the great variety of the B_s that differ in their morphology, size, and DNA content. Nevertheless, being different in the initial stage, they can probably converge on a certain stage of their evolution acquiring similar morphology. However, then B_s could pass through further rearrangements to the next stage of evolution. Besides insertions and DNA amplification, B_s are also prone to converting to biarmed chromosomes. Probably the break near centromeric region and incorrect DNA reparation can result in iso-B_s. Then each arm undergoes independent evolution. A similar mechanism of iso-B chromosome development was suggested for the B chromosome evolution in the Korean field mouse, *Apodemus peninsula* [81].

4.8. The Maintenance of B_s in Natural Populations

The existence of positive or negative natural selection against the specimens with B_s remains an open question. A strong positive selection in favor of the B_s should increase the frequency of specimens with the B_s in populations and probably increase their number per specimens. Strong negative selection should lead to the loss of the B_s. There are species without the B_s and the B_s are present in the majority of specimens [82]. However, in many species the specimens with B_s represent a minor part in most of populations. From our point of view, such maintenance of the B_s

requires at least two factors. For example, we can suggest weak negative natural selection against the B_s on one hand, and on the other, the preferential B chromosome transfer in oocyte or more efficient spermatocytes with the B_s. This drive will increase the probability of the B chromosome transfer to the next generation. We should also consider an alternative suggestion: positive natural selection for carriers of the B_s and decreased frequency of the B chromosome transfer to the next generation. Data on the genes present in the B_s of *E. plorans* and their transcriptional activity [71] stimulate intensive study, keeping in mind both suggestions. It is also possible that among different kinds of B_s (the B_s differ by their origin, DNA composition, influence on adaptiveness of their carriers, and being at different stages of their evolution) there are different modes of selection. For instance, the loss of euchromatic regions in B_s neutralize the mode of selection, but enlargement of their C-positive regions will cause negative selection. The development of B_s can lead to switching of the natural selection direction and converting the maintenance of the B_s in populations to a dynamic process providing the permanent wide diversity of the B_s that depends on particularities of basic genome and habitat versatility.

5. Conclusion

The possible mechanisms of the appearance and evolution of the Bs in grasshoppers were analyzed and discussed. We suggest that different mechanisms can be involved in these processes. One of them includes the stage of a small supernumerary chromosome with the following stages of DNA fragment incorporation and amplification. Another mechanism could be associated with the distribution of repeat clusters along the arm of the A chromosome and the loss of euchromatic regions located between them. Both autosomes and sex chromosomes can be involved in the B chromosome's appearance. It is probable that at least in some cases the process of the B chromosome evolution is similar to the neo-Y degradation described in Pamphagidae grasshoppers. The origination and evolution of the Bs depends on such particularities of the basic genome as a location of hotspots of chromosome rearrangements, mobility of various genomic elements, and a predisposition of some DNA fragments to their amplification. In some grasshoppers, rDNA or DNA fragments homologous to rDNA are prone to transpositions and amplification, leading in some species to enrichment of the B chromosome regions for DNA at least partly homologous to rDNA. The question of natural selection in favor or against the Bs is still open. Furthermore, we cannot rule out the switchover to different direction of natural selection during B chromosome evolution.

Further studies on the grasshopper Bs shall be devoted to identification of their euchromatic regions, and determination of their gene content and transcription activity, including its role and significance. The question that requires special attention concerns the determination of DNA content of mysterious C-negative B chromosome regions containing no or few repeats that are typical for euchromatic regions of As.

Author Contributions: Conceptualization, I.Y.J., A.G.B. and N.B.R.; Data curation, A.G.B.; Formal analysis, I.Y.J. and V.V.D.; Funding acquisition, A.G.B.; Investigation, I.Y.J., A.G.B., V.V.D. and N.B.R.; Methodology, A.G.B. and N.B.R.; Resources, A.G.B. and N.B.R.; Supervision, A.G.B. and N.B.R.; Visualization, I.Y.J.; Writing—original draft, I.Y.J. and N.B.R.; Writing—review and editing, I.Y.J., A.G.B. and N.B.R.

Funding: This research was funded by the project #0324-2018-0019 of The Federal Research Center Institute of Cytology and Genetics, the Siberian Branch of Russian Academy of Science and research by the Russian Foundation for Basic Research (grant #18-04-00192) and the Federal Fundamental Scientific Research Program for 2013–2020, grant N° VI.51.1.5 (AAAA-A16-116121410121-7).

Acknowledgments: We thanks Juan Pedro Martinez Camacho for donating material of *E. plorans* with the B24 chromosome; Olesya Buleu, for technical assistance in chromosome preparation and C-banding, Tatiana Karamysheva for assistance with *E. plorans* microdissected DNA library preparation; Kira Zadesenets for assistance with *P. sapporensis* microdissected DNA library preparation. The authors express gratitude to Haruki Tatsuta (Ryukyu University, Nishihara, Japan) and Elżbieta Warchałowska-Śliwa for their comprehensive assistance in Hokkaido, and Nadezhda Vorobyeva for English language proofreading.

Conflicts of Interest: The authors declare no conflict of interest. The funders had no role in the design of the study; in the collection, analyses, or interpretation of data; in the writing of the manuscript, or in the decision to publish the results.

References

1. Wilson, B.E. The supernumerary chromosomes of Hemiptera. *Sci. N. Y.* **1907**, *26*, 870–871.
2. Jones, N. New species with B chromosomes discovered since 1980. *Nucleus* **2017**, *60*, 263–281. [CrossRef]
3. D'Ambrosio, U.; Alonso-Lifante, M.P.; Barros, K.; Kovařík, A.; Mas de Xaxars, G.; Garcia, S. B-chrom: A database on B-chromosomes of plants, animals and fungi. *New Phytol.* **2017**, 635–642. [CrossRef] [PubMed]
4. Palestis, B.G.; Cabrero, J.; Trivers, R.; Camacho, J.P.M. Prevalence of B chromosomes in Orthoptera is associated with shape and number of A chromosomes. *Genetica* **2010**, *138*, 1181–1189. [CrossRef] [PubMed]
5. Camacho, J.P.M.; Sharbel, T.F.; Beukeboom, L.W. B-chromosome evolution. *Philos. Trans. R. Soc. B Biol. Sci.* **2000**, *355*, 163–178. [CrossRef] [PubMed]
6. Camacho, J.P.M. Chapter 4—B Chromosomes. In *The Evolution of the Genome*; Gregory, T., Ed.; Academic Press: Burlington, NJ, USA, 2005; pp. 223–286. ISBN 978-0-12-301463-4.
7. Blagojević, J.; Vujošević, M. B chromosomes and developmental homeostasis in the yellow-necked mouse, *Apodemus flavicollis* (Rodentia, Mammalia): Effects on nonmetric traits. *Heredity* **2004**. [CrossRef] [PubMed]
8. Jojić, V.; Blagojević, J.; Vujošević, M. B chromosomes and cranial variability in yellow-necked field mice (*Apodemus flavicollis*). *J. Mammal.* **2011**. [CrossRef]
9. Hewitt, G.M. The integration of supernumerary chromosomes into the orthopteran genome. *Cold Spring Harb. Symp. Quant. Biol.* **1974**, *38*, 183–194. [CrossRef] [PubMed]
10. Camacho, J.P.M.; Schmid, M.; Cabrero, J. B chromosomes and sex in animals. *Sex. Dev.* **2011**, *5*, 155–166. [CrossRef] [PubMed]
11. Ruiz-Ruano, F.J.; López-León, M.D.; Cabrero, J.; Camacho, J.P.M. High-throughput analysis of the satellitome illuminates satellite DNA evolution. *Nat. Publ. Gr.* **2016**, *6*, 128333. [CrossRef] [PubMed]
12. Palacios-Gimenez, O.M.; Dias, G.B.; de Lima, L.G.; Kuhn, G.C.E.S.S.; Ramos, É.; Martins, C.; Cabral-de-Mello, D.C. High-throughput analysis of the satellitome revealed enormous diversity of satellite DNAs in the neo-Y chromosome of the cricket *Eneoptera surinamensis*. *Sci. Rep.* **2017**, *7*, 6422. [CrossRef] [PubMed]
13. Ruiz-Ruano, F.J.; Castillo-Martínez, J.; Cabrero, J.; Gómez, R.; Camacho, J.P.M.; López-León, M.D. High-throughput analysis of satellite DNA in the grasshopper *Pyrgomorpha conica* reveals abundance of homologous and heterologous higher-order repeats. *Chromosoma* **2018**, *127*, 323–340. [CrossRef] [PubMed]
14. Cabrero, J.; Camacho, J.P.M. Cytogenetic studies in gomphocerine grasshoppers. I. Comparative analysis of chromosome C-banding pattern. *Heredity* **1986**, *56*, 365–372. [CrossRef]
15. Santos, J.L.; Arana, P.; Giráldez, R.; Girfildez, R.; Giráldez, R.; Girfildez, R. Chromosome C-banding patterns in Spanish Acridoidea. *Genetica* **1983**, *61*, 65–74. [CrossRef]
16. Abdelaziz, M.; Teruel, M.; Chobanov, D.; Camacho, J.P.M.; Cabrero, J. Physical mapping of rDNA and satDNA in A and B chromosomes of the grasshopper *Eyprepocnemis plorans* from a Greek population. *Cytogenet. Genome Res.* **2007**, *119*, 143–146. [CrossRef] [PubMed]
17. López-León, M.D.; Neves, N.; Schwarzacher, T.; Heslop-Harrison, J.S.; Hewitt, G.M.; Camacho, J.P.M. Possible origin of a B chromosome deduced from its DNA composition using double FISH technique. *Chromosom. Res.* **1994**, *2*, 87–92. [CrossRef]
18. Teruel, M.; Ruíz-Ruano, F.J.; Marchal, J.A.; Sánchez, A.; Cabrero, J.; Camacho, J.P.M.; Perfectti, F. Disparate molecular evolution of two types of repetitive DNAs in the genome of the grasshopper *Eyprepocnemis plorans*. *Heredity* **2014**, *112*, 531–542. [CrossRef] [PubMed]
19. Makunin, A.I.; Kichigin, I.G.; Larkin, D.M.; O'Brien, P.C.M.; Ferguson-Smith, M.A.; Yang, F.; Proskuryakova, A.A.; Vorobieva, N.V.; Chernyaeva, E.N.; O'Brien, S.J.; et al. Contrasting origin of B chromosomes in two cervids (Siberian roe deer and grey brocket deer) unravelled by chromosome- specific DNA sequencing. *BMC Genom.* **2016**, *17*. [CrossRef] [PubMed]
20. Makunin, A.I.; Rajičić, M.; Karamysheva, T.V.; Romanenko, S.A.; Druzhkova, A.S.; Blagojević, J.; Vujošević, M.; Rubtsov, N.B.; Graphodatsky, A.S.; Trifonov, V.A. Low-pass single-chromosome sequencing of human small supernumerary marker chromosomes (sSMCs) and Apodemus B chromosomes. *Chromosoma* **2018**, *127*, 301–311. [CrossRef] [PubMed]
21. Makunin, A.; Romanenko, S.; Beklemisheva, V.; Perelman, P.; Druzhkova, A.; Petrova, K.; Prokopov, D.; Chernyaeva, E.; Johnson, J.; Kukekova, A.; et al. Sequencing of Supernumerary Chromosomes of Red Fox and Raccoon Dog Confirms a Non-Random Gene Acquisition by B Chromosomes. *Genes* **2018**, *9*, 405. [CrossRef] [PubMed]

22. Loreto, V.; Cabrero, J.; López-León, M.D.; Camacho, J.P.M.; Souza, M.J. Possible autosomal origin of macro B chromosomes in two grasshopper species. *Chromosome Res.* **2008**, *16*, 233–241. [CrossRef] [PubMed]

23. Darlington, C.D.; la Cour, L.F. *The Handling of Chromosomes*, 5th ed.; George Allen And Unwin Ltd.: London, UK, 1969; ISBN 0045750130.

24. Bugrov, A.G.; Karamysheva, T.V.; Rubtsov, D.N.; Andreenkova, O.V. Comparative FISH analysis of distribution of B chromosome repetitive DNA in A and B chromosomes in two subspecies of *Podisma sapporensis* (Orthoptera, Acrididae). *Cytogenet. Genome Res.* **2004**, *106*, 284–288. [CrossRef] [PubMed]

25. Telenius, H.; Carter, N.; Bebb, C.E.; Nordensjold, M.; Ponder, B.A.; Tunnacliffe, A. Degenerate oligonucleotide-primed PCR: General amplification of targeted DNA by single degenerate primer. *Genomics* **1992**, *13*, 718–725. [CrossRef]

26. Rubtsov, N.B.; Karamisheva, T.V.; Astakhova, N.M.; Liehr, T.; Claussen, U.; Zhdanova, N.S. Zoo-FISH with region-specific paints for mink chromosome 5q: Delineation of inter-and intrachromosomal rearrangements in human, pig, and fox. *Cytogenet. Genome Res.* **2000**, *90*, 268–270. [CrossRef] [PubMed]

27. Jetybayev, I.Y.; Bugrov, A.G.; Ünal, M.; Buleu, O.G.; Rubtsov, N.B. Molecular cytogenetic analysis reveals the existence of two independent neo-XY sex chromosome systems in Anatolian Pamphagidae grasshoppers. *BMC Evol. Biol.* **2017**, *17*, 20. [CrossRef] [PubMed]

28. Jetybayev, I.E.; Bugrov, A.G.; Karamysheva, T.V.; Camacho, J.P.M.; Rubtsov, N.B. Chromosomal localization of ribosomal and telomeric DNA provides new insights on the evolution of *Gomphocerinae* grasshoppers. *Cytogenet. Genome Res.* **2012**, *138*, 36–45. [CrossRef] [PubMed]

29. Jetybayev, I.Y.; Bugrov, A.G.; Buleu, O.G.; Bogomolov, A.G.; Rubtsov, N.B. Origin and evolution of the neo-sex chromosomes in Pamphagidae grasshoppers through chromosome fusion and following heteromorphization. *Genes* **2017**, *8*, 323. [CrossRef] [PubMed]

30. Warchałowska-Śliwa, E.; Bugrov, A.G.; Tatsuta, H.; Akimoto, S.I. B chromosomes, translocation between B and autosomes, and C-heterochromatin polymorphism of the grasshopper *Podisma sapporensis Shir.* (Orthoptera, Acrididae) in Hokkaido, northern Japan. *Folia Biol.* **2001**, *49*, 63–76.

31. Bugrov, A.G.; Karamysheva, T.V.; Pyatkova, M.S.; Rubtsov, D.N.; Andreenkova, O.V.; Elzbieta, W.-S.; Rubtsov, N.B. B chromosomes of the *Podisma sapporensis Shir.* (Orthoptera, Acrididae) analysed by chromosome microdissection and FISH. *Folia Biol.* **2003**, *51*, 1–12.

32. Dzyubenko, V.V.; Karagyan, G.A.; Çiplak, B.; Rubtsov, N.B.; Bugrov, A.G. The B chromosome polymorphism in Armenian and Turkey populations of the grasshopper *Eyprepocnemis plorans plorans* (Charpentier) (Orthoptera, Acrididae, Eyprepocnemidinae). *Tsitologiya* **2006**, *48*, 1016–1022.

33. Liehr, T.; Claussen, U.; Starke, H. Small supernumerary marker chromosomes (sSMC) in humans. *Cytogenet. Genome Res.* **2004**, *107*, 55–67. [CrossRef] [PubMed]

34. Bugrov, A.G.; Karamysheva, T.V.; Perepelov, E.A.; Elisaphenko, E.A.; Rubtsov, D.N.; Warchałowska-Śliwa, E.E.; Tatsuta, H.; Rubtsov, N.B. DNA content of the B chromosomes in grasshopper *Podisma kanoi Storozh.* (Orthoptera, Acrididae). *Chromosom. Res.* **2007**, *15*, 315–325. [CrossRef] [PubMed]

35. Montiel, E.E.; Cabrero, J. Gypsy, RTE and Mariner transposable elements populate *Eyprepocnemis plorans* genome. *Genetica* **2012**, *140*, 365–374. [CrossRef] [PubMed]

36. Palacios-Gimenez, O.M.; Bueno, D.; Cabral-de-Mello, D.C. Chromosomal mapping of two *Mariner*-like elements in the grasshopper *Abracris flavolineata* (Orthoptera: Acrididae) reveals enrichment in euchromatin. *Eur. J. Entomol.* **2014**, *111*, 329–334. [CrossRef]

37. López-León, M.D.; Cabrero, J.; Dzyubenko, V.V.; Bugrov, A.G.; Karamysheva, T.V.; Rubtsov, N.B.; Camacho, J.P.M. Differences in ribosomal DNA distribution on A and B chromosomes between eastern and western populations of the grasshopper *Eyprepocnemis plorans plorans*. *Cytogenet. Genome Res.* **2008**, *121*, 260–265. [CrossRef] [PubMed]

38. Ruiz-Estévez, M.; Badisco, L.; Broeck, J.V.; Perfectti, F.; López-León, M.D.; Cabrero, J.; Camacho, J.P.M. B chromosomes showing active ribosomal RNA genes contribute insignificant amounts of rRNA in the grasshopper *Eyprepocnemis plorans*. *Mol. Genet. Genom.* **2014**, *289*, 1209–1216. [CrossRef] [PubMed]

39. López-León, M.D.; Cabrero, J.; Camacho, J.P.M. Unusually high amount of inactive ribosomal DNA in the grasshopper *Stauroderus scalaris*. *Chromosom. Res.* **1999**, *7*, 83–88. [CrossRef]

40. Bugrov, A.G.; Jetybayev, I.E.; Karagyan, G.H.; Rubtsov, N.B. Sex chromosome diversity in Armenian toad grasshoppers (Orthoptera, Acridoidea, Pamphagidae). *Comp. Cytogenet.* **2016**, *10*, 45–59. [CrossRef] [PubMed]

41. Plohl, M.; Meštrović, N.; Mravinac, B. Centromere identity from the DNA point of view. *Chromosoma* **2014**, *123*, 313–325. [CrossRef] [PubMed]

42. Ruiz-Ruano, F.J.; Cabrero, J.; López-León, M.D.; Sánchez, A.; Camacho, J.P.M. Quantitative sequence characterization for repetitive DNA content in the supernumerary chromosome of the migratory locust. *Chromosoma* **2018**, *127*, 45–57. [CrossRef] [PubMed]

43. Cabrero, J.; López-León, M.D.; Bakkali, M.; Camacho, J.P.M. Common origin of B chromosome variants in the grasshopper *Eyprepocnemis plorans*. *Heredity* **1999**, *83*, 435–439. [CrossRef] [PubMed]

44. Liehr, T.; Buleu, O.; Karamysheva, T.; Bugrov, A.; Rubtsov, N. New Insights into Phasmatodea Chromosomes. *Genes* **2017**, *8*, 327. [CrossRef] [PubMed]

45. Liehr, T. *Small Supernumerary Marker Chromosomes (sSMC)*; Springer: Berlin/Heidelberg, Germany, 2012; ISBN 978-3-642-20765-5.

46. Eichler, E.E. Masquerading Repeats: Paralogous Pitfalls of the Human Genome. *Genome Res.* **1998**, *8*, 758–762. [CrossRef] [PubMed]

47. Eichler, E.E.; Archidiacono, N.; Rocchi, M. CAGGG repeats and the pericentromeric duplication of the hominoid genome. *Genome Res.* **1999**, *9*, 1048–1058. [CrossRef] [PubMed]

48. Horvath, J.E.; Bailey, J.A.; Locke, D.P.; Eichler, E.E. Lessons from the human genome: Transitions between euchromatin and heterochromatin. *Hum. Mol. Genet.* **2001**, *10*, 2215–2223. [CrossRef] [PubMed]

49. Dorer, D.R.; Henikoff, S. Expansions of transgene repeats cause heterochromatin formation and gene silencing in *Drosophila*. *Cell* **1994**, *77*, 993–1002. [CrossRef]

50. Torres, E.M.; Williams, B.R.; Amon, A. Aneuploidy: cells losing their balance. *Genetics* **2008**, *179*, 737–746. [CrossRef] [PubMed]

51. Zadesenets, K.S.; Vizoso, D.B.; Schlatter, A.; Konopatskaia, I.D.; Berezikov, E.; Schärer, L.; Rubtsov, N.B. Evidence for karyotype polymorphism in the free-living flatworm, *Macrostomum lignano*, a model organism for evolutionary and developmental biology. *PLoS ONE* **2016**, *11*, e0164915. [CrossRef] [PubMed]

52. Zadesenets, K.S.; Schärer, L.; Rubtsov, N.B. New insights into the karyotype evolution of the free-living flatworm *Macrostomum lignano* (Platyhelminthes, Turbellaria). *Sci. Rep.* **2017**, *7*, 6066. [CrossRef] [PubMed]

53. Zadesenets, K.; Ershov, N.; Berezikov, E.; Rubtsov, N.; Zadesenets, K.S.; Ershov, N.I.; Berezikov, E.; Rubtsov, N.B. Chromosome Evolution in the Free-Living Flatworms: First Evidence of Intrachromosomal Rearrangements in Karyotype Evolution of *Macrostomum lignano* (Platyhelminthes, Macrostomida). *Genes* **2017**, *8*, 298. [CrossRef] [PubMed]

54. Hamid Al-Rikabi, A.B.; Pekova, S.; Fan, X.; Jancuskova, T.; Liehr, T. Small Supernumerary Marker Chromosome May Provide Information on Dosage-insensitive Pericentric Regions in Human. *Curr. Genom.* **2018**, *19*, 192–199. [CrossRef] [PubMed]

55. Hewitt, G.M.; John, B. Parallel Polymorphism for Supernumerary Segments in *Chorthippus parallelus* (ZETTERSTEDT) I. British Populations. *Chromosoma* **1968**, *25*, 319–342. [CrossRef] [PubMed]

56. Gosálvez, J.; López-Fernández, C. Extra heterochromatin in natural populations of *Gomphocerus Sibiricus* (Orthoptera: Acrididae). *Genetica* **1981**, *56*, 197–204. [CrossRef]

57. López-Fernández, C.; de la Vega García, C.; Gosálvez, J. Unstable B-Chromosomes in *Gomphocerus Sibiricus* (Orthoptera). *Caryologia* **1986**, *39*, 185–192. [CrossRef]

58. Gusachenko, A.M.; Vysotskaya, L.V.; Bugrov, A.G. Cytogeneic analysis of Siberian Grasshopper (Orthoptera, Acrididae) from Altay. *Dokl. Biol. Sci.* **1993**, *328*, 250–252.

59. Cella, D.M.; Ferreira, A. The Cytogenetics of *Abracris flavolineata* (Orthoptera, Caelifera, Ommatolampinae, Abracrini). *Rev. Bras. Genet.* **1991**, *14*, 315–329.

60. Henriques-Gil, N.; Santos, J.L.; Arana, P. Evolution of a complex B-chromosome polymorphism in the grasshopper *Eyprepocnemis plorans*. *Chromosoma* **1984**, *89*, 290–293. [CrossRef]

61. Bakkali, M.; Camacho, J.P.M. The B chromosome polymorphism of the grasshopper *Eyprepocnemis plorans* in North Africa: III. Mutation rate of B chromosomes. *Heredity* **2004**, *92*, 428–433. [CrossRef] [PubMed]

62. Camacho, J.P.M.; Carballo, A.R.; Cabrero, J. The B-chromosome system of the grasshopper *Eyprepocnemis plorans* subsp. *plorans* (Charpentier). *Chromosoma* **1980**, *80*, 163–176. [CrossRef]

63. Gallagher, A.; Hewitt, G.; Gibson, I. Differential Giemsa staining of heterochromatic B-chromosomes in *Myrmeleotettix maculatus* (Thunb.) (Orthoptera: Acrididae). *Chromosoma* **1972**, *40*, 167–172. [CrossRef]

64. Cabrero, J.; Viseras, E.; Camacho, J.P.M. The B-chromosomes of Locusta migratoria I. Detection of negative correlation between mean chiasma frequency and the rate of accumulation of the B's; a reanalysis of the available data about the transmission of these B-chromosomes. *Genetica* **1984**, *64*, 155–164. [CrossRef]

65. Bensasson, D.; Petrov, D.A.; Zhang, D.X.; Hartl, D.L.; Hewitt, G.M. Genomic gigantism: DNA loss is slow in mountain grasshoppers. *Mol. Biol. Evol.* **2001**, *18*, 246–253. [CrossRef] [PubMed]

66. Gregory, T.R. Animal Genome Size Database. Available online: http://www.genomesize.com (accessed on 18 October 2018).

67. Li, X.-J.; Zhang, D.-C.; Wang, W.-Q.; Zheng, J.-Y. The chromosomal C-banding karyotypes of two pamphagid species from China. *Chin. Bull. Entomol.* **2008**, *45*, 549–553.

68. Camacho, J.P.M.; Ruiz-Ruano, F.J.; Martín-Blázquez, R.; López-León, M.D.; Cabrero, J.; Lorite, P.; Cabral-de-Mello, D.C.; Bakkali, M. A step to the gigantic genome of the desert locust: Chromosome sizes and repeated DNAs. *Chromosoma* **2015**, *124*, 263–275. [CrossRef] [PubMed]

69. Jetybayev, I.; Karamysheva, T.; Bugrov, A.; Rubtsov, N. Cross-hybridization of repetitive sequences from the pericentric heterochromatic region of *Chorthippus apricarius* (L.) with the chromosomes of grasshoppers from the Gomphocerini tribe. *Eurasian Entomol. J.* **2010**, *9*, 433–436.

70. Eckert, W.A.; Plass, C.; Weith, A.; Traut, W.; Winking, H. Transcripts from amplified sequences of an inherited homogeneously staining region in chromosome 1 of the house mouse (*Mus musculus*). *Mol. Cell. Biol.* **1991**, *11*, 2229–2235. [CrossRef] [PubMed]

71. Navarro-Domínguez, B.; Ruiz-Ruano, F.J.; Cabrero, J.; Corral, J.M.; López-León, M.D.; Sharbel, T.F.; Camacho, J.P.M. Protein-coding genes in B chromosomes of the grasshopper *Eyprepocnemis plorans*. *Sci. Rep.* **2017**, *7*, 45200. [CrossRef] [PubMed]

72. Yudkin, D.V.; Trifonov, V.A.; Kukekova, A.V.; Vorobieva, N.V.; Rubtsova, N.V.; Yang, F.; Acland, G.M.; Ferguson-Smith, M.A.; Graphodatsky, A.S. Mapping of KIT adjacent sequences on canid autosomes and B chromosomes. *Cytogenet. Genome Res.* **2007**, *116*, 100–103. [CrossRef] [PubMed]

73. Kichigin, I.G.; Lisachov, A.P.; Giovannotti, M.; Makunin, A.I.; Kabilov, M.R.; O'Brien, P.C.M.; Ferguson-Smith, M.A.; Graphodatsky, A.S.; Trifonov, V.A. First report on B chromosome content in a reptilian species: The case of *Anolis carolinensis*. *Mol. Genet. Genom.* **2018**. [CrossRef] [PubMed]

74. Valente, G.T.; Conte, M.A.; Fantinatti, B.E.A.A.; Cabral-De-Mello, D.C.; Carvalho, R.F.; Vicari, M.R.; Kocher, T.D.; Martins, C. Origin and evolution of B chromosomes in the cichlid fish *Astatotilapia latifasciata* based on integrated genomic analyses. *Mol. Biol. Evol.* **2014**, *31*, 2061–2072. [CrossRef] [PubMed]

75. Hall, L.E.; Mitchell, S.E.; O'Neill, R.J. Pericentric and centromeric transcription: A perfect balance required. *Chromosom. Res.* **2012**, *20*, 535–546. [CrossRef] [PubMed]

76. Saksouk, N.; Simboeck, E.; Déjardin, J. Constitutive heterochromatin formation and transcription in mammals. *Epigenet. Chromatin* **2015**, *8*, 3. [CrossRef] [PubMed]

77. Bugrov, A.G.; Grozeva, S. Neo-XY chromosome sex determination in four species of the pamphagid grasshoppers (Orthoptera, Acridoidea, Pamphagidae) from Bulgaria. *Caryologia* **1998**, *51*, 115–121. [CrossRef]

78. Bugrov, A.G.; Warchałowska-Śliwa, E. Chromosome numbers and C-banding patterns in some Pamphagidae grasshoppers (Orthoptera, Acridoidea) from the Caucasus, Central Asia, and Transbaikalia. *Folia Biol.* **1997**, *45*, 133–138.

79. Bugrov, A.G.; Buleu, O.G.; Jetybayev, I.E. Chromosome polymorphism in populations of *Asiotmethis heptapotamicus* (Zub.) (Pamphagidae, Thrinchinae) from Kazakhstan. *Eurasian Entomol. J.* **2016**, *15*, 545–549.

80. Teruel, M.; Cabrero, J.; Perfectti, F.; Camacho, J.P.M. B chromosome ancestry revealed by histone genes in the migratory locust. *Chromosoma* **2010**, *119*, 217–225. [CrossRef] [PubMed]

81. Rubtsov, N.B.; Borissov, Y.M.; Karamysheva, T.V.; Bochkarev, M.N.; Rubtsov, N.B.; Borosov, Y.M.; Karamysheva, T.V.; Bochkarev, M.N.N. The mechanisms of formation and evolution of B chromosomes in Korean field mice *Apodemus peninsulae* (Mammalia, Rodentia). *Russ. J. Genet.* **2009**, *45*, 389–396. [CrossRef]

82. Rubtsov, N.B.; Kartavtseva, I.V.; Roslik, G.V.; Karamysheva, T.V.; Pavlenko, M.V.; Iwasa, M.A.; Koh, H.S. Features of the B chromosome in Korean field mouse *Apodemus peninsulae* (Thomas, 1906) from Transbaikalia and the Far East identified by the FISH method. *Russ. J. Genet.* **2015**, *51*, 278–288. [CrossRef]

© 2018 by the authors. Licensee MDPI, Basel, Switzerland. This article is an open access article distributed under the terms and conditions of the Creative Commons Attribution (CC BY) license (http://creativecommons.org/licenses/by/4.0/).

![genes logo] **genes**

MDPI

Review

Sequence Composition and Evolution of Mammalian B Chromosomes

Nikolay B. Rubtsov [1,2,*] **and Yury M. Borisov** [3]

1   The Federal Research Center Institute of Cytology and Genetics SB RAS, Lavrentiev Ave. 10, Novosibirsk 630090, Russia
2   Novosibirsk State University, Pirogova Str. 2, Novosibirsk 630090, Russia
3   Severtzov Institute of Ecology and Evolution, Russia Academy of Sciences, Leninsky Pr. 33, Moscow 119071, Russia; boriss-spb@yandex.ru
*   Correspondence: rubt@bionet.nsc.ru; Tel.: +7-913-941-5682

Received: 28 August 2018; Accepted: 1 October 2018; Published: 10 October 2018

Abstract: B chromosomes (Bs) revealed more than a hundred years ago remain to be some of the most mysterious elements of the eukaryotic genome. Their origin and evolution, DNA composition, transcriptional activity, impact on adaptiveness, behavior in meiosis, and transfer to the next generation require intensive investigations using modern methods. Over the past years, new experimental techniques have been applied and helped us gain a deeper insight into the nature of Bs. Here, we consider mammalian Bs, taking into account data on their DNA sequencing, transcriptional activity, positions in nuclei of somatic and meiotic cells, and impact on genome functioning. Comparative cytogenetics of Bs suggests the existence of different mechanisms of their formation and evolution. Due to the long and complicated evolvement of Bs, the similarity of their morphology could be explained by the similar mechanisms involved in their development while the difference between Bs even of the same origin could appear due to their positioning at different stages of their evolution. A complex analysis of their DNA composition and other features is required to clarify the origin and evolutionary history of Bs in the species studied. The intraspecific diversity of Bs makes this analysis a very important element of B chromosome studies.

Keywords: B chromosomes; karyotypes; genome evolution; interphase nucleus; mammals; genes; repetitive DNA; transcription of heterochromatin

1. Introduction

The story of studying and describing B chromosomes (Bs) dates back to 1907, when Edmund Wilson, working on hemipteran chromosomes, noticed those that appeared to be additional to the main karyotype and were present only in a fraction of individuals [1]. However, the term 'B chromosome' was only established 11 years later. In 1928, Lowell Fitz Randolph working on variation in maize chromosomes proposed to call stable chromosomes of the standard complement 'A chromosomes', and those coming additional to the standard complement and being variable in number and morphology, 'B chromosomes' [2]. This term has since become widely recognized and used, even though it causes certain issues. It is considered that Bs are the chromosomes that exist in addition to the chromosomes of the main karyotype and are present only in some of the individuals of any given species. They have been found in karyotypes of species in most of the large taxa of multicellular organisms. Bs have not yet been reported in birds whose chromosomes are similar to mammalian ones showing G and R-bands. Germline-restricted chromosomes revealed in zebra and Bengalese finches [3–5] are sometimes named Bs, however, they do not fulfill the requirements of the given definition since they are obligatory elements of germline cell karyotype. It is possible that Bs do occur in some avian species but happen to be camouflaged by numerous microchromosomes precluding an accurate count of chromosome number on metaphase plates. It cannot be excluded either that their absence in avian karyotypes is due to strict negative selection against 'superfluous' DNA in the evolution of small avian genome [6].

B chromosomes are not necessary for normal development or reproduction, although in some species they are present in most individuals' karyotypes [7,8]. Most DNA in Bs consists of repeated elements. Often, they can be arranged in arrays of tandem repeats. In some species, Bs contained also clusters of ribosomal DNA (rDNA) forming nucleolus organizer regions (NORs) detected with AgNOR silver staining [9] or regions enriched for repeat homologous to rDNA [10,11]. Sequences identified in Bs as being homologous to unique sequences of the main genome probably occur in multiple copies too [12,13]. As DNA composition and morphology of Bs are extremely diverse within a single species, population and even individual animals [8,14], it is very difficult to say how fast they are changing throughout generations. Nevertheless, divergence time was estimated for a few sequences revealed in Bs of some grasshoppers (0.75 Myr) [15] and in Bs of some rye species (1.1–1.3 Myr) [16].

Of special interest is the meiotic behavior of Bs and their effects on conjugation, chiasma formation and A chromosome crossover. It has been demonstrated that their presence can lead to changes in chiasma frequency and recombination rate variation [17,18]. The question as to whether Bs are genomic parasites or they are precious items in the evolutionary inventory has yet to be answered. Probably, some of Bs are just genomic parasites while other Bs could contain elements that are able to develop in the future to new genes or could modify some genetic processes. It has been hypothesized that variation in the number and morphology of Bs may help some species to adapt to special environments [19] or at least changes the phenotypic features [20–22].

While Bs in different species look similar, due to Bs sequencing and some other results obtained by modern experimental techniques we make an attempt to find and follow differences between these Bs by taking into account particularities of the main genome of their carriers. The comparative analysis of specific features of mammalian Bs and similar features of mammalian A chromosomes was performed with respect to specific distribution of genes in chromosome regions, their replication timing, transcriptional regulation, and preferred places they occupy in the nuclei of somatic and meiotic cells. Despite of impressive progress in B chromosome studies taking place during the last years, it should be mentioned that the knowledge of mammalian Bs remains scarce and their organization, significance and evolution should be discussed once again following the sequencing of new Bs, description of their transcriptional activity, and intensive studies of their other features.

2. Mammalian Genomes: Organization and Diversity

The genome size in mammals shows little interspecific variation [6,23]. There is no evidence for a whole-genome duplication event in mammalian genome evolution [24]. In mammals, only a limited number of tissues are composed of polyploid cells. The hypothesis that the vizcacha rat might be tetraploid [25] failed to be confirmed by later efforts [26]. For most of its part, variation in mammalian genome size is explained by differences in the volume of repeated DNA, although duplications and even amplifications of a small 'unique' portion of the genome are known to have taken place in various phylogenetic lines of mammals [27]. In this respect, Bs are special entities in mammalian genome evolution providing in some species large variation of constitutive heterochromatin volume [18].

In humans, size variation of constitutive heterochromatin (C-positive chromosome regions) in A chromosomes usually is not associated with developmental abnormalities and there is few data on the influence of the size of heterochromatic regions on adaptiveness or other phenotypical features. From work on this topic, we would like to mention a study devoted to comparison of the heterochromatic region size in chromosomes of individuals from human populations with a long record of living in contrasting climatic conditions (the Far North, the subtropics and tropics, lowlands and uplands) [28,29]. The results of these studies suggest that at least some of the size variation of heterochromatic regions may be of adaptive value as a species is exploring a new ecological niche to settle in. In some species, Bs provide considerable variation of the heterochromatin volume [18], nevertheless, there is only a few data on the influence of their number on phenotype of their carrier [20,21].

The euchromatic regions of mammalian chromosomes consist of G and R-bands. These bands differ in many ways including replication timing profiles and gene transcription activity. Being mostly heterochromatic, Bs for a long time had been considered as elements containing no genes and transcriptionally inert. However, DNA fragments homologous to different genes were revealed recently in Bs of some species despite their C-positive pattern [30,31]. Furthermore, transcription of some of them was shown [32]. We should also mention that most of Bs are highly heterochromatic but in some species G-banding of Bs was described [33–35].

DNA fragments of Bs homologous to genes of main genome could bring into total genome additional copies of 'unique' DNA fragments. The significance of copy number variations (CNV) of different DNA regions were studied in numerous investigations of patient genomes performed with whole genome sequencing and microarray-based comparative genome hybridization [36]. It was shown that the size of the regions varied in copy numbers could be strongly different in size. Some of them have been found to exert a considerable effect on the phenotype, while the others have not been found to be associated with any developmental pathology [37,38]. Taking together, these data encourage thorough comparative analysis of the regions containing homologous genes in A and B chromosomes. Application of Hi-C technology [39] for analysis of interaction of these B chromosome regions with other elements of main genome can be efficient approach for such comparative studies.

3. Mammalian B Chromosomes: Prevalence and DNA Content

Bs can add to the main genome a considerable amount of DNA homologous both to its repeated and unique sequences, including the introns and exons of various genes and sometimes probably entire genes together with their flanking sequences [13,30,31,40]. Up to now, Bs have been found in more than 70 mammalian species [7] and the number is growing constantly. Furthermore, because Bs can be found in different species in different numbers, it does not seem feasible to count how many mammalian species contain them. While some species have them in most individuals, others have them only in a few. Moreover, differences in the frequency of the Bs are known even between populations. For example, Bs are present in most (over 90%) Korean field mice dwelling in Siberia and the Russian Far East. A cell may contain up to thirty Bs [8,41,42]. At the same time, no Bs have been found in these mice's conspecifics inhabiting Sakhalin and Stenin Island [42]. That is, if the karyotyping of a small number of individuals from several populations fails to reveal Bs, it is, to say the least, precautious to conclude that Bs do not occur in this species at all.

A lot of relevance lies with the studies of small supernumerary marker chromosomes (SMCs) in humans. At the First B chromosome Conference held in 1993 in Madrid, the following definition of the B chromosome was given: "A dispensable supernumerary chromosome that does not recombine with the A chromosomes and follows its own evolutionary pathway" [10]. Small SMCs are dispensable supernumerary elements of human karyotype. In contrast to majority of Bs, human small SMCs are usually exact copy of the original chromosome region [43,44]. The question on their recombination with A chromosomes remains open but it is possible suggesting that the frequency of such recombination is at least diminished. It is impossible to say that they already "follow their own evolutionary pathway" but they are probably similar to Bs on the initial stage of their evolution. It cannot be excluded also that there are small SMCs that have undergone some additional changes and have thus become almost full-fledged Bs; however, the example of human small SMC with additional changes has yet to be found.

Thus, there is little wonder why the question as to whether small SMCs in humans should be regarded as Bs was addressed separately [45]. Like many Bs, human small SMCs normally contain pericentric C-positive regions of A chromosomes. The presence of additional euchromatic regions in them could lead to developmental abnormalities [44,46,47]. Composition of small SMCs derived from human chromosomes 15 and 22 is rather precisely determined, while the euchromatic part of other SMC composition often remains an open question. This inaccuracy in identifying breakpoint positions, and consequently, in determining DNA content of small SMCs makes it difficult to predict what clinical effects they may have [48]. Furthermore, SMCs containing small euchromatic regions next to pericentric heterochromatin do not cause development abnormalities [49]. We suppose that human small SMC might not be regarded as Bs but the mechanism of their formation could be similar to those one of proto-B chromosomes. The study of human small SMC can provide useful data to understand chromosome rearrangements leading to B chromosome formation and its DNA content on initial stage of their evolution.

Analysis of B chromosome composition usually reveals DNA homologous to the pericentric region of A chromosomes, this DNA being often multiply amplified forms of additional extended arms of Bs [8,9,14]. Sometimes a more detailed analysis of Bs reveals DNA that is atypical for pericentric regions. Bs were found to contain additional types of repeats [8,13,31] and sequences homologous to unique DNA fragments in the main genome, including gene fragments and probably entire genes accompanied with adjacent sequences [31,40]. Moreover, data suggest that Bs may contain extended DNA regions—up to several millions of base pairs in size—homologous to euchromatic regions in A chromosomes [31]. The first evidence that those regions can be found in Bs was discovered as a by-product of a comparative cytogenetic analysis performed on red fox and raccoon dog chromosomes, using fluorescence in situ hybridization (FISH) with DNA probes designed from canine bacterial artificial chromosomes containing *C-KIT* and its adjacent sequences [30]. Red fox (*Vulpes vulpes*) and raccoon dog (*Nyctereutes procyonoides*) diverged about 12.5 million years ago; however, Bs in both of them contain homologous DNA regions [40].

Do these Bs share a common ancestor, or did they emerge independently, with genes being present in them solely because of a high frequency of inclusions of the DNA fragments containing these genes in Bs? While searching for an answer, it is important to remember that the sequencing and analysis of the human genome have revealed a large number of clusters consisted of duplicated sequences. Some of these duplicons are large and may include whole genes together with their flanking regions [50]. Bs probably are excellent recipients for such duplicated sequences. Another explanation for the presence in Bs of homologous gene copies could include the occurrence in an ancestor of a chromosome containing the genes or their copies in the pericentric region and the chromosomal rearrangement hotspot distal to it. The presence of such A chromosome in karyotype could lead to a high frequency of proto-Bs containing this group of gene copies. Hypothetical mechanisms leading to arising of Bs with different DNA composition are shown in Figure 1. Considering these hypotheses, it is important to keep in mind a complex organization of Bs in different species. Verification of the suggested hypothesis requires the study of their DNA composition and gene content.

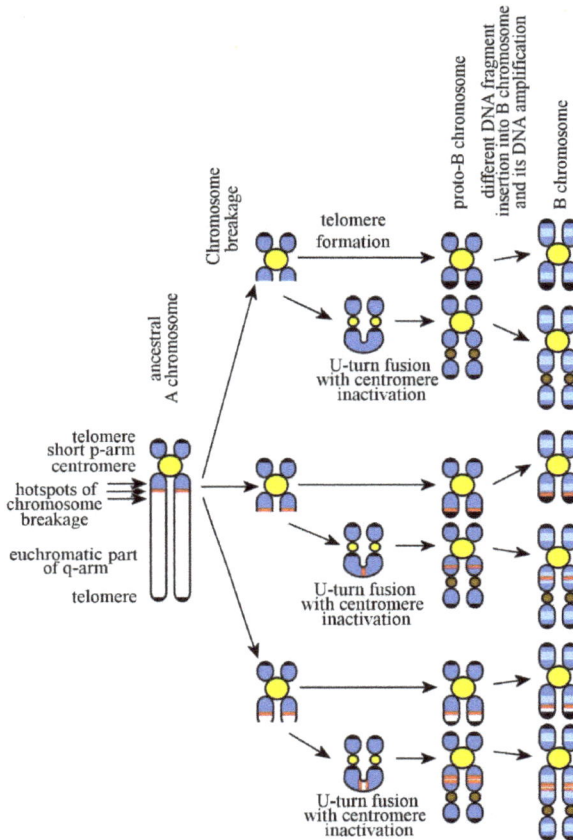

Figure 1. The B chromosomes (Bs) result from a chromosomal breakage in a pericentric heterochromatic region, a duplicon cluster, and a proximal euchromatic region. Euchromatic regions and insertions of euchromatic material in Bs are white; heterochromatic regions and insertions of heterochromatic material in Bs are blue; clusters of duplicons are red; active centromeres are yellow; inactive centromeres are brown; telomeres at the chromosome termini are black.

Breakthrough in this field is owing to the application of high-throughput sequencing of microdissected or flow-sorted DNA libraries obtained from Bs. This approach was used for studying Bs in six mammalian species [13,31,51,52] and remarkable differences between their gene content and DNA composition were found. The Siberian roe deer (*Capreolus pygargus*) Bs had two regions homologous to those in cattle chromosomes (1.42–1.98 Mbp in total). They contained three genes, while the gray brocket deer (*Mazama gouazoubira*) Bs had 26 such DNA regions (8.28–9.31 Mbp in total) with 34 whole genes and 21 gene fragments, including the proto-oncogenes *C-KIT* and *RET*, of which homologs had earlier been found in canid Bs [52]. There is a large number of mutations that distinguish the homologous sequences of the Bs in the Siberian roe deer, suggesting no strict selection has acted to keep the original DNA sequences in the Bs. By contrast, DNA in the Bs of the gray brocket deer was more homogeneous and more similar to DNA in A chromosomes [52]. Regrettably, the number of individual Bs of deer species that have been studied in detail is too low to discuss the intraspecific DNA heterogeneity in these Bs. We should note that DNA content of Bs in Siberian roe deer and brocket deer was revised recently and the total size of regions homologous to cattle chromosomes was increased from 1.96 Mbp to 2.36 Mbp in Bs of Siberian roe deer and from 9.31 Mbp to 10.46 Mbp in Bs of gray brocket deer [31].

As it was earlier shown, Bs in the raccoon dog (*Nyctereutes procyonoides*) contained at least two types of heterochromatin, interstitial telomere repeats, and DNA fragments homologous to rDNA [53–55]. Later, a more detailed analysis of Bs in two subspecies of raccoon dog and in red fox (*V. vulpes*) revealed DNA fragments homologous to C-KIT [30]. It was shown that these Bs also contained the flanking regions [40]. Sequencing of flow-sorted and microdissected DNA libraries of red fox and raccoon dog Bs revealed numerous sequences homologous to DNA fragments located in different A chromosomes of red fox and reference species [31] suggesting that numerous independent insertions could take place during evolution of these Bs or a large duplicon cluster derived from ancestral chromosome was included in initial proto-Bs of canid species. Two overlapping regions revealed in the fox and raccoon dog Bs we consider as argument in favor of the latter suggestion [31].

Other Bs that were intensively studied are the Bs in the Korean field mouse. The studies were started with a comparative analysis of Bs in mice captured at different locations of this species' habitat using various methods of cytogenetic analysis: From routine morphometry to microdissected DNA library generation followed by FISH on metaphase chromosomes [8,14,41].

The results obtained revealed a high diversity of Bs in number, size, repeated sequences, and ancestral chromosomes. Virtually all Bs attributes and properties studied varied in a wide range. The number of Bs in the Korean field mouse (*Apodemus peninsulae*) varies from 0 to 30 [41]. The size of Bs in this species varies from a dot-like chromosome to the largest chromosome in the karyotype [9]. Some of the Bs occur as acrocentrics and some as bi-armed chromosomes [8,41]. Natural populations of mice located in distant geographical regions appear also to be different in the frequency of Bs referred to different morphotypes [8,14].

Differences in the DNA of pericentric regions give reason to believe that some Bs in mice from Siberian populations originated from autosomes, while Bs in mice from Russian Far East are from a sex chromosome [8]. Analysis of chromatin folding has revealed both C-positive regions in Bs and regions that remain poorly condensed even in the late metaphase [9]. The latter regions cannot be visualized by Giemsa or 4′,6-Diamidine-2′-phenylindole dihydrochloride (DAPI) staining techniques; however, thanks to their specific DNA content, they are reliably visualized and identified by two-color FISH with microdissected DNA probes derived from micro-Bs and the pericentric region of large autosome of this species. As was found, each of the most micro-Bs are composed of a small pericentric region including DNA homologous to DNA in autosomal pericentric heterochromatin and a region(s) consisted of other repeats. The latter regions remain poorly condensed in mitosis. Poorly condensed regions, containing repeats such as these, have also been found in the distal regions of some macro-Bs [9]. These repeats have not been observed clustered in A chromosomes or in the interstitial regions of macro-Bs. Additionally, in some Bs rDNA was detected. In A chromosomes of Korean field mouse, rDNA clusters are in the terminal regions of the long arms of two pairs of autosomes, suggesting that rDNA was inserted into Bs at advance evolutionary stages [9]. The sequencing of microdissected DNA libraries of macro and micro-Bs also suggests multiple insertions of DNA fragments from A chromosomes to developed modern Bs [13]. In the mouse genome, the total size of DNA regions homologous to macro-B DNA was estimated at 9.4 Mbp (without repeated sequences totaling 7296 kbp length), while that of DNA regions homologous to micro-B DNA was 5.5 Mbp (without repeated sequences totaling 3935 kbp length). Bs included regions homologous to DNA in different mouse chromosomes: MMU1 (*Cntnap5a* and *Cntnap5b*), MMU5 (*Vmn2r84-Vmn2r87*), MMU7 (*NLR* genes), MMU9 (*Kif23*), MMU10 (the gene encoding the G protein-coupled receptors active in vomeronasal sensory neurons, and probably *Tespa1*), MMU12, MMU13, and MMU17 (*Cntnap5c*). Regions homologous to DNA in MMU1, MMU5, and MMU9 were present in both macro-B and micro-B, while the others, only in the macro-B. As was expected, the DNA of both Bs was found to be enriched for L1 long interspersed nuclear elements [13,31]. Considering the differences revealed with FISH using microdissected DNA probes on Korean field mouse metaphase chromosomes [8,9], the differences found later in the content of repeated sequences [13,31] were expected. A comparison of DNA content in the Bs of Korean field mouse and yellow-necked mouse (*Apodemus flavicollis*) showed that the Bs of both *Apodemus* species contain DNA homologous to the *Vrk1* gene (vaccinia related kinase 1) in mice [31,56]. However, the margins of *Vrk1* region differed in Bs of the studied species, suggesting independent insertions of *Vrk1*-containing DNA fragments into these Bs [13].

It should be noted that the genes homologous to sequences found in the Bs of the Korean field mouse had previously been associated with evolutionary breakpoint regions in the porcine genome [57]. The same is probably true for Bs of other species [12]. Taking all data together, it is possible to suggest the different reasons for B chromosome enrichment for copies of certain genes: (*i*) these genes are often involved in duplicon cluster formation; (*ii*) they are located in the vicinities of evolutionary breakpoints; (*iii*) there exist positive natural selection for the functional activity of their copies located in Bs; (*iv*) a combination of the above reasons; (*v*) unknown evidence.

To date, gene content in Bs has been determined with high-throughput sequencing of flow-sorted or microdissected libraries in six mammalian species belonging to canids, ruminants, and rodents [31]. Comparative analysis of the list of these genes allowed suggesting that this list is enriched for cell-cycle-related genes, development-related genes, and genes functioning in synapses. Genes belonging to these groups were also found in Bs of non-mammalian lineages: In Bs of cichlid *Lithochromis rubripinnis* morphogene *Ihhb* (Indian hedgehog b) [22]; in cichlid *Astatotilapia latifiscata* genes associated with cell division [58]; in rye *Secale cereal* pseudogenes and regulatory genes [26,59]; in a grasshopper *Eyprepocnemis plorans* five genes involved in cell division [60].

Ribosomal DNA or DNA partially homologous to it is a usual component of many Bs. There are multiple examples of species in which Bs are enriched by DNA fragments homologous to rDNA with or without nucleolus organizer region formation [9,11]. Change of rDNA location within and between the chromosomes of even closely related species has been found in many phylogenetic lines of mammals [61]. It is possible that Bs are recipients of rDNA in transposition and offer good conditions for the amplification of inserted copies.

Virtually any detailed study of DNA content in Bs using high-throughput sequencing of the DNA libraries of these Bs and other techniques have revealed sequences homologous to gene fragments and quite extended A chromosomal regions [13,31]. To answer the question on similarity or diversity of Bs within species or B chromosome presence in different species, special importance should be given to more extended studies of their DNA content.

4. Transcriptional Activity of DNA in Mammalian B Chromosomes

The finding that Bs contain DNA sequences homologous to genes of the main genome raised a question about their transcriptional activity. In relation to Bs in various species, this question was addressed in detail in recent review [62], allowing us to focus on mammalian Bs. Probably, it would be useful to divide the discussion of this problem in two parts: transcription of DNA homologous to genes of main genome and transcription of repetitive DNA. Gene transcription from Bs can be reliably detected due to differences between the B chromosome gene sequences and homologous A chromosome gene sequences. With reliance on these differences, the transcriptional activity of genes found in Bs of the Siberian roe deer has been demonstrated [32]. Considering the size variation of DNA inserts in Bs and the diversity of their flanking regions, we would like to speculate that the transcriptional activity of genes in Bs may vary considerably. This is consistent with data on the transcriptional activity of genes in Bs in plants and insects [62]. Most mammalian Bs contain extended heterochromatic regions [54,63]. If some genes in B chromosome are close to these regions, their transcriptional activity can be partially or fully suppressed. In our opinion, some data on gene transcription in Bs may represent a record of low-level transcription, which has no effect on normal cell function. This is supported by data from patients with human small SMCs that contain small euchromatic regions next to pericentric heterochromatin [49]. Healthy carriers of small SMCs with euchromatic centromere-near (ECN) imbalances in small (0.3–5 Mbp) euchromatic regions have been revealed. However, the matter of B chromosome gene transcription is far from being clear. There are Bs containing extended C-negative regions. Some of their examples are Bs of the yellow-necked mouse, *A. flavicollis* [64]. Differential display reverse transcription-polymerase chain reaction (DDRT-PCR) was used for comparative analysis of gene expression in these animals with and without B chromosome. The following three complementary DNA (cDNA) fragments with differential expression were revealed: Chaperonin containing TCP-1 subunit 6b (zeta) (*CCT6B*), fragile histidine triad gene (*FHIT*), and a hypothetical gene XP transcript. Their differential expression was confirmed by real-time PCR. The study of DNA content of five Bs in *A. flavicollis* [13,31] revealed DNA fragments homologous to 38 genes. Twenty-nine of them may form parts of functional clusters. The largest number of the genes revealed were those encoding microtubule-associated and cell-cycle gene proteins: *Cenpe* (centromere protein E), *Dync1i2* (dynein cytoplasmic 1 intermediate chain 2), *Mns1* (meiosis-specific nuclear structural protein 1), and *Mapre1* (microtubule-associated protein, RP/EB family, member 1). It should be noted that four of the studied Bs were from the Serbian populations and one was from Russian. The one from the Russian population contained a DNA fragment homologous to a ~300 kbp sequence of house mouse chromosome 9, which the Bs of the animals captured in Serbia did not. The intraspecific differences found in DNA content between Bs complicate the study on transcription of the genes in Bs. Data about the copy numbers of studied genes, DNA methylation and histone modification, regions flanking genes, and distance from heterochromatic regions are required for the correct interpretation of revealed transcriptional activity.

It would be appropriate to look back at the transcriptional activity of repetitive DNA in C-positive regions. Transcription of pericentric satellite repeats have been observed in a large number of species and have been associated with various processes, including cell proliferation, ontogenesis, cellular differentiation, ageing, stress response and cellular transformation [65–67]. Transcription of gene copies located in heterochromatic Bs could be a result of their involvement in the general process of repeat transcription [66,68,69], probably taking place in the Bs, or it could represent a specific process of major importance for perfect ontogenesis.

Unfortunately, little is known about the transcription of repeated sequences in heterochromatic regions of mammalian chromosomes. Transcription of pericentric heterochromatin in mouse embryonic fibroblasts is one of the examples of repetitive DNA studied in detail [70]. Here, the first wave of major satellite transcription was observed in late G_1 phase, peaking in early S phase. The transcripts were heterogeneous, varying in length from 1000 to more than 8000 bp. It is possible that this stage-dependent transcription is associated with the preparation of DNA in pericentric heterochromatin for replication, which also makes this DNA accessible for transcription. Of no less interest is the next wave of pericentric transcription, which is confined to mitosis and characterized with smaller transcripts (~200 bp). Features of pericentric satellite transcription, specific for ontogenetic stages, cell ageing, cellular stress, cancers and other diseases have recently been reviewed [65–67]. These reviews consider features of histone modifications, DNA-protein interactions, and DNA methylation in pericentric heterochromatin.

More is known about the matter in fission yeast. Transcription in fission yeast proceeds during S phase of a cell cycle [71–80]. However, mechanisms of repeated DNA transcription and its regulation in mammals including repeats of Bs remain poorly studied. At least in some cases gene transcription in Bs might be a part of total heterochromatin transcription. It is also possible that by leading to a considerably increased share of heterochromatin in the genome, Bs can play a role in the regulation of both transcription of repeated sequences in A and B chromosomes and structural organization of chromatin in the nucleus. Unfortunately, our knowledge on transcription in Bs is limited. Data on Bs mostly derived from transcription analysis of a few isolated B chromosome genes and from gene composition of the Bs that were studied recently [31,32].

5. Where Do Mammalian B Chromosomes Reside in the Interphase Nucleus?

In interphase, mammalian chromosomes are not fixed at certain positions; however, in most cell types, chromosomal regions or even entire chromosomes show preference either at the internal or external compartment of the nucleus as well as to certain positions relative to the nuclear lamina and the nucleolus—depending on the number and transcriptional activity of the genes in these chromosomal regions or chromosomes [81–84]. C-positive regions are located in the nucleus peripherally and near nucleoli, usually in contact with the nuclear lamina. G-band material is also found closer to the nuclear periphery and the nuclear lamina, while R-band DNA most commonly is located in the internal compartment and makes no close contact with the nuclear lamina. The number of studies seeking to locate Bs in an interphase nucleus is not high. One worked with the raccoon dog and fox Bs [85], and another with Bs in fibroblasts and spermatocytes of the Korean field mouse [18].

The fox and raccoon dog Bs contained the *C-KIT* gene but had different sizes and preferences for locations in the interphase nuclei of the fibroblasts. Small-sized Bs of the fox were mostly found in the internal compartment of the nucleus, while their larger counterparts in the raccoon dog preferred peripheral locations. According to the authors [85], their data is consistent with the hypothesis that a chromosome is located in a nucleus depending on its size which is a statement from the chromosome size-dependent theory. However, a more detailed consideration indicates that this data is also consistent with evidence that transcriptionally active chromosomal regions show preference for the internal compartment of the nucleus, while transcriptionally inactive chromosomal regions for its periphery [83,86,87]. If a chromosome is small enough, this rule applies to the entire chromosome. A classic example is the contrasting manner in which human chromosomes 18 and 19, similar in size, are localized in the nucleus [88]. These chromosomes are contrasting in the number of transcriptionally active genes they contain. Chromosome 18 consisting mostly of G-bands is most commonly localized peripherally; while chromosome 19, mostly composed of R-bands, normally resides in the internal compartment. Localization data on canid Bs are difficult to interpret due to a lack of detailed information about DNA content in particular Bs. It is possible that small-sized Bs in the fox and medium-sized Bs in the raccoon dog have small transcriptionally active regions similar in size, and the difference in size that they still have could be due to a much larger heterochromatic block in the Bs of the raccoon dog. In this case, differences in the location of Bs can be accounted for by different ratios of transcriptionally active to transcriptionally inactive chromatin. It leaves no doubt that a detailed description of Bs appearing in the studies of interphase nuclear organization will be valuable for interpreting any relevant data obtained.

In a work analyzing the localization of Bs in ten the Korean field mouse specimens, all 103 Bs being studied were characterized by FISH with microdissected DNA probes, allowing the authors to assess the size of the regions composed of repeated sequences and the diversity of their repeats [18]. The number of Bs in a single individual's cells varied from 3 to 19. The size, too, varied from a dot-like B to a large autosome. Regions homologous to particular chromosomal regions in the house mouse have been identified in two of these Bs [13], but no data on their transcriptional activity is available. In the fibroblasts, Bs were localized on the periphery of the interphase nuclei in associations with C-positive regions of A chromosomes. However, the distribution of the Bs in these associations was non-random and varied across individuals [18]. It is possible that the observed differences were due to different content of repeated sequences in the Bs that were present in the mice studied.

It is worth noting that the presence of supernumerary chromosomes leads to DNA increase in the genome. The amount of B material has been estimated to vary from 4% to 32% of the haploid genome of the Korean field mouse [18]. Furthermore, the additional material appeared as C-positive heterochromatin. In animals with a large number of macro-Bs, the amount of additional heterochromatin was substantially higher than the total amount of heterochromatin in A chromosomes [18]. Nevertheless, this variation of heterochromatin volume in the genome did not affect the normal function of the genetic machinery of the Korean field mouse cells in any way. The association of Bs with the C-positive regions of A chromosomes was likely to have helped preserve the architecture of the internal compartment of the nucleus, where most transcriptionally active genes reside. Apparently, the nucleus volume grew with the genome increasing [87,89,90]. However, because of denser heterochromatin folding, the former should have been increased to a lesser degree than the latter. Finding Bs on the nuclear periphery in associations with C-positive regions of A chromosomes is indicative of the need for an additional volume of the external compartment of the nucleus [18]. However, when it comes to change, the volume of the external compartment will be growing at a faster pace than that of the nucleus in its entirety. The ratio of the volumes of the external to the internal compartment can also be corrected for by a minor change of nuclear morphology. This correction is likely to take place naturally, because the formation of the nuclear envelope starts from the formation of the nuclear membrane on the heterochromatic regions of chromosomes [87,91].

In our opinion, the role of Bs in building the architecture of the interphase nucleus mostly depends on the composition of these chromosomes. Bs containing a small heterochromatic region and actively transcribed regions tend to have preference to the internal compartment of the nucleus, while the presence of large heterochromatic regions in Bs appears to promote their peripheral localization, in associations with heterochromatin of A chromosomes. This pattern of B chromosome localization will help maintain the optimal infrastructure of the internal compartment of the nucleus even if the number of Bs will be high.

6. Mammalian B Chromosomes: Origin and Evolution

Addressing the question as to the emergence and evolution of Bs can be staged as follows: (*i*) Emergence of a B chromosome or its ancestor (proto-B chromosome); (*ii*) changes in its DNA content; (*iii*) its behavior in mitosis and meiosis; and (*iv*) action of natural selection. In theory, several scenarios can be proposed to explain the emergence of the B chromosome or its ancestor. Due to the structural similarities between mammalian Bs and human small SMCs, it is possible to suggest that the mechanism of their formation is also similar [43] (Figure 1). Additionally, it cannot be excluded either that Bs arose in result of a gradual degradation of the ancestral A chromosome [10] (Figure 2a) or developed through insertions of foreign DNA (Figure 2b).

If most proto-Bs really arose similarly to human small SMCs, it should be expected that Bs arose more frequently in the species with acrocentrics in their karyotypes. This tendency was earlier noticed [92,93]. The probability of a break in the long arm of an acrocentric chromosome is higher than that of two breaks at once in the proximal regions of two arms of a bi-armed chromosome. Moreover, the formation of a small SMC as an inverted duplication [43,94,95] (Figure 1) removes the problem of its protection from digestion with exonuclease. All these propositions make the suggestion of B chromosome formation mostly from acrocentric chromosomes very attractive. However, Bs were also found in species with karyotype containing mostly bi-armed chromosomes [7].

It is also possible that Bs may have arisen during interspecific hybridization, of which the role in speciation appears to be underestimated [96]. No proved case of mammalian Bs originated from interspecies hybrids has been published yet. However, we should keep in mind that traces of ancient interspecific hybridization could be erased during long time evolution.

The possibility of the emergence of Bs by a gradual degradation of the X-chromosome [10] and a gradual degradation of neo-Y chromosome [97] in grasshoppers make us consider one additional mechanism of B chromosome formation. The mechanism of gradual A chromosome degradation may, at least in part, be similar to that leading to small heterochromatic neo-Y-chromosome in some grasshoppers [97]. A hypothesis was put forward, which proposed the following sequence of events (Figure 2a): (*i*) Formation of the intercalary heterochromatic blocks in the euchromatic part of the original chromosome's arm, (*ii*) reduction in the transcriptional activity of the euchromatic regions between heterochromatic blocks, (*iii*) increase of meiotic aberrations because of the presence of additional C-positive chromosomal regions, (*iv*) trisomy of this chromosome (one of the chromosomes with additional intercalary heterochromatic blocks) and (*v*) loss of euchromatic regions between heterochromatic blocks [97]. There is no described example of B chromosome formed by this mechanism, but it cannot be excluded that at least some mammalian Bs could have arisen in this way.

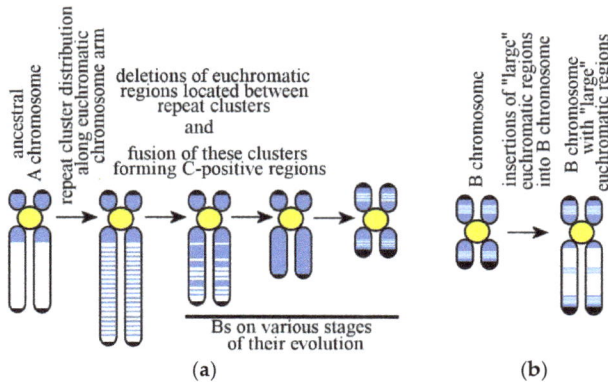

Figure 2. B chromosome arising through cluster repeats distributed along the euchromatic part of the chromosome arm of ancestral chromosome followed by the loss of the euchromatic region located between them (**a**) and Bs development through insertions of large euchromatic regions of A chromosomes or foreign DNA (**b**). Euchromatic regions and insertions of euchromatic material in Bs are white; heterochromatic regions and insertions of heterochromatic material in Bs are blue; active centromeres are yellow; inactive centromeres are brown; and telomeres on chromosome termini are black.

To add to this, there are mammalian Bs that consist mostly of C-negative regions [64]. Five Bs of yellow-necked mouse were studied by generating micro-dissected DNA libraries followed by their sequencing. Those Bs were large and composed predominantly of C-negative material. They contained C-negative regions showing G-banding [33] and DNA homologous to chromosomal regions in the house mouse, totaling a few millions of base pairs [13,31]. The organization of these chromosomes remains a mystery. They were devoid of regions homologous to extended regions of the reference genome but showed larger C-negative regions. In total, DNA in these Bs is homologous to about 3.5 Mbp of the euchromatic portion of mouse genome. It may be that they contain a large cluster of duplicons multiplied many times over. These chromosome regions might probably remain C-negative, despite the homology of their DNA to a relatively small portion of euchromatin in the reference genome. However, if this assumption is true, then they have not resulted from a gradual degradation of A chromosomes. Today, the suggestion on insertions into Bs rather large euchromatic regions from A chromosomes (Figure 2b) should remain among the hypothesis set about B chromosome origin and their further evolution.

Data on B chromosome sequencing provided additional information on particularities of the B chromosome organization that requested for their explanation the development of the general theory B chromosome evolution. The sequencing of micro-dissected DNA libraries obtained from Bs demonstrated that Bs usually contain fragments of DNA from different A chromosomes [31], suggesting that most of them could be inserted into Bs at advanced stages of their evolution or could be included in the proto-Bs as duplicon cluster of an ancestral chromosome. To distinguish these two ways of B chromosome formation and further development, it is necessary to analyze DNA in regions between the centromere and the chromosomal hotspot breakpoint involved in the rearrangement provided proto-Bs.

Analysis of homology between the pericentric regions of Bs and A chromosomes turned out to be intricate as well. In many species, pericentric heterochromatic regions contain—in addition to their specific sequences—a considerable number of sequences with homology to DNA in the pericentric regions of other chromosomes. The evolutionary rates of DNA in pericentric regions raise doubts as to the reliability of using these DNAs for identification of the origin of Bs [98–100]. These DNA markers might be reliable in some cases, but those cases are rare. For example, in the Korean field mouse, DNA sequences in the pericentric regions of autosomes and sex chromosomes are quite different [9]. FISH with microdissected DNA probes prepared from the pericentric region of an autosome painted the pericentric regions of all autosomes but gave no signal in the pericentric regions of the sex chromosomes. Most Bs of individuals from Siberian populations have in their pericentric regions DNA homologous to DNA in the pericentric regions of autosomes, while the Bs of specimens from the Russian Far East are devoid of this DNA in their pericentric regions [8]. It is possible that chromosomal rearrangement hotspots occur in different chromosomes in mice from Siberia and in mice from the Russian Far East. In the Siberian mice, they could occur in one or more autosomes; while in the Far East mice in a sex chromosome. The sequencing and a comparative analysis of Bs from inhabitants of these geographical regions could confirm their different origin and probably different ways of gaining even homologous sequences. Independent insertion of identical DNA fragments in Bs of different origin may be indicative of a high frequency of their transposition across the genome or positive selection in favor of the Bs that contain, such as DNA fragments. For example, detection of different margins of *Vrk1* region in Bs of two *Apodemus* species was considered as an indication to independent recruitment of the DNA fragments from the same genomic region [13].

Considering possible mechanisms by which Bs will be evolving, three points should be addressed: (*i*) changing of original DNA of proto-Bs; (*ii*) DNA amplification in Bs; and (*iii*) rearrangements involving large regions of Bs. Unfortunately, to date the sequencing of B chromosomal DNA has been performed for a small number of species and a small number of Bs [31]. Nevertheless, the differences between the homologous DNA insertions discovered in Bs of *Apodemus* mice captured from different populations suggested that the insertion of new DNA fragments from the main genome could take place also at advance stage of the B chromosome evolution [13].

In most cases, Bs grow in size due to amplification of DNA fragments. The formation of amplicons and their amplification lead to the emergence of B chromosomal regions composed of tandem repeats [51]. This amplification may probably involve both repeated and unique DNA fragments of the main genome [31,51]. In the course of evolution, Bs may undergo rearrangements affecting their large regions. In a study on Bs in the Korean field mouse using FISH of microdissected DNA probes derived from individual Bs and their regions, regions along the chromosome arms were differentiated, and thus iso-Bs were revealed, and thus were Bs that have evolved from them [8,9]. The presence of iso-Bs in the karyotypes studied is consistent with the hypothesis stating that the proximal region of the ancestral A chromosome contained a chromosomal rearrangement hotspot and that more rearrangements involving the same hotspot occurred at more advanced stages of B chromosome evolution. Those later rearrangements led to an iso-B chromosome appearing in the form of an inverted duplication of one of its arms. The last rearrangement resembles the human small SMCs derived from the chromosome 15. Later on, differences in DNA amplification in the arms of this iso-B would lead to differences between these arms in size and DNA composition (Figure 3).

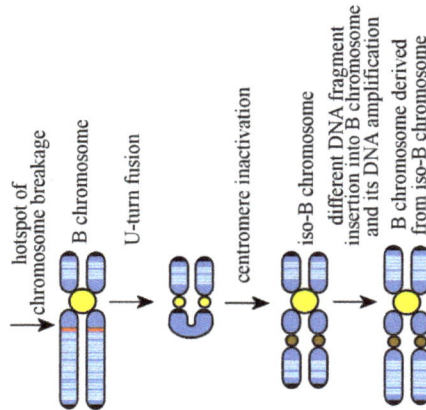

Figure 3. Iso-B chromosome formation through chromosome breakage at a hotspot of chromosomal rearrangements in a proximal region of a B chromosome followed by U-turn fusion, one centromere inactivation, and further B chromosome development. Euchromatic regions and insertions of euchromatic material in Bs are white; heterochromatic regions and insertions of heterochromatic material in Bs are blue; clusters of duplicons are red; active centromeres are yellow; inactive centromeres are brown; and telomeres on chromosome termini are black.

Analysis of a large number of the Korean field mouse specimens living in geographically well-spaced populations and displaying clear-cut differences in the sets of their Bs allowed the assessment of impact of known molecular genetic processes on evolution of the B sets in those populations [14]. We assumed that the observed differences in the morphology and DNA content of Bs is caused by different frequencies of chromosomal rearrangements in hotspots located in various chromosomes as well as variable intensities of some molecular genetic processes responsible for insertion of DNA fragments into proto-B chromosomes and amplification of their DNA like in formation of homogeneously staining regions (HSRs) [50,101]. Leaving open the question of whether natural selection has played a role in the formation of various sets of Bs in the Korean field mouse, we propose that different sets of Bs in isolated populations may have evolved without such selection.

Analysis of mammalian B chromosome evolution must include an analysis the meiotic behavior of Bs. Analysis of the 3D organization of the nucleus in the Korean field mouse spermatocytes showed that Bs tend to be localized in the immediate vicinity of the sex bivalent [18]. These data agree well with the results of an analysis of chromosome spreads prepared from the cells at the pachytene stage [17,18]. The occurrence of Bs with both similar and contrasting DNA content could account for both paired and unpaired Bs in pachytene spreads. Some of the unpaired Bs, especially those without synaptonemal complex protein 1 (SCP1), were associated with asynapsed chromatin and the XY bivalent. The other unpaired Bs could form a separate association with asynapsed chromatin without the XY bivalent [17,18]. A and B chromosomes were distinguished in meiosis based on the absence or presence of phosphorylated histone H2A.X at different pachytene substages [18]. Some Bs still have H2A.X on them, while double-strand break repair in A chromosomes has been completed and phosphorylated histone H2A.X was not detected on them [18]. As is known, histone H2A.X phosphorylation is involved in meiotic sex chromosome inactivation (MSCI) and transcriptional silencing of unpaired chromatin (MSUC) in autosomes, which may lead to pachytene checkpoint activation and apoptosis [102,103]. However, it should be noted that a low degree of asynapsis can be ignored, as has been demonstrated for mice and humans [104–108]. It is possible that the pairing of Bs with each other or the pairing of different regions within a B chromosome and their association with the transcriptionally inactive XY bivalent allows the cell to avoid apoptosis and to accomplish meiosis.

How Bs behave during female meiosis and whether their behavioral features may have impact on the probability of these chromosomes being transferred to an oocyte has yet to be elucidated.

7. Conclusions

A comparative analysis of B chromosomes in different mammalian species suggests that the emergence and evolution of these chromosomes must be due to a relatively small number of molecular and cellular mechanisms as well as some features of the main genome. They include different localization of chromosomal rearrangement hotspots in the proximal regions of chromosomes, proto-B chromosomes resulting from double-strand breaks at hotspots followed by errors in DNA repair, insertion of DNA fragments of the main genome into B chromosomes, and DNA amplification in B chromosomes. Differences in the frequency of these events and in the intensity of repair, transposition and amplification of DNA fragments result in different variants of B chromosomes, and their specific sets for particular species and even populations. When isolated from the main genome, DNA in B chromosomes may be evolving at increased rates. We do not exclude that natural selection favors B chromosomes with valuable genes, and although no evidence that genetic material moves from B chromosomes back to A chromosomes is available, it cannot be excluded.

Author Contributions: N.B.R. wrote original Draft Preparation and prepared its further review; Y.M.B. took part in text discussion.

Funding: This research was funded by the project #0324-2018-0019 of The Federal Research Center Institute of Cytology and Genetics, the Siberian Branch of Russian Academy of Science and research grant from the Russian Foundation for Basic Research #16-04-01185a.

Acknowledgments: We would like to thank Vladimir Filonenko for helping with the English.

Conflicts of Interest: The authors declare no conflict of interest.

References

1. Wilson, E. The supernumerary chromosomes of Hemiptera. *Science* **1907**, *26*, 870–873. [CrossRef]
2. Randolph, L.F. Types of supernumerary chromosomes in maize. *Anat. Rec.* **1928**, *41*, 102.
3. Pigozzi, M.I.; Solari, A.J. Germ cell restriction and regular transmission of an accessory chromosome that mimics a sex body in the zebra finch, *Taeniopygia guttata*. *Chromosom. Res.* **1998**, *6*, 105–113. [CrossRef]
4. Biederman, M.K.; Nelson, M.M.; Asalone, K.C.; Pedersen, A.L.; Saldanha, C.J.; Bracht, J.R. Discovery of the first germline-restricted gene by subtractive transcriptomic analysis in the zebra finch, *Taeniopygia guttata*. *Curr. Biol.* **2018**, *28*, 1620–1627. [CrossRef] [PubMed]
5. Del Priore, L.; Pigozzi, M.I. Histone modifications related to chromosome silencing and elimination during male meiosis in Bengalese finch. *Chromosoma* **2014**, *123*, 293–302. [CrossRef] [PubMed]
6. Kapusta, A.; Suh, A.; Feschotte, C. Dynamics of genome size evolution in birds and mammals. *Proc. Natl. Acad. Sci. USA* **2017**, *114*, E1460–E1469. [CrossRef] [PubMed]
7. D'Ambrosio, U.; Alonso-Lifante, M.P.; Barros, K.; Kovařík, A.; Mas de Xaxars, G.; Garcia, S. B-chrom: A database on B-chromosomes of plants, animals and fungi. *New Phytol.* **2017**, *635*–642. [CrossRef] [PubMed]
8. Rubtsov, N.B.; Kartavtseva, I.V.; Roslik, G.V.; Karamysheva, T.V.; Pavlenko, M.V.; Iwasa, M.A.; Koh, H.S. Features of the B chromosome in Korean field mouse *Apodemus peninsulae* (Thomas, 1906) from Transbaikalia and the Far East identified by the FISH method. *Russ. J. Genet.* **2015**, *51*, 278–288. [CrossRef]
9. Rubtsov, N.B.; Karamysheva, T.V.; Andreenkova, O.V.; Bochkaerev, M.N.; Kartavtseva, I.V.; Roslik, G.V.; Borissov, Y.M. Comparative analysis of micro and macro B chromosomes in the Korean field mouse *Apodemus peninsulae* (Rodentia, Murinae) performed by chromosome microdissection and FISH. *Cytogenet. Genome Res.* **2004**, *106*, 289–294. [CrossRef] [PubMed]
10. Camacho, J.P.M.; Sharbel, T.F.; Beukeboom, L.W. B-chromosome evolution. *Philos. Trans. R. Soc. B Biol. Sci.* **2000**, *355*, 163–178. [CrossRef] [PubMed]
11. Stitou, S.; Díaz De La Guardia, R.; Jiménez, R.; Burgos, M. Inactive ribosomal cistrons are spread throughout the B chromosomes of *Rattus rattus* (Rodentia, Muridae). Implications for their origin and evolution. *Chromosom. Res.* **2000**, *8*, 305–311. [CrossRef]

12. Duke Becker, S.E.; Thomas, R.; Trifonov, V.A.; Wayne, R.K.; Graphodatsky, A.S.; Breen, M. Anchoring the dog to its relatives reveals new evolutionary breakpoints across 11 species of the Canidae and provides new clues for the role of B chromosomes. *Chromosom. Res.* **2011**, *19*, 685–708. [CrossRef] [PubMed]

13. Makunin, A.I.; Rajičić, M.; Karamysheva, T.V.; Romanenko, S.A.; Druzhkova, A.S.; Blagojević, J.; Vujošević, M.; Rubtsov, N.B.; Graphodatsky, A.S.; Trifonov, V.A. Low-pass single-chromosome sequencing of human small supernumerary marker chromosomes (sSMCs) and *Apodemus* B chromosomes. *Chromosoma* **2018**, 1–11. [CrossRef] [PubMed]

14. Rubtzov, N.B.; Borissov, Y.M.; Karamysheva, T.V.; Bochkarev, M.N. The mechanisms of formation and evolution of B chromosomes in Korean field mice *Apodemus peninsulae* (Mammalia, Rodentia). *Russ. J. Genet.* **2009**, *45*, 389–396. [CrossRef]

15. Teruel, M.; Cabrero, J.; Perfectti, F.; Camacho, J.P.M. B chromosome ancestry revealed by histone genes in the migratory locust. *Chromosoma* **2010**, *119*, 217–225. [CrossRef] [PubMed]

16. Martis, M.M.; Klemme, S.; Banaei-Moghaddam, A.M.; Blattner, F.R.; Macas, J.; Schmutzer, T.; Scholz, U.; Gundlach, H.; Wicker, T.; Simkova, H.; et al. Selfish supernumerary chromosome reveals its origin as a mosaic of host genome and organellar sequences. *Proc. Natl. Acad. Sci. USA* **2012**, *109*, 13343–13346. [CrossRef] [PubMed]

17. Kolomiets, O.L.; Borbiev, T.E.; Safronova, L.D.; Borisov, Y.M.; Bogdanov, Y.F. Synaptonemal complex analysis of B-chromosome behavior in meiotic prophase I in the east-asiatic mouse *Apodemus peninsulae* (Muridae, Rodentia). *Cytogenet. Genome Res.* **1988**, *48*, 183–187. [CrossRef] [PubMed]

18. Karamysheva, T.V.; Torgasheva, A.A.; Yefremov, Y.R.; Bogomolov, A.G.; Liehr, T.; Borodin, P.M.; Rubtsov, N.B. Spatial organization of fibroblast and spermatocyte nuclei with different B-chromosome content in Korean field mouse, *Apodemus peninsulae* (Rodentia, Muridae). *Genome* **2017**, *60*, 815–824. [CrossRef] [PubMed]

19. Plowman, A.B.; Bougourd, S.M. Selectively advantageous effects of B chromosomes on germination behaviour in *Allium schoenoprasum* L. *Heredity* **1994**, *72*, 587–593. [CrossRef]

20. Blagojević, J.; Vujošević, M. B chromosomes and developmental homeostasis in the yellow-necked mouse, *Apodemus flavicollis* (Rodentia, Mammalia): Effects on nonmetric traits. *Heredity* **2004**, *93*, 249–254. [CrossRef] [PubMed]

21. Jojić, V.; Blagojević, J.; Vujošević, M. B chromosomes and cranial variability in yellow-necked field mice (*Apodemus flavicollis*). *J. Mammal.* **2011**, *92*, 396–406. [CrossRef]

22. Yoshida, K.; Terai, Y.; Mizoiri, S.; Aibara, M.; Nishihara, H.; Watanabe, M.; Kuroiwa, A.; Hirai, H.; Hirai, Y.; Matsuda, Y.; et al. B chromosomes have a functional effect on female sex determination in Lake Victoria cichlid fishes. *PLoS Genet.* **2011**, *7*, e1002203. [CrossRef]

23. Graphodatsky, A.S.; Trifonov, V.A.; Stanyon, R. The genome diversity and karyotype evolution of mammals. *Mol. Cytogenet.* **2011**, *4*, 1–16. [CrossRef] [PubMed]

24. Moriyama, Y.; Koshiba-Takeuchi, K. Significance of whole-genome duplications on the emergence of evolutionary novelties. *Brief. Funct. Genom.* **2018**, *17*, 329–338. [CrossRef] [PubMed]

25. Gallardo, M.H.; González, C.A.; Cebrián, I. Molecular cytogenetics and allotetraploidy in the red vizcacha rat, *Tympanoctomys barrerae* (Rodentia, Octodontidae). *Genomics* **2006**, *88*, 214–221. [CrossRef] [PubMed]

26. Evans, B.J.; Upham, N.S.; Golding, G.B.; Ojeda, R.A.; Ojeda, A.A. Evolution of the Largest Mammalian Genome. *Genome Biol. Evol.* **2017**, *9*, 1711–1724. [CrossRef] [PubMed]

27. Perry, G.H.; Dominy, N.J.; Claw, K.G.; Lee, A.S.; Fiegler, H.; Redon, R.; Werner, J.; Villanea, F.A.; Mountain, J.L.; Misra, R.; et al. Diet and the evolution of human amylase gene copy number variation. *Nat. Genet.* **2007**, *39*, 1256–1260. [CrossRef] [PubMed]

28. Ibraimov, A.I.; Mirrakhimov, M.M.; Nazarenko, S.A.; Axenrod, E.I.; Akbanova, G.A. Human chromosomal polymorphism. I. Chromosomal Q polymorphism in Mongoloid populations of central Asia. *Hum. Genet.* **1982**, *60*, 1–7. [CrossRef] [PubMed]

29. Ibraimov, A.I.; Mirrakhimov, M.M.; Axenrod, E.I.; Kurmanova, G.U. Human chromosomal polymorphism. IX. Further data on the possible selective value of chromosomal Q-heterochromatin material. *Hum. Genet.* **1986**, *73*, 151–156. [CrossRef] [PubMed]

30. Graphodatsky, A.S.; Kukekova, A.V.; Yudkin, D.V.; Trifonov, V.A.; Vorobieva, N.V.; Beklemisheva, V.R.; Perelman, P.L.; Graphodatskaya, D.A.; Trut, L.N.; Yang, F.; et al. The proto-oncogene *C-KIT* maps to canid B-chromosomes. *Chromosom. Res.* **2005**, *13*, 113–122. [CrossRef]

31. Makunin, A.; Romanenko, S.; Beklemisheva, V.; Perelman, P.; Druzhkova, A.; Petrova, K.; Prokopov, D.; Chernyaeva, E.; Johnson, J.; Kukekova, A.; et al. Sequencing of supernumerary chromosomes of red fox and raccoon dog confirms a non-random gene acquisition by B chromosomes. *Genes* **2018**, *9*, 405. [CrossRef] [PubMed]

32. Trifonov, V.A.; Dementyeva, P.V.; Larkin, D.M.; O'Brien, P.C.M.; Perelman, P.L.; Yang, F.; Ferguson-Smith, M.A.; Graphodatsky, A.S. Transcription of a protein-coding gene on B chromosomes of the Siberian roe deer (*Capreolus pygargus*). *BMC Biol.* **2013**, *11*, 1. [CrossRef] [PubMed]

33. Vujosevic, M.; Zivkovic, S. Numerical chromosome polymorphism in *Apodemus flaviocollis* and *A. sylvaticus* (Mammalia: Rodentia) caused by supernumerary chromosomes. *Acta Vet.* **1987**, *37*, 115–122.

34. Vujošević, M.; Blagojević, J. B chromosomes in populations of mammals. *Cytogenet. Genome Res.* **2004**, *106*, 247–256. [CrossRef] [PubMed]

35. Yonenaga-Yassuda, Y.; Maia, V.; L'Abbate, M. Two tandem fusions and supernumerary chromosomes in *Nectomys squamipes* (Cricetidae, Rodentia). *Caryologia* **1988**, *41*, 25–39. [CrossRef]

36. Liehr, T. *Benign and Pathological Chromosomal Imbalances: Microscopic and Submicroscopic Copy Number Variations (CNVs) in Genetics and Counseling*; Academic Press: Cambridge, MS, USA, 2014; ISBN 9780124046313.

37. McCarroll, S.A.; Altshuler, D.M. Copy-number variation and association studies of human disease. *Nat. Genet.* **2007**, *39*, S37–S42. [CrossRef] [PubMed]

38. Nowakowska, B. Clinical interpretation of copy number variants in the human genome. *J. Appl. Genet.* **2017**, *58*, 449–457. [CrossRef] [PubMed]

39. Sati, S.; Cavalli, G. Chromosome conformation capture technologies and their impact in understanding genome function. *Chromosoma* **2017**, *126*, 33–44. [CrossRef] [PubMed]

40. Yudkin, D.V.; Trifonov, V.A.; Kukekova, A.V.; Vorobieva, N.V.; Rubtsova, N.V.; Yang, F.; Acland, G.M.; Ferguson-Smith, M.A.; Graphodatsky, A.S. Mapping of *KIT* adjacent sequences on canid autosomes and B chromosomes. *Cytogenet. Genome Res.* **2007**, *116*, 100–103. [CrossRef] [PubMed]

41. Borisov, I.M.; Afanas'ev, A.G.; Lebedev, T.T.; Bochkarev, M.N. Multiplicity of B microchromosomes in a Siberian population of mice *Apodemus peninsulae* (2n = 48 + 4–30 B chromosomes). *Russ. J. Genet.* **2010**, *46*, 505–711. [CrossRef]

42. Kartavtseva, I.V.; Roslik, G.V. A complex B chromosome system in the Korean field mouse, *Apodemus peninsulae. Cytogenet. Genome Res.* **2004**, *106*, 271–278. [CrossRef] [PubMed]

43. Liehr, T.; Claussen, U.; Starke, H. Small supernumerary marker chromosomes (sSMC) in humans. *Cytogenet. Genome Res.* **2004**, *107*, 55–67. [CrossRef] [PubMed]

44. Liehr, T.; Weise, A. Frequency of small supernumerary marker chromosomes in prenatal, newborn, developmentally retarded and infertility diagnostics. *Int. J. Mol. Med.* **2007**, *9*, 719–731. [CrossRef]

45. Liehr, T.; Mrasek, K.; Kosyakova, N.; Ogilvie, C.; Vermeesch, J.; Trifonov, V.; Rubtsov, N. Small supernumerary marker chromosomes (sSMC) in humans; are there B chromosomes hidden among them. *Mol. Cytogenet.* **2008**, *1*, 12. [CrossRef] [PubMed]

46. Armanet, N.; Tosca, L.; Brisset, S.; Liehr, T.; Tachdjian, G. Small supernumerary marker chromosomes in human infertility. *Cytogenet. Genome Res.* **2015**, *146*, 100–108. [CrossRef] [PubMed]

47. Hashemzadeh-Chaleshtori, M.; Teimori, H.; Ghasemi-Dehkordi, P.; Jafari-Ghahfarokhi, H.; Moradi-Chaleshtori, M.; Liehr, T. Small supernumerary marker chromosomes and their correlation with specific syndromes. *Adv. Biomed. Res.* **2015**, *4*, 140. [CrossRef] [PubMed]

48. Reddy, K.S.; Aradhya, S.; Meck, J.; Tiller, G.; Abboy, S.; Bass, H. A systematic analysis of small supernumerary marker chromosomes using array CGH exposes unexpected complexity. *Genet. Med.* **2013**, *15*, 3–13. [CrossRef] [PubMed]

49. Hamid Al-Rikabi, A.B.; Pekova, S.; Fan, X.; Jancuskova, T.; Liehr, T. Small supernumerary marker chromosome may provide information on dosage-insensitive pericentric regions in human. *Curr. Genom.* **2018**, *19*, 192–199. [CrossRef] [PubMed]

50. Horvath, J.E. Lessons from the human genome: transitions between euchromatin and heterochromatin. *Hum. Mol. Genet.* **2001**, *10*, 2215–2223. [CrossRef] [PubMed]

51. Makunin, A.I.; Dementyeva, P.V.; Graphodatsky, A.S.; Volobouev, V.T.; Kukekova, A.V.; Trifonov, V.A. Genes on B chromosomes of vertebrates. *Mol. Cytogenet.* **2014**, *7*, 1–10. [CrossRef] [PubMed]

52. Makunin, A.I.; Kichigin, I.G.; Larkin, D.M.; O'Brien, P.C.M.; Ferguson-Smith, M.A.; Yang, F.; Proskuryakova, A.A.; Vorobieva, N.V.; Chernyaeva, E.N.; O'Brien, S.J.; et al. Contrasting origin of B chromosomes in two cervids (Siberian roe deer and grey brocket deer) unravelled by chromosome-specific DNA sequencing. *BMC Genom.* **2016**, *17*, 1–14. [CrossRef] [PubMed]

53. Wurster-Hill, D.H.; Ward, O.G.; Davis, B.H.; Park, J.P.; Moyzis, R.K.; Meyne, J. Fragile sites, telomeric DNA sequences, B chromosomes, and DNA content in raccoon dogs, *Nyctereutes procyonoides*, with comparative notes on foxes, coyote, wolf, and raccoon. *Cytogenet. Genome Res.* **1988**, *49*, 278–281. [CrossRef] [PubMed]

54. Trifonov, V.A.; Perelman, P.L.; Kawada, S.I.; Iwasa, M.A.; Oda, S.I.; Graphodatsky, A.S. Complex structure of B-chromosomes in two mammalian species: *Apodemus peninsulae* (Rodentia) and *Nyctereutes procyonoides* (Carnivora). *Chromosom. Res.* **2002**, *10*, 109–116. [CrossRef]

55. Szczerbal, I.; Switonski, M. B chromosomes of the Chinese raccoon dog (*Nyctereutes procyonoides procyonoides* Gray) contain inactive NOR-like sequences. *Caryologia* **2003**, *56*, 213–216. [CrossRef]

56. Bugarski-Stanojevic, V.; Stamenković, G.; Blagojević, J.; Liehr, T.; Kosyakova, N.; Rajičić, M.; Vujošević, M. Exploring supernumeraries—A new marker for screening of B-chromosomes presence in the yellow necked mouse *Apodemus flavicollis*. *PLoS ONE* **2016**, *11*, 1–18. [CrossRef] [PubMed]

57. Groenen, M.A.M.; Archibald, A.L.; Uenishi, H.; Tuggle, C.K.; Takeuchi, Y.; Rothschild, M.F.; Rogel-Gaillard, C.; Park, C.; Milan, D.; Megens, H.J.; et al. Analyses of pig genomes provide insight into porcine demography and evolution. *Nature* **2012**, *491*, 393–398. [CrossRef] [PubMed]

58. Valente, G.T.; Conte, M.A.; Fantinatti, B.E.A.; Cabral-De-Mello, D.C.; Carvalho, R.F.; Vicari, M.R.; Kocher, T.D.; Martins, C. Origin and evolution of B chromosomes in the cichlid fish *Astatotilapia latifasciata* based on integrated genomic analyses. *Mol. Biol. Evol.* **2014**, *31*, 2061–2072. [CrossRef] [PubMed]

59. Banaei-Moghaddam, A.M.; Meier, K.; Karimi-Ashtiyani, R.; Houben, A. Formation and expression of pseudogenes on the B chromosome of rye. *Plant Cell* **2013**, *25*, 2536–2544. [CrossRef] [PubMed]

60. Navarro-Domínguez, B.; Ruiz-Ruano, F.J.; Cabrero, J.; Corral, J.M.; López-León, M.D.; Sharbel, T.F.; Camacho, J.P.M. Protein-coding genes in B chromosomes of the grasshopper *Eyprepocnemis plorans*. *Sci. Rep.* **2017**, *7*, 1–12. [CrossRef]

61. Sochorová, J.; Garcia, S.; Gálvez, F.; Symonová, R.; Kovařík, A. Evolutionary trends in animal ribosomal DNA loci: Introduction to a new online database. *Chromosoma* **2018**, *127*, 141–150. [CrossRef] [PubMed]

62. Banaei-Moghaddam, A.M.; Martis, M.M.; Macas, J.; Gundlach, H.; Himmelbach, A.; Altschmied, L.; Mayer, K.F.X.; Houben, A. Genes on B chromosomes: Old questions revisited with new tools. *Biochim. Biophys. Acta* **2015**, *1849*, 64–70. [CrossRef] [PubMed]

63. Karamysheva, T.V.; Andreenkova, O.V.; Bochkaerev, M.N.; Borissov, Y.M.; Bogdanchikova, N.; Borodin, P.M.; Rubtsov, N.B. B chromosomes of Korean field mouse *Apodemus peninsulae* (Rodentia, Murinae) analysed by microdissection and FISH. *Cytogenet. Genome Res.* **2002**, *96*, 154–160. [CrossRef] [PubMed]

64. Tanić, N.; Vujošević, M.; Dedović-Tanić, N.; Dimitrijević, B. Differential gene expression in yellow-necked mice *Apodemus flavicollis* (Rodentia, Mammalia) with and without B chromosomes. *Chromosoma* **2005**, *113*, 418–427. [CrossRef] [PubMed]

65. Hall, L.E.; Mitchell, S.E.; O'Neill, R.J. Pericentric and centromeric transcription: A perfect balance required. *Chromosom. Res.* **2012**, *20*, 535–546. [CrossRef] [PubMed]

66. Saksouk, N.; Simboeck, E.; Déjardin, J. Constitutive heterochromatin formation and transcription in mammals. *Epigenet. Chromatin* **2015**, *8*, 1–17. [CrossRef] [PubMed]

67. Wang, J.; Jia, S.T.; Jia, S. New Insights into the regulation of heterochromatin. *Trends Genet.* **2016**, *32*, 284–294. [CrossRef] [PubMed]

68. Enukashvily, N.I.; Ponomartsev, N.V. Mammalian satellite DNA: A speaking dumb. *Adv. Protein Chem. Struct. Biol.* **2013**, *90*, 31–65. [CrossRef] [PubMed]

69. Eymery, A.; Horard, B.; el Atifi-Borel, M.; Fourel, G.; Berger, F.; Vitte, A.L.; Van den Broeck, A.; Brambilla, E.; Fournier, A.; Callanan, M.; et al. A transcriptomic analysis of human centromeric and pericentric sequences in normal and tumor cells. *Nucleic Acids Res.* **2009**, *7*, 6340–6354. [CrossRef] [PubMed]

70. Lu, J.; Gilbert, D.M. Proliferation-dependent and cell cycle-regulated transcription of mouse pericentric heterochromatin. *J. Cell Biol.* **2007**, *179*, 411–421. [CrossRef] [PubMed]

71. Chen, E.S.; Zhang, K.; Nicolas, E.; Cam, H.P.; Zofall, M.; Grewal, S.I.S. Cell cycle control of centromeric repeat transcription and heterochromatin assembly. *Nature* **2008**, *451*, 734–737. [CrossRef] [PubMed]

72. Kloc, A.; Zaratiegui, M.; Nora, E.; Martienssen, R. RNA interference guides histone modification during the S phase of chromosomal replication. *Curr. Biol.* **2008**, *8*, 490–495. [CrossRef] [PubMed]

73. Motamedi, M.R.; Verdel, A.; Colmenares, S.U.; Gerber, S.A.; Gygi, S.P.; Moazed, D. Two RNAi complexes, RITS and RDRC, physically interact and localize to noncoding centromeric RNAs. *Cell* **2004**, *119*, 789–802. [CrossRef] [PubMed]

74. Sugiyama, T.; Cam, H.; Verdel, A.; Moazed, D.; Grewal, S.I.S. From the cover: RNA-dependent RNA polymerase is an essential component of a self-enforcing loop coupling heterochromatin assembly to siRNA production. *Proc. Natl. Acad. Sci. USA* **2005**, *102*, 152–157. [CrossRef] [PubMed]

75. Verdel, A.; Jia, S.; Gerber, S.; Sugiyama, T.; Gygi, S.; Grewal, S.I.S.; Moazed, D. RNAi-mediated targeting of heterochromatin by the RITS complex. *Science* **2004**, *303*, 672–676. [CrossRef] [PubMed]

76. Bayne, E.H.; Bijos, D.A.; White, S.A.; de Lima Alves, F.; Rappsilber, J.; Allshire, R.C. A systematic genetic screen identifies new factors influencing centromeric heterochromatin integrity in fission yeast. *Genome Biol.* **2014**, *15*, 481. [CrossRef] [PubMed]

77. Hong, E.J.E.; Villén, J.; Gerace, E.L.; Gygi, S.P.; Moazed, D. A cullin E3 ubiquitin ligase complex associates with Rik1 and the Clr4 histone H3-K9 methyltransferase and is required for RNAi-mediated heterochromatin formation. *RNA Biol.* **2005**, *2*, 106–111. [CrossRef] [PubMed]

78. Jia, S.; Kobayashi, R.; Grewal, S.I.S. Ubiquitin ligase component Cul4 associates with Clr4 histone methyltransferase to assemble heterochromatin. *Nat. Cell Biol.* **2005**. [CrossRef] [PubMed]

79. Horn, P.J.; Bastie, J.N.; Peterson, C.L. A Rik1-associated, cullin-dependent E3 ubiquitin ligase is essential for heterochromatin formation. *Genes Dev.* **2005**, *19*, 1705–1714. [CrossRef] [PubMed]

80. Noma, K.I.; Sugiyama, T.; Cam, H.; Verdel, A.; Zofall, M.; Jia, S.; Moazed, D.; Grewal, S.I.S. RITS acts in *cis* to promote RNA interference-mediated transcriptional and post-transcriptional silencing. *Nat. Genet.* **2004**, *36*, 1174–1180. [CrossRef] [PubMed]

81. Cremer, T.; Cremer, C. Chromosome territories, nuclear architecture and gene regulation in mammalian cells. *Nat. Rev. Genet.* **2001**, *2*, 292–301. [CrossRef] [PubMed]

82. Cremer, T.; Cremer, M. Chromosome Territories. *Cold Spring Harb. Perspect. Biol.* **2010**, *2*, a003889. [CrossRef] [PubMed]

83. Lanctôt, C.; Cheutin, T.; Cremer, M.; Cavalli, G.; Cremer, T. Dynamic genome architecture in the nuclear space: Regulation of gene expression in three dimensions. *Nat. Rev. Genet.* **2007**, *8*, 104–115. [CrossRef] [PubMed]

84. Fritz, A.J.; Barutcu, A.R.; Martin-Buley, L.; Van Wijnen, A.J.; Zaidi, S.K.; Imbalzano, A.N.; Lian, J.B.; Stein, J.L.; Stein, G.S. Chromosomes at work: Organization of chromosome territories in the interphase nucleus. *J. Cell. Biochem.* **2016**, *117*, 9–19. [CrossRef] [PubMed]

85. Kociucka, B.; Sosnowski, J.; Kubiak, A.; Nowak, A.; Pawlak, P.; Szczerbal, I. Three-dimensional positioning of B chromosomes in fibroblast nuclei of the red fox and the Chinese raccoon dog. *Cytogenet. Genome Res.* **2013**, *139*, 243–249. [CrossRef] [PubMed]

86. Tanabe, H.; Küpper, K.; Ishida, T.; Neusser, M.; Mizusawa, H. Inter- and intra-specific gene-density-correlated radial chromosome territory arrangements are conserved in Old World monkeys. *Cytogenet. Genome Res.* **2005**, *108*, 255–261. [CrossRef] [PubMed]

87. Solovei, I.; Thanisch, K.; Feodorova, Y. How to rule the nucleus: divide et impera. *Curr. Opin. Cell Biol.* **2016**, *40*, 47–59. [CrossRef] [PubMed]

88. Tanabe, H.; Habermann, F.A.; Solovei, I.; Cremer, M.; Cremer, T. Non-random radial arrangements of interphase chromosome territories: Evolutionary considerations and functional implications. *Mutat. Res.* **2002**, *504*, 37–45. [CrossRef]

89. Olmo, E. Nucleotype and cell size in vertebrates: A review. *Basic Appl. Histochem.* **1983**, *27*, 227–256. [PubMed]

90. Gillooly, J.F.; Hein, A.; Damiani, R. Nuclear DNA content varies with cell size across human cell types. *Cold Spring Harb. Perspect. Biol.* **2015**, *7*, 1–28. [CrossRef] [PubMed]

91. Hetzer, M.W. The Nuclear Envelope. *Cold Spring Harb. Perspect. Biol.* **2010**, *2*, a000539. [CrossRef] [PubMed]

92. Palestis, B.G.; Burt, A.; Jones, R.N.; Trivers, R. B chromosomes are more frequent in mammals with acrocentric karyotypes: Support for the theory of centromeric drive. *Proc. R. Soc. B Biol. Sci.* **2004**, *271*, 22–24. [CrossRef] [PubMed]

93. Palestis, B.G.; Trivers, R.; Burt, A.; Jones, R.N. The distribution of B chromosomes across species. *Cytogenet. Genome Res.* **2004**, *106*, 151–158. [CrossRef] [PubMed]
94. Murmann, A.E.; Conrad, D.F.; Mashek, H.; Curtis, C.A.; Nicolae, R.I.; Ober, C.; Schwartz, S. Inverted duplications on acentric markers: Mechanism of formation. *Hum. Mol. Genet.* **2009**, *15*, 2241–2256. [CrossRef] [PubMed]
95. Sheth, F.; Ewers, E.; Kosyakova, N.; Weise, A.; Sheth, J.; Patil, S.; Ziegler, M.; Liehr, T. A neocentric isochromosome Yp present as additional small supernumerary marker chromosome—Evidence against U-type exchange mechanism? *Cytogenet. Genome Res.* **2009**, *125*, 115–116. [CrossRef] [PubMed]
96. Pennisi, E. Shaking up the tree of life. *Science* **2016**, *354*, 817–821. [CrossRef] [PubMed]
97. Jetybayev, I.Y.; Bugrov, A.G.; Buleu, O.G.; Bogomolov, A.G.; Rubtsov, N.B. Origin and evolution of the neo-sex chromosomes in pamphagidae grasshoppers through chromosome fusion and following heteromorphization. *Genes* **2017**, *8*, 323. [CrossRef] [PubMed]
98. Karamysheva, T.V.; Bogdanov, A.S.; Kartavtseva, I.V.; Likhoshvay, T.V.; Bochkarev, M.N.; Kolcheva, N.E.; Marochkina, V.V.; Rubtsov, N.B. Comparative FISH analysis of C-positive blocks of centromeric chromosomal regions of pygmy wood mice *Sylvaemus uralensis* (Rodentia, Muridae). *Russ. J. Genet.* **2010**, *46*, 712–724. [CrossRef]
99. Rubtsov, N.B.; Karamysheva, T.V.; Bogdanov, A.S.; Likhoshvay, T.V.; Kartavtseva, I.V. Comparative FISH analysis of C-positive regions of chromosomes of wood mice (Rodentia, Muridae, Sylvaemus). *Russ. J. Genet.* **2011**, *47*, 1096–1110. [CrossRef]
100. Rubtsov, N.B.; Karamysheva, T.V.; Bogdanov, A.S.; Kartavtseva, I.V.; Bochkarev, M.N.; Iwasa, M.A. Comparative analysis of DNA homology in pericentric regions of chromosomes of wood mice from genera *Apodemus* and *Sylvaemus*. *Russ. J. Genet.* **2015**, *51*, 1233–1242. [CrossRef]
101. L'Abbate, A.; Macchia, G.; D'Addabbo, P.; Lonoce, A.; Tolomeo, D.; Trombetta, D.; Kok, K.; Bartenhagen, C.; Whelan, C.W.; Palumbo, O.; et al. Genomic organization and evolution of double minutes/homogeneously staining regions with *MYC* amplification in human cancer. *Nucleic Acids Res.* **2014**, *42*, 9131–9145. [CrossRef] [PubMed]
102. Turner, J.M.A.; Mahadevaiah, S.K.; Fernandez-Capetillo, O.; Nussenzweig, A.; Xu, X.; Deng, C.X.; Burgoyne, P.S. Silencing of unsynapsed meiotic chromosomes in the mouse. *Nat. Genet.* **2005**, *37*, 41–47. [CrossRef] [PubMed]
103. Baarends, W.M.; Wassenaar, E.; van der Laan, R.; Hoogerbrugge, J.; Sleddens-Linkels, E.; Hoeijmakers, J.H.J.; de Boer, P.; Grootegoed, J.A. Silencing of Unpaired Chromatin and Histone H2A Ubiquitination in Mammalian Meiosis. *Mol. Cell. Biol.* **2005**, *25*, 1041–1053. [CrossRef] [PubMed]
104. Sciurano, R.; Rahn, M.; Rey-Valzacchi, G.; Solari, A.J. The asynaptic chromatin in spermatocytes of translocation carriers contains the histone variant γ-H2AX and associates with the XY body. *Hum. Reprod.* **2007**, *22*, 142–150. [CrossRef] [PubMed]
105. Sciurano, R.B.; Rahn, M.I.; Rey-Valzacchi, G.; Coco, R.; Solari, A.J. The role of asynapsis in human spermatocyte failure. *Int. J. Androl.* **2012**, *35*, 541–549. [CrossRef] [PubMed]
106. Naumova, A.K.; Fayer, S.; Leung, J.; Boateng, K.A.; Camerini-Otero, R.D.; Taketo, T. Dynamics of Response to Asynapsis and Meiotic Silencing in Spermatocytes from Robertsonian Translocation Carriers. *PLoS ONE* **2013**, *8*, e75970. [CrossRef] [PubMed]
107. Burgoyne, P.S.; Mahadevaiah, S.K.; Turner, J.M.A. The consequences of asynapsis for mammalian meiosis. *Nat. Rev. Genet.* **2009**, *10*, 207–216. [CrossRef] [PubMed]
108. Manterola, M.; Page, J.; Vasco, C.; Berríos, S.; Parra, M.T.; Viera, A.; Rufas, J.S.; Zuccotti, M.; Garagna, S.; Fernández-Donoso, R. A high incidence of meiotic silencing of unsynapsed chromatin is not associated with substantial pachytene loss in heterozygous male mice carrying multiple simple Robertsonian translocations. *PLoS Genet.* **2009**, *5*, e1000625. [CrossRef] [PubMed]

© 2018 by the authors. Licensee MDPI, Basel, Switzerland. This article is an open access article distributed under the terms and conditions of the Creative Commons Attribution (CC BY) license (http://creativecommons.org/licenses/by/4.0/).

genes

MDPI

Review

B Chromosomes in Populations of Mammals Revisited

Mladen Vujošević *, Marija Rajičić and Jelena Blagojević

Institute for Biological Research "Siniša Stanković", Department of Genetic Research, University of Belgrade, Bulevar despota Stefana 142, Belgrade 11060, Serbia; marija.rajicic@ibiss.bg.ac.rs (M.R.); jelena.blagojevic@ibiss.bg.ac.rs (J.B.)
* Correspondence: mladenvu@ibiss.bg.ac.rs

Received: 29 August 2018; Accepted: 3 October 2018; Published: 9 October 2018

Abstract: The study of B chromosomes (Bs) started more than a century ago, while their presence in mammals dates since 1965. As the past two decades have seen huge progress in application of molecular techniques, we decided to throw a glance on new data on Bs in mammals and to review them. We listed 85 mammals with Bs that make 1.94% of karyotypically studied species. Contrary to general view, a typical B chromosome in mammals appears both as sub- or metacentric that is the same size as small chromosomes of standard complement. Both karyotypically stable and unstable species possess Bs. The presence of Bs in certain species influences the cell division, the degree of recombination, the development, a number of quantitative characteristics, the host-parasite interactions and their behaviour. There is at least some data on molecular structure of Bs recorded in nearly a quarter of species. Nevertheless, a more detailed molecular composition of Bs presently known for six mammalian species, confirms the presence of protein coding genes, and the transcriptional activity for some of them. Therefore, the idea that Bs are inert is outdated, but the role of Bs is yet to be determined. The maintenance of Bs is obviously not the same for all species, so the current models must be adapted while bearing in mind that Bs are not inactive as it was once thought.

Keywords: supernumerary chromosomes; additional chromosomes; chromosome polymorphism; evolution

1. Introduction

The presence of supernumerary or B chromosomes (Bs) is the oldest known chromosome polymorphism [1], and yet, after more than a century of research, the biological importance of Bs is still to be better determined. The knowledge about Bs in mammals is more recent and dates since 1965 when they were found in the greater glider, *Petauroides* (*Schoinobates*) *volans* by Hayman and Martin [2] and in the red fox, *Vulpes vulpes* by Moore and Elder [3].

A complex collection of diverse chromosomes, such as Bs, is difficult to describe. Yet, Bs are defined as dispensable supernumerary chromosomes which do not recombine with members of the basic A chromosome set (As), and do not follow the rules of Mendelian segregation law [4]. This definition assembles a pool of various chromosomes that do not share a complete set of features but only the mentioned dispensability, which alludes that a regular growth and development take place with or without Bs. A typical B chromosome is seen as a supernumerary, heterochromatic chromosome, smaller and morphologically different from chromosomes of the standard set, that does not evoke visible phenotypic effects. Nevertheless, the Bs that do not fit either partly or entirely into this picture are far from being atypical. In reality, when it comes to Bs, being out of the ordinary is considered to be a rule.

The earlier Beukeboom's estimate that 15% of all species carry Bs seems to be too high. The more accurate calculation stating that only 3% of karyologically studied extant species, across the majority of taxonomic groups carry Bs, was given by D'Ambrosio et al. [5]. Although it was thought that species with Bs in mammals are many times less frequent than in plants, it seems that this is not well grounded. According to the data that Jones [6] summarized, there are 1252 plant species with Bs that make about 2.4% of karyotipically studied plant species [7]. In the first review of Bs in mammals Volobujev [8] listed 14 species, but the next year he expanded list to 25 species [9]. Vujošević [10] increased the list to 34 species, and in 2004 we recorded fifty-five species carrying Bs [11]. There are nearly 70 species with Bs that were mentioned by Trifonov et al. [12], but the very list of species was not presented. As it can be seen in Table 1, the number of mammalian species carrying Bs has increased to 85. At the same time, the list of documented mammalian species has also increased from 4629 [13] to 6399 extant ones [14]. It appears that ~1.9% of 4380 karyotipically studied mammalian species (according to chromosome number database [15]) are featured by presence of Bs. We gave all species proper names according to the list of Burgin et al. [14], but even in such a detailed list, some species remain questionable. Besides adding new species to the list, we also removed some due to either being listed multiple times under different names, or incorrectly mentioned as species with Bs, such as the pocket gopher, *Thomomys umbrinus* [16].

Despite the vast body of knowledge on Bs within mammalian species, the question of what factors determine the distribution of Bs across different species is yet to be answered. Why are Bs present in some species and not in others? Is there some innate property of the genome, or karyotype, which determines whether a species is likely to carry Bs or not? As passed two decades witnessed huge progress in application of molecular techniques, we decided to re-examine the data on Bs in mammals, and to suggest the future directions of the research.

2. Morphological Characteristics and Size of B Chromosomes

B chromosomes were sorted in three categories [8,10,11] based on their size in relation to chromosomes from the standard set (Table 1). The most frequent Bs are of the same size (II) as the chromosomes from A set (52 species, 65.0%), so they cannot be recognized using standard cytogenetic techniques. In this category, the size of Bs often corresponds to the size of the smallest chromosome in the genome. Furthermore, about less than half (39 species, 48.8%) of species have Bs that are smaller than the smallest chromosomes of the standard set. They belong to category I. Ungulates and bats have only Bs from this category (Table 1). There are two primate species that possess micro Bs, but there is still a debate if human small supernumerary marker chromosomes (sSMC) could be considered as B chromosomes [17].

The rarest (3.5%) of the species are the ones with Bs either larger or the same size as the largest chromosomes from the standard complement (III): *Uromys caudimaculatus* [18], *Holochilus brasiliensis* [19] and *Apodemus peninsulae* [20]. Additionally, the presence of different types of Bs in the same genome contributes to a large variability of Bs. Eleven species from Table 1 show variation in size and morphology of Bs, so there are different types that are recognised. This variation is well studied in *A. peninsulae* where five classes with different morphological types of Bs are present.

Among Bs from type II, metacentrics and submetacentrics are more present (62.7%) than acrocentrics. Hewitt [21] noted that large Bs tend to be mitotically stabile while the small ones have an opposite tendency. It means that the intraindividual variability rises with the decrease in size of Bs (see more details in Section 3).

Table 1. List of species with B chromosomes.

ORDER / Species	Common Name [◊]	2n	NFa	X/Y	No. Bs	Bs Morphology		References
						Size *	Cent. Position †	
PERAMELEMORPHIA								
Echymipera kalubu	Common Echymipera	13–14 XX/X0	26	M/A	1-5	I	M	[22]
DIPROTODONTIA								
Petauroides (Schoinobates) volans	Greater glider	22	38	M/A	1-8	I	mi	[2,23]
INSECTIVORA								
Crocidura leucodon	Bicolored shrew	28	52	SM/SM, A	1	II	A	[24]
Crocidura malayana	Malayan shrew	38	62	SM/M	1-2	II	M	[25]
Crocidura suaveolens	Lesser shrew	40	46	M/A	1	II	A	[26]
Sorex bedfordiae	Lesser stripe-backed shrew	24	44	A/A	1-2	II	M	[27]
CHIROPTERA								
Myotis macrodactylus	Big-footed Myotis	44	56	M/A	1	I	mi	[28]
Nyctalus leisleri	Lesser Noctule	44	54	M/A	1-3	I	mi	[29]
Pipistrellus tenuis (mimus)	Least Pipistrelle	38	50	M/A	2-4	I	mi	[30]
PRIMATES								
Alouatta seniculus	Red howler monkey	46	64	M/A	1-3	I	A	[31,32]
A.seniculus maccomelli		47–49			1-3			[33]
Homo sapiens	Human	46	78	SM/A	2	I	mi	[34,35]
CARNIVORA								
Atelocynus microtis	Short-eared dog	74	72	SM/SM	2	I	mi	[36]
Chrysocyon brachyurus	Maned wolf	76	72	SM/SM	1	II	A	[37]
Nyctereutes p. procyonoides	Raccoon dog	54	62	M/M	1-4	II	A, SM	[38]
Nyctereutes p. viverrinus		38	62	M/M	1-5	II	A	[39]
Vulpes (Alopex) lagopus	Arctic fox	50	92	M/A	1	II	M	[40]
Vulpes bengalensis	Bengal fox	60	68	M/A		I	mi	[41]
Vulpes pallida	Pale fox							
Vulpes vulpes (fulvus)	Red fox	34	64	M/A	1-10	I	A, M	[3,43]
ARTIODACTYLA								
Capreolus pygargus	Siberian roe deer	70	72	SM/A	1-14	I	mi	[44,45]
Mazama americana	Red brocket	42–53	42–52	SM/A	2-5	I	mi, A	[44,46]
Mazama bororo	Small red brocket	34	46	M	4-6	I	mi, A	[47,48]
Mazama gouazoubira	Gray brocket	69–70	68–69	M, A/A, mi	1-2	I	mi, A	[47,49,50]
Mazama nana	Brazilian dwarf brocket	70	68	A/M	1-3	I		
Mazama nemorivaga	Amazonian brown brocket	36	54	M/mi	1-6	I	mi, A	[47,51]
Moschus moschiferus (sibiricus)	Siberian musk deer	67–69	69–72	SM/A,M	2-7	I	mi	[52]
RODENTIA		58	56	A/A	1-2	-	-	[53]

Table 1. Cont.

ORDER / Species	Common Name (◊)	2n	NFa	X/Y	No. Bs	Bs Morphology Size *	Bs Morphology Cent. Position †	References
Acomys ngurui	Spiny mouse	59–61	68	M/A,SM	1	II	SM	[54,55]
Acomys spinosissimus		59–61		A/SM	1	II	A	[54]
Akodon mollis	Soft grass mouse	22	42	M/A	1	II	M	[56]
Akodon montensis (arviculoides)	Montane Akodont	24	42	A/A	1–3	II	SM	[57,58]
Apodemus agrarius	Striped field mouse	48	54	A/A	1	I,II	mi, A	[59]
Apodemus argenteus	Small Japanese field mouse	46	48	A/A	1	I,II	mi, SM	[60]
Apodemus flavicollis	Yellow-necked field mouse	48	46	A/A	1–9	II	A	[61,62]
Apodemus mystacinus	Eastern broad-toothed field mouse	48	50	A/A	2	-	-	[63]
Apodemus peninsulae (giliacus)	Korean field mouse	48	46	A/A	1–30	I,II,III	mi, A, SM, M	[64,65]
Apodemus sylvaticus	Long-tailed field mouse	48	46	A/A	1–3	II	A	[66]
Bandicota indica	Greater bandicoot rat	44/45, XX/XO	84	SM/A	1–3	II	SM	[67]
Bandicota savilei	Savile's bandicoot rat	43	58	SM	1	I	SM	[68]
Berylmys berdmorei	Berdmore's Berylmys	40	62	A	1	II	M	[68]
Blarinomys breviceps	Brazilian shrew mouse	29–50	50	A	2	II	M	[69]
Chaetodipus (Perognathus) baileyi	Bailey's pocket mouse	46	64	M/M	1–10	II	M	[70]
Dasymys rufulus	West African shaggy rat	36, 38, 39, 40	42–50	A,SM,M/A,SM,M	1–3	II	M	[71]
Dasyprocta fuliginosa	Black Agouti	64	118	M/SM	1	II	SM	[72]
Dasyprocta leporina	Red-rumped Agouti	64	118	M/M	1	I	M	[72]
Dasyprocta prymnolopha (nigriclunis)	Black-rumped Agouti	64	118	M/SM	1	II	M	[72]
Dasyprocta sp.		64	118	M/SM	1	II	M	[72]
Dicrostonyx groenlandicus (kilangmiutak)	Northern collared lemming	47–50	48	M/SM	1–3	I,II	A, M	[73]
Dicrostonyx torquatus	Palearctic collared lemming	44	56	A,M/A,SM	1–8	II	M	[74,75]
Golunda ellioti	Indian bush rat	54	54	A,SM/A	1–42	II	SM, M	[76]
Grammomys (Thamnomys) dolichurus	Woodland thicket rat	54	68	SM/A	1–4	II	A	[77]
Grammomys macmillani (Thamnomys gazellae)	Macmillan's thicket rat	48	70	SM/A	4–7	I	A, M	[78]
Holochilus brasiliensis	Web-footed marsh rat	48–56	58	A/A	2–17	II,III	mi	[79]
Holochilus chacarius	Chaco marsh rat	44	56–60	A/A	1–2	II	SM, M	[80]
Holochilus venezuelae		36	56	A/A	1–2	II	M	[81]
Holochilus vulpinus		48	58	SM/SM	1	II	A	[19]
Mastacomys fuscus	Broad-toothed rat	48	56	SM/SM	1–3	II	A	[82]
Mastomys erythroleucus	Guinea multimammate mouse	38	54	M/A	1	II	SM	[83]
Mastomys natalensis	Natal multimammate mouse	32	54	A/A	2	II	mi, A, SM, M	[84]
Melomys burtoni	Grassland Melomys	48	50	A/A	1–8	I,II	SM, A	[85]
Melomys capensis	Cape York Melomys	48	50	A/A	3–6	-	A	[85]
Melomys cervinipes	Fawn-footed Melomys	48	50	A/A	4–13	I,II	M	[82]
Microtus gregalis	Narrow-headed vole	36	50	M/A	1–4	II	A	[86]
Microtus longicaudus	San bernardino long-tailed vole	56	84	M/A	1–14	I	M	[87]
Mus cookii	Ryley's spiny mouse	40	38	A/A	1	I,II	A, M	[68]

Table 1. Cont.

ORDER Species	Common Name (◊)	2n	NFa	X/Y	No. Bs	Bs Morphology Size *	Cent. Position †	References
Mus shortridgei	Shortridge's mouse	46	46	A/A	1–3	I, II	A, M	[88]
Nannospalax (Spalax) leucodon	Lesser blind mole rat	60	74	SM/M	1–3	I	mi, A	[89]
Nectomys rattus	Common water rat	52	50	A, SM/A, SM	1–3	II	A, SM, M	[90]
Nectomys squamipes	South American water rat	56	54	A, SM/A, SM, M	1–3	II	A, SM	[91]
Oecomys concolor	Natterer's Oecomys	60	62	SM	1–2	I	SM	[92]
Oligoryzomys (Oryzomys) fornesi	Fornes colilargo	62–66	64	SM/SM	1–2	I	A	[93]
Oligoryzomys flavescens	Yellow pygmy rice rat	64	64	SM/SM	1–2	I	mi	[94]
Olomys irroratus	Southern African vlei rat	28	44	M/SM	2–4	II	SM, M	[95]
Praechimys sp.		26			1	I	mi	[96]
Rattus fuscipes	Bush rat	38	58		1–3	II	M	[92,97]
Rattus norvegicus	Brown rat	42	60		1	II	A	[98]
Rattus rattus	House rat	42	60–64		1–3	II	M	[99]
Rattus r. diardii		42			1–4	II	M	[100]
Rattus r. frugivorus		38			1–3	II	M	[101]
Rattus r. kandianus		40			1	II	M	[102]
Rattus r. talmezumi		42			1	II	M	[102]
Rattus r. thai		42			1–6			[103]
Rattus tunneyi	Pale field rat	42	60	A/A	1	II	M	[104]
Reithrodontomys megalotis	Southern marsh harvest mouse	42	72	M/A, SM	1–7	I	mi	[105]
Reithrodontomys montanus	Plains harvest mouse	36	50	A/M	1	I	SM	[106]
Sigmodon hispidus	Hispid cotton rat	52			3–4	-		[107]
Scooretamys angouya (Oryzomys angouya, O. buccinatus, O. ratticeps)		58	60	A/A	2	I, II	mi, SM	[108–110]
Thallomys nigricauda	Black-tailed tree rat	48	60			-		[111]
Thomomys bottae	Animas mountains pocket gopher	76	130	SM/mi	6–12		mi	[112]
Trinomys (Proechimys) iheringi	Ihering's spiny rat	60	116	SM/SM	1–6	I	mi	[113,114]
Tscherskia (Cricetulus) triton	Greater long-tailed hamster	28	30	A/M	1–2	II	A	[112,113]
Uromys caudimaculatus	White-tailed giant rat	46	50	A/A	2–12	II, III	A, SM, M	[18,115]

2n—diploid number, NFa—fundamental number of autosomes, X/Y—morphology of sex chromosomes; * Category: I—Bs smaller than chromosomes from A set, II—same as A, III—larger than A; † mi—micro Bs, M—metacentric, SM—submetacentric, A—acrocentric chromosomes; (◊) Mammal Diversity Database [116].

3. Frequency of B Chromosomes

A large variability of Bs in mammalian species is displayed on all levels: intra-individual, intra- and inter-populational. The most common for mammals is frequent appearance of intraindividual variability that can feature the same tissue or appear between different tissues. Mosaicism for the number of Bs was scored in *Echymipera kalubu* [22], *Rattus rattus* [103], *V. vulpes* [117], *Myotis macrodactylus* [28], *A. peninsulae* [118], *Dicrostonix torquatus* [119], *Trinomys iheringi* [113], *Nictereutes procionides viverinus, Capreolus pygargus* [44], *Alouatta seniculus* [32], *Dasyprocta fuliginosa, Dasyprocta leporine, Dasyprocta prymnolopha* [72], *Apodemus flavicollis* [11], *Nictereutes procionides procyonides* [120], *Mazama nana* [51], *Mazama americana* [51], *Grammomys macmillani* [78], *Acomys ngurui, Tscherskia triton* [121] and *Mazama nemorivaga* [52].

The mosaicism for the number of Bs is extensively studied in Korean field mouse, *A. peninsulae*, first noticed in early studies [122] and then confirmed in different areas of species' wide distribution [123,124]. The frequency of mosaics extends from 0.05 in South Korea [125] to even 0.85 in populations from Primorskii region and Hokkaido [123]. Furthermore, it has been found that the variability of B chromosome numbers is higher in the group of mosaics [8,124,126].

The great variability produced by intraindividual mosaicism is also characteristic for genus *Mazama*. In this genus, Bs appear in five out of eight species [51]. For instance, in *M. americana*, Abril et al. [127] found Bs in all 18 studied animals with intraindividual variability from 0–6 Bs. The same occurrence happened in *M. nana* [51] and in *M. nemorivaga* [52] where all studied animals had 0–7 Bs. Besides Bs, genus *Mazama* is featured with other kinds of chromosome polymorphisms, involving autosomes and sex chromosomes. This is also the case with *Acodon montensis* [128]. On the other hand, genus *Apodemus* with one third of species with Bs is karyotypically very stable.

A presence of one B chromosome is the most common situation, but the number of Bs per animal can vary widely. The highest number of Bs, which counted 42, was found in *D. torquatus* [75]. Up to 30 Bs in a single animal have been detected in *A. peninsulae* [65], while there have been 17 Bs identified in *Thamnomys gazellae* (now *Grammomys macmillani*) [78]. The average maximal number of Bs per specimen found in 85 mammalian species is 4.88 (Table 1).

There are some species with Bs whose populations cover wide geographic areas. The yellow-necked field mouse, *A. flavicollis*, common in the Western Palearctic region, has showed Bs presence almost everywhere through its range with frequencies ranging from 0.07 to 0.94 [62,129–138]. The frequency of animals with Bs in different geographic regions varies, but rules cannot be established easily. The variation in frequency of Bs that is generally present in *A. flavicollis* is also characteristic for small areas. For instance, we studied Bs presence in 40 populations from Serbia and the frequencies ranged from 0.11 to 0.67 [66,133,139–142]. Generally, the frequency of animals with Bs increases with altitude towards harsher climatological conditions [141,143]. However, this trend was not confirmed in the samples from Poland [138].

Shellhammer [144] suggested that the most reasonable explanation of great B frequency variation in southern marsh harvest mouse, *Reithrodontomys megalotis*, is a general increase in genetic variability towards the periphery of species distribution. The same was proposed for Bailey's pocket mouse, *Chaetodipus baileyi* [70], while Boyeskorov et al. [145] found the highest B frequency in *A. flavicollis* (0.81) in a peripheral area of its distribution. A north-to-south increase in frequency of Bs was found in grassland Melomys, *Melomys burtoni* [85].

Besides being found in almost all studied populations, Bs in *A. peninsulae* are often present in all individuals. For instance, in the populations in East Asia, the frequency of animals with Bs vary from 0 to 1.0, while in the Siberian populations, from 0.99 to 1.0 [123,146]. So far, the only exceptions are Sakhalin Island and Stenina Island, where Bs repeatedly have not been found [146,147]. The distribution of Bs varies significantly between populations [62,148–152], however, these differences are still largely unexplained. The difference in the maximal number of Bs between regions is also evident varying from 30 in Siberia [65], to 6 in South Korea [123]. Roslik and Kartavtseva [124] established variability in modal number of Bs. Each population is characterized by a certain modal number of Bs. This number is also specific for regions. Roslik and Kartavtseva [153] documented the presence of clinal decreasing in frequency of rare B morphotypes from East to Northwest in the studied area.

Contrary to such high frequencies of Bs in *A. peninsulae* and *A. flavicollis*, Zima and Macholán [62] found that the frequency of animals with Bs in populations of long-tailed field mouse, *A. sylvaticus*, is very low (2.4%). Such sporadic occurrence of Bs is characteristic for another species from the same genus, the striped field mouse, *Apodemus agrarius* [59]. While *A. peninsulae* and *A. flavicollis* are typical forest-dwelling species, *A. sylvaticus* is limited to the edges of forests and *A. agrarius* is a typical field mouse.

4. Structure and Composition of B Chromosomes

The newly discovered facts about Bs are mostly concerning their structure. Bs were for a long time seen as chromosomes without genes or, at least without active ones, due to a prevailing absence of their visible phenotypic effects. Although the counterevidence was repeatedly suggested, they were generally ignored until recently when technological advances [154] in genome analysis and sequencing armed investigators with a variety of new technical approaches to shake this dogmatic view. Molecular studies represent Bs as assemblage of various repeated sequences originating from one or more A chromosomes [155–157] or even from all [158]. Non-coding repetitive sequences or mobile elements present in both A and B chromosomes prevail but some of them are more frequent in Bs [159]. Some paralogs of genes located on A chromosomes could be found on Bs as intact or as degenerate sequences [154]. Unique sequences specific for Bs are rarely found [23,160,161]. Yet, thanks to the new technology, the list of genes identified on Bs is promisingly increasing.

The previous studies on mammalian Bs that were based mostly on differential staining revealed that 60% of them are C positive [11]. Those studies showed that when different types of Bs are present they could be C positive or C negative, such as in *A. peninsulae* and *M. nana* [51,162]. Furthermore, the analyses of molecular DNA composition of Bs in *A. peninsulae* [163] showed a presence of two specific forms of chromatin with presumed autonomous origin. Besides that, homology to the heterochromatic region of sex chromosomes and pericentromeric DNA of autosomes was established [163–165].

Molecular composition of Bs for 19 mammalian species is presented in Table 2. A presence of ribosomal genes (rDNA) was detected in 5 species by using silver staining and fluorescent in situ hybridization (FISH). Telomeric repeats are most frequently found on Bs (12 species), but centromeric were detected in only three cases. The presence of molecular markers specific for Bs was found in *P. volans* [23] and *A. flavicollis* [160].

Table 2. Current data on molecular composition of B chromosomes in mammalian species.

Species	Found on B Chromosome	Method	References
Petauroides volans	centromeric regions, B specific regions	FISH, PCR	[23]
Nyctereutes procyonoides procyonoides	interstitial telomeric sequences	FISH	[166]
	rDNA (NOR)	FISH, silver staining	[167]
	C-KIT	FISH	[168]
	Kdr, RPL23A pseudogene	FISH, PCR	[169]
	rDNA	PRINS (primed in situ DNA synthesis)	[170]
	Lrig1	FISH	[171]
	Ret		
	Lrig1 Ret	FISH	
	C-KIT (no transcriptional activitiy)	PCR, RT-PCR	[172]
	100 sequences located on B, homologous to genes involved in cell proliferation, differentiation, neuron sinapse, cell junction	sequencing of microdissected B	[173]
Nyctereutes procyonoides viverrinus	interstitial telomeric sequences	FISH	[166]
	three types of B-specific heterochromatin	FISH	[173]
	C-KIT	FISH	[168,169,171]
	Kdr RPL23A pseudogene	FISH, PCR	[169]
Vulpes vulpes	C-KIT	FISH	[168,171]
	RPL23A pseudogene	PCR	[169]
	Mdn1, Cfndd2	FISH	[171]
	49 sequences located on B, homologous to genes associated with cell division machinery, cell cycle control functions, microtubule, centrosomes, cell differentiation, proliferation	sequencing of microdissected B	[173]
Capreolus pygargus	Tnni3k, Fpgt, Lrriq3	FISH, flow-sorted DNA libraries derived from Bs	[174,175]
	9 genes located on B	re-analyzed data from [175]	[173]

Genes **2018**, *9*, 487

Table 2. *Cont.*

Species	Found on B Chromosome	Method	References
Mazama gouazoubira	55 sequences located on B, homologous to genes associated with functional clusters associated with ATP-binding/kinase, mitochondria, cell cycle, Zn-ion binding/Zn-finger, membrane, cell proliferation/differentiation, positive regulation of protein kinase activity	sequencing of microdissected B	[175]
Acomys sp.	107 sequences located on B homologous to genes	re-analyzed data from [175]	[173]
	telomeric repeat	FISH	[54]
Akodon montensis (arviculoides)	rDNA (NOR)	silver staining	[57,108]
	telomeric repeat, rDNA (NOR)	FISH, silver staining	[176]
Apodemus flavicollis	rDNA (NOR)	silver staining	[177]
	B specific regions	AP-PCR RT-PCR	[160] [178]
	rDNA	RT-PCR	[179]
	Vrk1	ISSR-PCR, sequencing	[180]
	38 sequences located on B, homologous to genes associated with microtubule, cell cycle proteins, and less significant nucleotide-binding, membrane and metal binding proteins. Satellite repeats, MurSatRep1, ERVL (MaLR), ERVK LTRs and transposable elements.	sequencing of microdissected B	[181]
	101 sequences located on B homologous to genes	re-analyzed data from [181]	[173]
Apodemus peninsulae	telomeric repeat, two types of B arm-specific repeats	FISH	[164] [163]
	two types of B-specific chromatin	FISH	[181]
	repetitive elements	FISH	[182]
	centromeric repeats, 32 sequences located on B homologous to genes associated with cell division machinery, cell cycle control, nucleotide-binding, laminin and EGF-like domain-containing, cytoskeleton and ion-bindings proteins, LINE L1 elements, centromeric repeats, satellite repeats MurSatRep1, ERVK and ERVL (MaLR) LTRs.	sequencing of microdissected Bs	[181]
	152 sequences located on Bs homologous to genes	re-analyzed data from [181]	[173]
Blarinomys breviceps	telomeric repeats, ITSs	FISH	[69]

Table 2. *Cont.*

Species	Found on B Chromosome	Method	References
Holochilus brasiliensis	OSHR, telomeric repeats	FISH	[183]
Nanospalax leucodon	telomeric repeat	FISH	[89]
Nectomys sp.	ITBs	FISH	[184]
Nectomys rattus	OSHR	FISH	[183]
Nectomys squamipes	OSHR, ITS	FISH	[183]
Rattus rattus	rDNA	FISH	[185]
	telomeric repeat	FISH	[186]
Reithrodontomys megalotis	telomeric repeat, LINE elements, centromeric repeats	FISH	[187]
Scoretamys angouya	rDNA (NOR)	silver staining	[108]
Trinomys iheringi	telomeric repeats	FISH	[114]

FISH—fluorecent in situ hybridization; RT-PCR—real time-PCR; rDNA (NOR)—ribosomal DNA (nucleolus organizer region); AP-PCR—arbitrarily primed-PCR; ISSR-PCR—inter simple sequence repeat-PCR; EGF—epidermal growth factor; LINE—long interspersed nuclear element; ERVK—endogenous retrovirus-K; ERVL—endogenous retroviruses-related; LTRs—long terminal repeats; ITS—interstitial telomeric sequences; OSHR—Oryzomyini shared heterochromatin region; ITBs—interstitial telomeric bands.

119

The first autosomal gene found on Bs of mammals was proto-oncogene receptor tyrosine kinase (*C-KIT*). It was found in three unrelated species, the red fox, *V. vulpes*, the Chinese and Japanese raccoon dog, *N. procyonoides* [168,169] and *M. gouazoubira* [175] but not in *A. flavicollis* [179]. Another gene (*Vrk1*) was found in two *Apodemus* species. By using ISSR-PCR, Bugarski-Stanojević et al. [180] found a part of *Vrk1* gene on Bs of *A. flavicollis*. The presence of this gene was confirmed upon isolation by microdissection [181] and additional 37 genes or parts of genes were found on Bs of this species. The Bs in *A. flavicollis* have similar structure as pericentromeric region of sex chromosomes [188]. Through comparison of gene groups in Bs of six mammalian species from different families, Makunin et al., [173] confirmed enrichment with genes related to cell-cycle, development and genes functioning in the neuron synapse. They pointed that the presence of genes on Bs involved in cell-cycle regulation and tissue differentiation could be important for the B chromosome formation.

There are also findings that propose the existence of regulatory interactions between coding sequences of A and B-chromosomes. Bugno-Poniewierska et al. [189], from studies of Bs in Chinese raccoon dogs and red fox determined that DNA methylation may maintain the transcriptional inactivation of DNA sequences situated on Bs. This could be the way to avoid some negative effects of Bs presence. Trifonov et al. [174] found, for the first time, the protein coding sequences on Bs of the Siberian roe deer, *C. pygargus*, which are not fully inactivated. Earlier, the gene expression in *A. flavicollis* showed elevated expression of three DNA fragments in the presence of Bs [178]. So, B'chromosome could be seen as a repository of various information which could be used depending on the selection pressure that a B carrier faces.

5. Origin of Bs in Mammals

There are several hypotheses proposed to explain the route of B chromosomes appearance [174,185,190]. In general, the source of Bs are chromosomes of the standard set, both autosomes and sex chromosomes, yet their origin from interspecies hybrids has also been proven in certain cases [191,192], but not in mammals. Whatever the source of their origin is, all proto-Bs must instantaneously pass through inactivation to avoid synapsis with the source chromosome. At present, a series of molecular processes are known as good candidates to achieve this condition, for instance mechanisms of sex chromosomes inactivation and epigenetic mechanisms. Bs can follow the same process operating in meiotic sex chromosome inactivation (MSCI) during the meiotic prophase I. Vujošević and Blagojević [11] proposed that B chromosomes are absent in birds due to genome reduction. Moreover, it appears that the sex chromosome specific silencing is absent in birds, although not yet been completely elucidated [193,194]. The same situation is found in egg laying monotremes [195] that also lack Bs. What is frequently overlooked in attempts to explain the initial steps of Bs origin is the possibility of simultaneous origin of proto-Bs in a population [11] that is far more probable in mammals due to their social organisation and population dynamics. This could promote spread of proto-Bs in populations.

The origin of Bs in mammalian species was based rather on presumptions than on facts. The circumstantial evidence come and is expected from molecular studies of Bs DNA contant. Sex chromosomes are proposed as a source of Bs in *E. kalubu* [22], *Dycrostonix groenlandicus* [196] and *Apodemus argenteus* [60]. In *A. peninsulae* a homology of heterochromatic region of Bs, sex chromosomes and autosomes was established [163–165]. Upon generation of microdissected DNA probes followed by FISH on metaphase chromosomes, another study found that Bs in *A. flavicollis* originate from pericentromeric region of sex chromosomes [188]. While there are five different types of Bs [197] with different origin (including sex chromosomes) in *A. peninsulae*, it was shown that all Bs in *A. flavicollis* have the same DNA contant regardless of their number or geographical distance which indicates a common origin from sex chromosomes [188]. Furthermore, whenever two or more types of Bs are present in one species, it appears that they do not have the same origin. A multiple origin of Bs in *A. peninsulae* was suggested by Matsubara et al., [198] based on the presence of 18S/28S rRNA genes only on meta- or submetacentric Bs. Some recent findings offer evidences for single origin of Bs in this species also [181]. A different origin for two types of Bs was also found in the harvest mouse, *R. megalotis*, by Peppers et al. [187].

Based on comparative cytogenetic studies [67,199] it was suggested that Bs in *Bandicota indica*, *R. rattus* and *Rattus fuscipes* originated before the divergence of these species occurred. Different origin was assumed for Bs in two species of Carnivora, *N. procyonides* and *V. vulpes* [163,173] based on molecular data.

A two-step appearance of Bs was proposed for *A. peninsulae* [164]. The first step is the destabilization of pericentromeric regions, produced by the invasion of DNA sequences from euchromatic parts of A chromosomes, which leads to a formation of microchromosomes in high frequency, and thus make proto-Bs. The second step is the insertion and amplification of new DNA sequences. Similar steps were proposed by Rubtsov et al. [200] that assumed that the origin of Bs start with a loss of a greater part of q arm of an ancestor autosome followed by subsequent evolution of Bs that includes additional constitutional rearrangements. Makunin et al. [175], by using sequencing of isolated Bs of two mammalian species, showed that Bs originate as segmental duplications of specific genomic regions, and subsequently passes through pseudogenization and a repeat accumulation.

Presently, it seems that the new data describing the molecular composition of Bs incites more questions than suggests answers to the old ones.

6. Behaviour of B Chromosomes during Meiosis and their Transmission

The number of species whose meiotic behaviour of Bs was studied increased just slightly in last 15 years but there are new details for some already studied species. Currently, the meiotic behaviour is known for 25 species and univalent Bs are present in all of them. Besides univalents, bivalents appear in 13 species, and multivalents in 7, while assimetrical bivalents are present in 6 species. There are 5 species (*A. peninsulae*, *D. groenlandicus*, *N. procyonides*, *C. baileyi* and *V. vulpes*) where all four mentioned types of configurations are found.

When a different type of Bs is characteristic for the same species, their meiotic behaviour is often type dependent. So Hyata [64] found that both paring and non-paring among Bs occur in *A. peninsulae*. He showed that small macro- and microchromosomes in most cases do not follow Mendelian inheritance, yet other types of supernumeraries do follow it. Further meiotic studies in this species [197] showed that Bs are able to form axial elements and synaptonemal complexes in prophase of the first meiotic division. The same authors found that univalents of dot like Bs of different morphology are obviously not homologous, while metacentric Bs showed a partial homology. Univalent Bs are commonly associated with sex bivalent. Ishak et al. [201] noticed an absence of transcriptional activity in Bs of this species during pachytene. Karamysheva et al. [202], through the use of 2D analysis of pachytene in *A. peninsulae*, found three types of configurations: synapsed bivalents, univalents, and univalents that contain the foldback structure. During meiosis, Bs in *A. flavicollis* appear as univalents, bivalents or, depending on number, combinations of both, but never as a pair with the members of A set [130,203]. In the same species, Banaszek and Jadwiszczak [204] found that Bs behave in non-Mendelian fashion during meiosis I of males.

In *N. squamipes* analysis of the synaptonemal complex revealed auto-pairing of univalent Bs [184]. In the Northen collared lemming, *D. groenlandicus*, it was found that, besides univalents, bivalents and trivalents Bs can make synaptic associations with the Y chromosome [196]. Studies of synaptonemal complex in *M. americana* by Aquino el al. [46] revealed the presence of both univalents and bivalents. Univalents appear in two forms: as autopaird or just univalents.

When Bs appear as univalents in the silver fox, they show a folding-back behaviour that ends as intrachromosomal pairing [205], which indicates the presence of repeated DNA sequences. Sosnowski et al. [206] conducted experiments with spermatocytes of the red fox and the Chinese raccoon dog, and found that Bs that conjugate together form diverse structures, such as bivalents, trivalents, and tetravalents. Sosnowski et al. [206] also concluded that the increase in the number of Bs in spermatocytes of the Chinese raccoon dog corresponds with the lack of conjugation more frequent. Basheva et al. [43] studied A- and B-chromosome pairing and recombination in the silver fox using electron and immunofluorescent microscopy. They found the same distribution of the foci along B- and A-bivalents and proved, for the first time, that meiotic recombination occurs in mammalian B chromosomes

The accumulation of B chromosomes in mammals appears to be a rarer event than expected. One of the reasons for sure is the lack of studies. Furthermore, Bs in some cases could maintain themselves without the apparent drive. In males, the evidences for accumulation of Bs were found in *C. baileyi* [207], *V. vulpes* [208], *A. peninsulae* [162,197,209], and in the greater long tailed hamster, *T. triton* [121]. In the latter, Bs were found in males only and the increase in number of Bs in germline cells was observed.

In lemmings, univalent Bs were eliminated from the polar body and incorporated into secondary oocytes [210,211]. The evidence of accumulation of Bs has been obtained in females of *R. rattus* [212,213] and *R. fuscipes* [97] by means of controlled crosses. In the case of experimental crosses done by Stitou et al. [213] in *R. rattus*, males showed Mendelian transmission rates, while only a slight accumulation of Bs happened in females.

Palestis et al. [214], following the theory of centromeric drive, based on a different ability of the two meiotic poles for capturing centromeres [215], showed that Bs in mammals are more common in species with acrocentric chromosomes. Since then, the number of mammalian species with Bs increased, therefore this theory is not valid anymore. The number of species with Bs with predominantly acrocentric chromosomes in standard set is just slightly larger, so it seems that such explanation for origin of Bs is reasonable only in proven cases.

Karamysheva et al. [202] studied nuclear organization of Bs in *A. peninsulae*. They showed that additional volume of heterochromatic regions of chromosomes and extra centromeres modify 3D architecture of interphase nuclei. The location of Bs in meiosis appeared not to be random, and unpaired Bs had a tendency to form a common compartment with unpaired part of the sex bivalent, and thus avoided pachytene check point.

7. The Effects of B-Chromosomes

Apart from few exceptions, Bs do not cause visible phenotypic manifestations at individual level. This makes the search for observable effects in mammals rather difficult. But even such a small amount of data, together with new findings of genes on Bs, raises objections to the idea of Bs genetic inertness. It has been found that Bs presence influences cell division, degree of recombination, development, some quantitative characteristics, host-parasite interactions and behaviour.

A new and interesting data came from three-dimensional studies of Bs behaviour during division in both somatic and germ cells. Kociucka et al. [216] studied three-dimensional positioning of B chromosomes in fibroblast nuclei of the red fox and the Chinese raccoon dog, and found that small Bs of the red fox are dominantly positioned in the interior of the nucleus, while the medium-sized Bs of the Chinese raccoon dog are in the peripheral area of the nucleus as well as in intermediate and interior locations. The data was in agreement with the chromosome size dependent theory [216]. But in the nuclei of the Korean field mouse all Bs, irrespective of their size, were located on nuclear periphery in common compartments with C-positive regions of A chromosomes [202]. They suppose that, at least for the Bs of the Korean field mouse, the DNA content is more important parameter that determines where Bs will be located inside the nucleus.

As we pointed earlier, Basheva et al. [43] proved that in the silver fox recombination occurs between B chromosomes which increases variability in the specimens that carry them. In earlier studies, the presence of B chromosomes was associated with increased chiasma frequency in Bailey's pocket mouse, *C. baileyi* [207] and *R. fuscipes* [199]. In both species this increase is not influenced by the number of Bs yet only by their presence.

Gileva [211] recorded a reduction of body and skull sizes in *D. torquatus* that carry numerous Bs, and proposed that this could be reflected as a negative selective value in extreme climate conditions.

Positive correlations between number of Bs and body weight were established in males of two species: *N. p. viverinus* [217] and *A. flavicollis* [62,218]. Effects of Bs presence are extensively studied in *A. flavicollis* and, in general, it was found that they influence the development of some morphometric characters, mostly cranial ones [140]. One of the two regions of the mandible shows almost a triple increase in intensity of integration in B carriers [219]. Furthermore, the maintenance of Bs in the same species was studied by examining their effects on 3 components of cranial variability: canalization, developmental stability, and morphological integration. It was suggested that B carriers follow different developmental pathway for generating covariations of cranial traits [220]. This specific developmental pathway is more sensitive to modifications caused by natural selection, which could be beneficial to B carriers under variable environmental conditions. It was previously established that reaction of animals with Bs to environmental changes differ from those without them [221]. Nonmetric traits analyses show that the population density influences, at the same time, both the variation in the frequency of specimens with Bs and the developmental homeostasis.

Adnađević et al. [222], by analysing effect of recorded endoparasites and parasite life-cycle stages in *A. flavicollis* on expression levels of genes *MHC II-DRB*, *IL-10* and *Tgf-β*, found that the presence of Bs is associated with lower expression level of *Tgf-β* gene. Although the influence of host genetic background on parasite infection has already been well established, this is the first study in mammals that correlates presence of Bs with immune response. Curiously enough, the presence of Bs in this species plays an important role in infrapopulations of their certain endoparasites by shifting sex balance to higher proportion of males [223].

Shellhammer [144] from studies on *R. megalotis* was the first to propose that Bs could have an effect on behaviour. The behaviour and the presence of Bs were connected in foxes through a series of experimental crosses [224–226]. It appeared that groups of foxes selected for specific behaviour differs significantly in frequency of mosaics for Bs.

8. Maintenance of B Chromosomes

The mostly discussed question about Bs during a century of research was the way they are maintained in populations through time. Two schools of thought grouped around two models giving opposite explanations of the way Bs are retained in natural populations. Both models assume that the frequency of specimens with Bs in population is at equilibrium but the explanations how this equilibrium is reached and kept are different. The model firstly named parasitic and then selfish [227] claims that Bs are maintained by balance of accumulation and elimination due to detrimental effects. Contrary to this, the heterotic model [228] suggests that, in the absence of mechanism of accumulation, a small number of Bs could offer an adaptive advantage to carriers, while a large number could be harmful. Currently, the parasitic model is predominant, mostly because the search for adaptive significance of Bs was mostly ineffective. Furthermore, the convergence of this paradigm partly comes from the popular theory of selfish or parasitic DNA, irresistible to some scientists. The number of cases with proved Bs accumulation, which is prerequisite for parasitic model, although larger than the number of cases without accumulation, is still very small in comparison with the known number of species with Bs. For instance, the accumulation of Bs was studied in about 70 plant species and among them 42 (60%) manifested accumulation mechanism [4] which makes only 3.4% of plant species with Bs. Furthermore, detailed studies are largely directed on commercially important species that possess Bs, like maize and rye, and pests such as grasshoppers, so the number of extensively studied species groups is rather small. In attempt to include species without mechanism of Bs accumulation into parasitic model, Camacho et al. [229] proposed that all Bs are initially parasitic, and later on, through arm race with A genome, may become neutral. From this stage they can disappear or become parasitic again. One of their arguments against heterotic Bs is that it is unexpected that Bs could be beneficial in the first step, so a drive is necessary to establish themselves in population. Yet, if Bs appear simultaneously in population, these arguments are not plausible. Therefore, when models are assessed it is not good to stay frozen within a particular paradigm.

Temporal analyses of B frequency and transmission in mammals are scarce. The frequency of animals with Bs was the same in two successive years in *Rattus rattus diardi* [100]. During 8 years of study, equilibrium frequencies of Bs in populations of *A. flavicollis* at one locality were maintained in spite of fluctuations in population density [11,230]. Zima and Macholan [62] and Wojcik et al. [138] also found the equilibrium during a three-year study in the same species. Contrary to stable frequency from year to year, seasonal changes in frequency of animals with Bs could escape from equilibrium in stress situations [231] or could keep it when there is no tough competition present [232]. Thus, though frequencies of Bs could significantly differ through a year, their values stay the same between years [230].

B chromosome frequencies in *A. peninsulae* show temporal variation. Comparison of Bs from the population from Altai Republic, trapped in the 1980 and 2002 showed that a mean number of Bs in this population increased almost threefold in the period of 22 years [200]. This increase was mainly due to the rise of numbers of small and large bi-armed Bs (by factors of 7.0 and 5.3, respectively) and a slight increase in the number of medium-sized biarmed B chromosomes (by a factor of 1.6). Nonetheless, Borisov et al. [233] found that the number of Bs and theirs morphotypes were stable over the period of 30 years in certain populations.

Direct or indirect evidences for B drive in mammals are provided for seven species only: *C. baileyi* [208], *V. vulpes* [209], *R. rattus* [212,213], *D. torquatus* [211], *R. fuscipes* [97], *A. peninsulae* [197], and *T. triton* [121]. In 3 of them, B drive is operating in females thus supporting the theory of centromeric drive [214]. Thomson [97] showed that the maintenance of Bs in *R. fuscipes* supports the parasitic model very well.

The maintenance of Bs in populations can be explained in terms of their contribution to overall genetic diversity of the species possessing them, and it might be arguable under the heterotic model [142]. The increased variability widens the probability that species will survive in changing environmental situations. In *A. flavicollis*, an increased frequency of animals with Bs is found in more extreme climatic conditions [141]. Frequency of B chromosomes and quality of habitat are negatively correlated indicating that B chromosomes in this species are mentioned due to the effects that they exert at the level of populations [143]. Possible adaptive effects of Bs were also postulated by Blagojević et al. [234] upon comparison of head morphology in three populations of this species that have Bs at different frequencies. Adnađević et al. [235], by using amplified fragment length polymorphism (AFLP) markers, made a comparison of populations of *A. flavicollis* settled in ecologically distinct habitats differing in frequency of Bs, and found that the greatest genetic diversity is in the population settled in optimal conditions for this species featured by the lowest frequency of animals with Bs. The majority of loci that are subject of directional selection, feature either population with lower or with a higher frequency of Bs. They suggested that the different frequency of B carriers in populations is related to adaptive differentiation to diverse habitats. Tokarskaia et al. [45] found that the presence of Bs is positively correlated with heterozygosity for random amplification of polymorphic DNA (RAPD) loci, in populations of the Siberian roe deer, *C. pygargus*, thus indicating influence of Bs on the genetic variation of the species. All these findings support the heterotic model of Bs maintenance.

Theoretically, inbreeding is harmful to parasitic Bs but beneficial to mutualistic ones. Social organization of rodent populations and some other mammalian groups supports inbreeding which opens new possibilities for the existence of beneficial Bs.

Extensive population studies of two species of the same genus *Apodemus* best illustrate that the present models do not exclude each other but rather call for further adjustments. If we try to fit the maintenance of Bs in *A. peninsulae* and *A. flavicollis* into the current models, *A. peninsulae* will follow the parasitic (or selfish) model, while Bs in *A. flavicollis* will better fit into heterotic model. But when we go into details, it seems that neither *A. peninsulae* nor *A. flavicollis* fit quite well into proposed models. *A. peninsule* do not have populations at equilibrium and tolerance for Bs is so great that it is not easy to say when Bs become detrimental. Furthermore, five different types of Bs present in this species, have different outcomes. Some types are inherited in almost a Mendelian fashion. On the other hand, Bs in *A. flavicollis* brings adaptive advantage in certain situations and in some environments. In other situations (and environments) they could be neutral or deleterious. Therefore, it could be hypothesized that the adaptive advantage of these Bs is not general, but it is dependent on events through which the individual or population is passing. The existing models need to be very much adjusted, but the adjustment must be based on detailed and intensive studies in natural populations.

9. Conclusions

After more than a century, it appears that B chromosomes research suffers from an unbalanced approach. That is also true for research of Bs in mammals. Population studies are a very difficult task and are still largely avoided. Even rarer attempts are made to resolve effects of Bs in different species that carry them. While molecular breaks into DNA composition of Bs are rapidly increasing, the number of species included in them is still scarce. Namely, a more detailed molecular composition is known for only six mammalian species. Although the confirmed presence of genes on Bs, in all cases, disproved the claims that Bs are inert, the gathered knowledge and data are not sufficient to explain the significance of Bs to their carriers. Are the paths of evolution of As and Bs opposite, or do the lanes of the same highway promise a greater success in adapting to environmental changes? This is yet to be resolved, but the answer seems to be inclining towards the latter statement.

Author Contributions: Conceptualization, M.V. and J.B.; Writing—original draft preparation, M.V. and J.B.; Writing—review and editing, M.V.; M.R. and J.B.; visualization, M.V.; M.R. and J.B.; supervision, M.V. and J.B.; project administration, M.V.; funding acquisition, M.V.

Funding: This research was funded by Ministry of Education, Science and Technological Development, Republic of Serbia (http://www.mpn.gov.rs/nauka), grant number 173003.

Acknowledgments: We thank the Editors and three anonymous reviewers for their helpful and constructive comments that greatly contributed to improving the final version of the paper.

Conflicts of Interest: The authors declare no conflict of interest. The funders had no role in the design of the study; in the collection, analyses, or interpretation of data; in the writing of the manuscript, or in the decision to publish the results.

References

1. Wilson, E.B. Studies on chromosomes. V. The chromosomes of Metapodius. A contribution to the hypothesis of the genetic continuity of chromosomes. *J. Exp. Zool.* **1906**, *6*, 147–205. [CrossRef]
2. Hayman, D.L.; Martin, P.G. Supernumerary chromosomes in the marsupial *Schoinobates volans* (Ker). *Aust. J. Biol. Sci.* **1965**, *18*, 1081–1082. [CrossRef] [PubMed]
3. Moore, J.W.; Elder, R.L. Chromosome of the fox. *J. Hered.* **1965**, *56*, 142–143. [CrossRef]
4. Jones, R.N. B chromosomes in plants. *New Phytol.* **1995**, *131*, 411–434. [CrossRef]
5. D'Ambrosio, U.; Alonso-Lifante, M.P.; Barros, K.; Kovařík, A.; Mas de Xaxars, G.; Garcia, S. B-chrom: A database on B-chromosomes of plants, animals and fungi. *New Phytol.* **2017**. [CrossRef] [PubMed]
6. Jones, N. New species with B chromosomes discovered since 1980. *Nucleus* **2017**, *60*, 263–281. [CrossRef]
7. Rice, A.; Glick, L.; Abadi, S.; Einhorn, M.; Kopelman, N.M.; Salman-Minkov, A.; Mayzel, J.; Chay, O.; Mayrose, I. The chromosome counts database (CCDB)—A community resource of plant chromosome numbers. *New Phytol.* **2015**, *206*, 19–26. [CrossRef] [PubMed]
8. Volobujev, V.T. The B-chromosome system of mammals. *Genetica* **1980**, *52/53*, 333–337. [CrossRef]
9. Volobujev, V.T. The B-chromosome system of the mammals. *Caryologia* **1981**, *34*, 1–23. [CrossRef]
10. Vujošević, M. B-chromosomes in mammals. *Genetika* **1993**, *25*, 247–258.
11. Vujošević, M.; Blagojević, J. B chromosomes in populations of mammals. *Cytogenet. Genome Res.* **2004**, *106*, 247–256. [CrossRef] [PubMed]
12. Trifonov, V.A.; Dementyeva, P.V.; Beklemisheva, V.R.; Yudkin, D.V.; Vorobieva, N.V.; Graphodatsky, A.S. Supernumerary chromosomes, segmental duplications, and evolution. *Russ. J. Genet.* **2010**, *46*, 1094–1096. [CrossRef]
13. Wilson, D.E.; Reeder, D.M. *Mammal Species of the World: A Taxonomic and Geographic Reference*, 2nd ed.; Smithsonian Institutions: Washington, DC, USA; London, UK, 1993; pp. 1–12, ISBN 1-56098-217-9.

14. Burgin, C.J.; Colella, J.P.; Kahn, P.L.; Upham, N.S. How many species of mammals are there? *J. Mammal.* **2018**, *99*, 1–14. [CrossRef]

15. Diploid Numbers of Mammalia. Available online: http://www.bionet.nsc.ru/labs/chromosomes/mammalia.htm (accessed on 25 July 2018).

16. Patton, J.L.; Sherwood, S.W. Genome evolution in pocket gophers (Genus *Thomomys*) I. Heterochromatin variation and speciation potential. *Chromosoma* **1982**, *85*, 149–162. [CrossRef] [PubMed]

17. Liehr, T.; Mrasek, K.; Kosyakova, N.; Ogilvie, C.M.; Vermeesch, J.; Trifonov, V.; Rubtsov, N. Small supernumerary marker chromosomes (sSMC) in humans; are there B chromosomes hidden among them. *Mol. Cytogenet.* **2008**, *1*, 12. [CrossRef] [PubMed]

18. Baverstock, P.R.; Wats, C.H.S.; Hogarth, J.T. Heterochromatin variation in Australian rodent *Uromys caudimaculatus*. *Chromosoma* **1976**, *59*, 397–403. [CrossRef]

19. Nachman, M.W. Geographic patterns of chromosomal variation in South American marsh rats, *Holochilus brasiliensis* and H. vulpinus. *Cytogenet. Cell Genet.* **1992**, *61*, 10–17. [CrossRef] [PubMed]

20. Kartavtseva, I.V.; Pavlenko, M.V. Chromosome Variation in the Striped Field Mouse *Apodemus agrarius* (Rodentia, Muridae). *Rus. J. Genet. C/C Genet.* **2000**, *36*, 162–174.

21. Hewitt, G.M. Orthoptera. In *Animal cytogenetics. 3. Insecta 1 Orthoptera*; John, B., Ed.; Gebr. Borntrager: Berlin, Germany, 1979; pp. 1–170.

22. Hayman, D.L.; Martin, P.G.; Waller, P.F. Parallel mosaicism of supernumerary chromosomes and sex chromosomes in *Echymipera kalabu* (Marsupialia). *Chromosoma* **1969**, *27*, 371–380. [CrossRef] [PubMed]

23. McQuade, L.R.; Hill, R.J.; Francis, D. B-chromosome systems in the greater glider, *Petauroides volans* (Marsupialia: Pseudocheiridae). *Cytogenet. Genome Res.* **1994**, *66*, 155–161. [CrossRef] [PubMed]

24. Atanassov, I.N.; Chassovnikarov, T.G. Karyotype characteristics of *Crocidura leucodon* Herman, 1780 and *Crocidura suaveolens* Pallas, 1811 (Mammalia: Insectivora: Soricidae) in Bulgaria. *Acta Zool. Bulg.* **2008**, *2*, 71–78.

25. Ruedi, M.; Maddalena, T.; Yong, H.; Vogel, P. The Crocidura fuliginosa species complex (Mammalia: Insectivora) in peninsular Malaysia: Biological, karyological and genetical evidence. *Biochem. Syst. Ecol.* **1990**, *18*, 573–581. [CrossRef]

26. Meylan, A.; Hausser, J. Position citotaxonomique de quelques museragines du genere *Crocidura* au Tessisn (Mammalia: Insectivora). Origine du dessin dentaire "*Apodemus*" (Rodentia, Mammalia). *CR Acad. Sci. Paris* **1974**, *264*, 711.

27. Motokawa, M.; Wu, Y.; Harada, M. Karyotypes of six Soricomorph species from Emei Shan, Sichuan Province, China. *Zool. Sci.* **2009**, *26*, 791–797. [CrossRef] [PubMed]

28. Obara, Y.; Tomyiasu, T.; Saitoh, K. Chromosome studies in the Japanese vespertilionid bats. I. Karyotypic variation in *Myotis macrodactylus* Temmink. *Jpn. J. Genet.* **1976**, *51*, 201–206. [CrossRef]

29. Volleth, M. Comparative analysis of the banded karyotypes of the European *Nyctalus* species (Vespertilionidae, Chiroptera). In *Prague Studies in Mammology*; Horáček, I., Vohralík, V., Eds.; Charles University Press: Prag, Czech Republic, 1992; pp. 221–226.

30. Bhatnagar, V.S.; Srivastava, M.D.L. Somatic chromosomes of four common bats of Allahabad. *Cytologia* **1974**, *39*, 327–334. [CrossRef] [PubMed]

31. Yunis, E.J.; De Caballero, O.M.T.; Ramirez, C.; Ramirez, Z.E. Chromosomal variations in the Primate *Alouatta seniculus seniculus*. *Folia Primatol.* **1976**, *25*, 215–224. [CrossRef] [PubMed]

32. Vassart, M.; Guedant, A.; Vie, J.C.; Keravec, J.; Seguela, A.; Volobouev, V.T. Chromosomes of *Alouatta seniculus* (Platyrrhini, Primates) from French Guiana 331–334. *J. Hered.* **1996**, *87*, 331–334. [CrossRef] [PubMed]

33. Oliveira, E.H.C.; De Lima, M.M.C.; Sbaloqueuro, I.J.; Dasvila, A.F. Analysis of polimorphic NORs in *Alouatta* species (Primates, Atelidae). *Caryologia* **1999**, *52*, 169–175. [CrossRef]

34. Walzer, S.M.D.; Breau, G.; Gerald, P.S. A chromosome survey of 2400 normal new born. *J. Pediatr.* **1969**, *74*, 438–448. [CrossRef]

35. Huang, B.; Crolla, J.A.; Christian, S.L.; Wolf-Ledbetter, M.E.; Macha, M.E.; Papenhausen, P.N.; Ledbetter, D.H. Refined molecular characterization of the breakpoints in small inv dup(15) chromosomes. *Hum. Genet.* **1997**, *99*, 11–17. [CrossRef] [PubMed]

36. Hsu, T.C.; Benirschke, K. *An Atlas of Mammalian Chromosomes 4*; Springer: Berlin, Germany, 1970; p. 178.

37. Pieńkowska-Schelling, A.; Schelling, C.; Zawada, M.; Yang, F.; Bugno, M.; Ferguson-Smith, M. Cytogenetic studies and karyotype nomenclature of three wild canid species: Maned wolf (*Chrysocyon brachyurus*), bat-eared fox (*Otocyon megalotis*) and fennec fox (*Fennecus zerda*). *Cytogenet. Genome Res.* **2008**, *121*, 25–34. [CrossRef] [PubMed]

38. Makinen, A.; Fredga, K. Banding analyses of the somatic chromosomes of raccoon dogs, *Nyctereutes procyonides*, from Finland. In Proceedings of the 4th Colloquium on the Cytogenetics of Domestic Animals, Uppsala, Sweden, 10–13 June 1980; pp. 420–430.

39. Ward, O.G. Chromosomes studies in Japanese raccoon dogs: X chromosomes, supernumeraries, and heteromorphism. *MCN* **1984**, *25*, 34.

40. Mäkinen, A.; Lohi, O.; Juvonen, M. Supernumerary chromosome in the chromosomally polymorphic blue fox, *Alopex lagopus*. *Hereditas* **1981**, *94*, 277–279. [CrossRef]

41. Bhatnagar, V.S. Microchromosomes in the somatic cells of *Vulpes bengalensis* Shaw. *Chromosom. Inf. Serv.* **1973**, *15*, 32.

42. Chiarelli, A.B. The chromosomes of the Canidae. In *Wild Canids Their Systematic Behaviour Ecology and Evolution*; Fox, M.W., Ed.; Van Nonstrand Reinhold: New York, NY, USA, 1975; pp. 40–53.

43. Basheva, E.A.; Torgasheva, A.A.; Sakaeva, G.R.; Bidau, C.; Borodin, P.M. A- and B-chromosome pairing and recombination in male meiosis of the silver fox (*Vulpes vulpes* L., 1758, Carnivora, Canidae). *Chromosom. Res.* **2010**, *18*, 689–696. [CrossRef] [PubMed]

44. Neitzel, H. Chromosome evolution of Cervidae: Karyotypic and molecular aspects. In *Cytogenetics*; Obe, G.B.A., Ed.; Springer: Berlin, Germany, 1987; pp. 90–112.

45. Tokarskaia, O.N.; Efremova, D.A.; Kan, N.G.; Danilkin, A.A.; Sempere, A.; Petrosian, V.G.; Semenova, S.K.; Ryskov, A.P. Variability of multilocus DNA markers in populations of the Siberian (*Capreolus pygargus* Pall.) and European (*C. capreolus* L.) roe deer. *Genetika* **2000**, *36*, 1520–1530. [PubMed]

46. Aquino, C.I.; Abril, V.V.; Duarte, J.M.B. Meiotic pairing of B chromosomes, multiple sexual system, and Robertsonian fusion in the red brocket deer *Mazama americana* (Mammalia, Cervidae). *Genet. Mol. Res.* **2013**, *12*, 3566–3574. [CrossRef] [PubMed]

47. Duarte, J.M.B.; Jorge, W. Chromosomal polymorphism in several populations of deer (Genus Mazama) from Brazil. *Arch. Zootec.* **1996**, *45*, 281–287.

48. Duarte, J.M.B.; Jorge, W. Morphologic and cytogenetic description of the small red brocket (*Mazama bororo* Duarte, 1996) in Brazil. *Mammalia* **2003**, *67*, 403–410. [CrossRef]

49. Valeri, M.P.; Tomazella, I.M.; Duarte, J.M.B. Intrapopulation Chromosomal Polymorphism in *Mazama gouazoubira* (Cetartiodactyla; Cervidae): The Emergence of a New Species? *Cytogenet. Genome Res.* **2018**. [CrossRef] [PubMed]

50. Rossi, R.V.; Bodmer, R.; Duarte, J.M.B.; Trovati, R.G. Amazonian brown brocket deer *Mazama nemorivaga* (Cuvier 1817). In *Neotropical Cervidology: Biology and Medicine of Latin American Deer*; Duarte, J.M.B., Gonzalez, S., Eds.; Funep: Jaboticabal, Brazil; IUCN: Gland, Switzerland, 2010; pp. 202–210, ISBN 978-85-7805-046-7.

51. Abril, V.V.; Duarte, J.M.B. Chromosome polymorphism in the Brazilian dwarf brocket deer, *Mazama nana* (Mammalia, Cervidae). *Genet. Mol. Biol.* **2008**, *31*, 53–57. [CrossRef]

52. Fiorillo, B.F.; Sarria-Perea, J.A.; Abril, V.V.; Duarte, J.M.B. Cytogenetic description of the Amazonian brown brocket *Mazama nemorivaga* (Artiodactyla, Cervidae). *Comp. Cytogenet.* **2013**, *7*, 25–31. [CrossRef] [PubMed]

53. Sokolov, V.E.; Prikhodko, I.V. Taxonomy of the musk deer moschus-moschiferus (Artiodactyla, Mammalia). *Izv. Akad. Nauk SSSR. Ser. Biol.* **1998**, *1*, 37–46.

54. Castiglia, R.; Makundi, R.; Corti, M. The origin of an unusual sex chromosome constitution in *Acomys sp.* (Rodentia, Muridae) from Tanzania. *Genetica* **2007**, *131*, 201–207. [CrossRef] [PubMed]

55. Castiglia, R.; Annesi, F. Cytotaxonomic considerations on the sex chromosome variation observed within *Acomys ngurui* Verheyen et al. 2011 (Rodentia Muridae). *Zootaxa* **2012**, *3493*, 35–38.

56. Lobato, L.; Cantos, G.; Araujo, B.; Bianchi, N.O.; Merani, S. Cytogenetics of the South American akodont rodents (Cricetidae) X. *Akodon mollis*: A species with XY females and B chromosomes. *Genetica* **1982**, *57*, 199–205.

57. Yonenaga-Yassuda, Y.; Assis, M.F.L.; Kasahara, S. Variability of the nucleolus organizer regions and the presence of the rDNA genes in the supernumerary chromosome of *Akodon aff. arviculoides* (Cricetidae, Rodentia). *Caryologia* **1992**, *45*, 163–174. [CrossRef]

58. Hass, I.; Soares, A.; Balieiro, P.; Miranda, M.L.P.; Alegri, C.K.; Dornelles, S.S.; Marques, D.; Sbalqueiro, I.J. *Akodon montensis* karyotype variability, in a sample of São Francisco do Sul—SC, arising 0–3 supernumerary chromosomes. In Proceedings of the 60 °Congresso Brasileiro de Genética, Guarujá, Brazil, 26–29 August 2014.

59. Kartavtseva, I.V. Description of B-chromosomes in karyotype of field mouse *Apodemus agrarius*. *Tsitol. Genet.* **1994**, *28*, 96–97. [PubMed]

60. Obara, Y.; Sasaki, S. Fluorescent approaches on the origin of B chromosomes of *Apodemus argenteus* hokkaidi. *Chrom. Sci.* **1997**, *1*, 1–5.

61. Soldatović, B.; Savić, I.; Seth, P.; Reichstein, H.; Tolksdorf, M. Comparative karyological study of the genus *Apodemus*. *Acta Vet.* **1975**, *25*, 1–10.

62. Zima, J.; Macholán, M. B chromosomes in the wood mice (genus *Apodemus*). *Acta Theriol.* **1995**, *40* (Suppl. 3), 75–86. [CrossRef]

63. Belcheva, R.G.; Topashka-Ancheva, M.N.; Atanassov, N. Karyological studies of five species of mammals from Bulgarian fauna. *Comptes Rendus l'Académie Bulg. des Sci.* **1988**, *42*, 125–128.

64. Hayata, J. Chromosomal polymorphism caused by supernumerary chromosomes in field mouse, *Apodemus giliacus*. *Chromosoma* **1973**, *42*, 403–414. [CrossRef] [PubMed]

65. Borisov, Y.M.; Afanasev, A.G.; Lebedev, T.T.; Bochkarev, M.N. Multiplicity of B microchromosomes in a Siberian population of mice *Apodemus peninsulae* (2n = 48 + 4–30 B chromosomes). *Russ. J. Genet.* **2010**, *46*, 705–711. [CrossRef]

66. Vujošević, M.; Živković, S. Numerical chromosome polymorphism in *Apodemus flavicollis* and *A. sylvaticus* (Mammalia: Rodentia) caused by supernumerary chromosomes. *Acta Vet.* **1987**, *37*, 81–92.

67. Gadi, I.K.; Sharma, T.; Raman, R. Supernumerary chromosomes in *Bandicota indica nemorivaga* and a female individual with XX/X0 mosaicism. *Genetica* **1982**, *58*, 103–108. [CrossRef]

68. Badenhorst, D.; Herbreteau, V.; Chaval, Y.; Pagès, M.; Robinson, T.J.; Rerkamnuaychoke, W.; Morand, S.; Hugot, J.P.; Dobigny, G. New karyotypic data for Asian rodents (Rodentia, Muridae) with the first report of B-chromosomes in the genus Mus. *J. Zool.* **2009**, *279*, 44–56. [CrossRef]

69. Ventura, K.; Sato-Kuwabara, Y.; Fagundes, V.; Geise, L.; Leite, Y.L.R.; Costa, L.P.; Silva, M.J.J.; Yonenaga-Yassuda, Y.; Rodrigues, M.T. Phylogeographic structure and karyotypic diversity of the Brazilian shrew mouse (*Blarinomys breviceps*, Sigmodontinae) in the Atlantic forest. *Cytogenet. Genome Res.* **2012**, *138*, 19–30. [CrossRef] [PubMed]

70. Patton, J.L. A complex system of chromosomal variation in the pocket mouse, *Perognathus baileyi* Merriam. *Chromosoma* **1972**, *36*, 241–255. [CrossRef] [PubMed]

71. Volobouev, V.T.; Sicard, B.; Aniskin, V.M.; Gautun, J.C.; Granjon, L. Robertsonian polymorphism, B chromosomes variation and sex chromosomes heteromorphism in the african water rat *Dasymys* (Rodentia, Muridae). *Chromosome Res.* **2000**, *8*, 689–697. [CrossRef] [PubMed]

72. Ramos, R.S.L.; Vale, W.G.; Assis, F.L. Karyotypic analysis in species of the genus *Dasyprocta* (Rodentia: Dasyproctidae) found in Brazilian Amazon. *An. Acad. Bras. Cienc.* **2003**, *75*, 55–69. [CrossRef] [PubMed]

73. Van Wynsberghe, N.R.; Engstrom, M.D. Chromosomal variation in collared lemmings (*Dicrostonyx*) from the western Hudson Bay region. *Musk-Ox* **1992**, *39*, 203–209.

74. Gileva, E.A. B-chromosomes, unusual inheritance of sex chromosomes and sex ratio in the varying lemming, *Dicrostonyx torquatus* Pall. (1779). *Proc. Natl. Acad. Sci. USA* **1973**, *213*, 952–955.

75. Chernyavsky, F.B.; Kozlovsky, A.I. Species status and history of the Arctic lemming (*Dicrostonyx*, Rodentia) of Wrangel Island. *Zoo. Zhurnal* **1980**, *59*, 266–273. (In Russian)

76. Rao, K.S.; Aswathanrayana, N.V.; Prakash, K.S. Supernumerary (B) chromosomes in Indian bush rat *Golunda ellioti* (Gray). *MCN* **1979**, *20*, 79.

77. Roche, J.; Capanna, E.; Civitelli, M.; Ceraso, A. caryotypes des rongeurs de Somalie. *Monit. Zool. Ital. Suppl.* **1984**, *19*, 259–277. [CrossRef]

78. Civitelli, M.V.; Consentino, P.; Capanna, E. Inter- and intra-individual chromosome variability in *Thamnomys* (*Grammomys*) *gazellae* (Rodentia, Muridae) B-chromosomes and structural heteromorphisms. *Genetica* **1989**, *79*, 93–105. [CrossRef] [PubMed]

79. Freitas, T.R.O.; Mattevi, M.S.; Oliveira, L.F.B.; Souza, M.J.; Yonenaga-Yassuda, Y.; Salzano, F.M. Chromosome relationship in three representatives of the genus *Holochilus* (Rodentia, Cricetidae) from Brazil. *Genetica* **1983**, *61*, 13–20. [CrossRef]

80. Vidal, O.R.; Riva, R.; Baro, N. Los cromosomas. del genero *Holochilus*. 1. Polimorfismo en *H. chacarius* Thomas (1906). *Phys. (B. Aires) Ser. C* **1976**, *35*, 75–85.

81. Sangines, N.; Aguilera, M. Chromosome polymorphism in *Holochilus venezuelae* (Rodentia: Cricetidae): C- and G-bands. *Genome* **1991**, *34*, 13–18. [CrossRef]

82. Baverstock, P.R.; Wats, C.H.S.; Hogarth, J.T. Chromosome evolution in Australian Rodents. I. The Pseudomyinae, the Hydrmyinae and the *Uromys/Melomys* group. *Chromosoma* **1977**, *61*, 95–125. [CrossRef] [PubMed]

83. Denys, C.; Lalis, A.; Aniskin, V.; Kourouma, F.; Soropogui, B.; Sylla, O.; Doré, A.; Koulemou, K.; Beavogui, Z.B.; Sylla, M.; et al. New data on the taxonomy and distribution of Rodentia (Mammalia) from the western and coastal regions of Guinea West Africa. *Ital. J. Zool.* **2009**, *76*, 111–128. [CrossRef]

84. Denys, C.; Lalis, A.; Lecompte, É.; Cornette, R.; Moulin, S.; Makundi, R.H.; Machang'u, R.S.; Volobouev, V.; Aniskine, V.M. A faunal survey in Kingu Pira (south Tanzania), with new karyotypes of several small mammals and the description of a new Murid species (Mammalia, Rodentia). *Zoosystema* **2011**, *33*, 5–47. [CrossRef]

85. Baverstock, P.R.; Watts, C.H.S.; Adams, M.; Gelder, M. Chromosomal and electrophorectic studies of Australian *Melomys* (Rodentia:Muridae). *Aust. J. Zool.* **1980**, *28*, 553–574. [CrossRef]

86. Kovalskaja, Y.M. B-chromosomes and XO-males in narrow-skulled vole, *Microtus/Stenocranius/gregalis* from North Mongolia. In Proceedings of the 7th All-Union Conference, Nalchik, Sverdlovsk, Russia; 1988; Volume 1, pp. 73–74.

87. Judd, S.R.; Cross, S.P. Chromosomal variation in *Microtus longicaudus* (Merriam). *Murrelet* **1980**, *61*, 2–5. [CrossRef]

88. Gropp, A.; Marshall, J.; Markvong, A. Hriomosomal findings in the spiny mice of Thailand (genus *Mus*) and occurrence of a complex intraspecific variation in *M. shortridgei*. *Z. Saugetierkd.* **1973**, *38*, 159–168.

89. Ivanitskaya, E.; Sözen, M.; Rashkovetsky, L.; Matur, F.; Nevo, E. Discrimination of 2n = 60 *Spalax leucodon* cytotypes (Spalacidae, Rodentia) in Turkey by means of classical and molecular cytogenetic techniques. *Cytogenet. Genome Res.* **2008**, *122*, 139–149. [CrossRef] [PubMed]

90. Maia, V.; Yonenaga-Yassuda, Y.; Freitas, T.R.O.; Kasahara, S.; Sune-Mattevi, M.; Oliviera, L.F. Supernumerary chromosomes, Robertsonian rearrangement and variability of the sex chromosomes in scaly-footed water rat *Nectomys squamipes* (Cricetidae, Rodentia). *Genetica* **1984**, *63*, 121–128. [CrossRef]

91. Yonenaga-Yassuda, Y.; do Prado, C.R.; Mello, D.A. Supernumerary chromosomes in *Holochilus brasiliensis* and comparative cytogenetic analysis with *Nectomys squamipes* (Cricetidae, Rodentia). *Rev. Bras. Genet.* **1987**, *X*, 209–220.

92. Andrades-Miranda, J.; Oliveira, L.F.B.; Zanchin, N.I.T.; Mattevi, M.S. Chromosomal description of the rodent genera *Oecomys* and *Nectomys* from Brazil. *Acta Theriol.* **2001**, *46*, 269–278. [CrossRef]

93. Myers, P.; Carleton, M.D. The species *Oryzomys* (*Oligoryzomys*) in Paraguay and the identity of Azara's "Rat sixime ou Rat Tarse Noir". *Publ. Mus. Univ. Mich.* **1981**, *161*, 1–141.

94. Sbalqueiro, I.J.; Mattevi, M.S.; Oliveira, L.F.B.; Solano, M.J.V. B chromosome system in populations of *Oryzomys flavescens* (Rodentia, Cricetidae) from southern Brazil. *Acta Theriol. (Warsz)* **1991**, *36*, 193–199. [CrossRef]

95. Contrafatto, G.; Meester, J.; Bronner, G.; Taylor, P.J.; Willan, K. Genetic variation in the african rodent sub-family Otomyinae (Muridae). iv: Chromosome G-banding analysis of *Otomys irroratus* and *O. angoniensis*. *Isr. J. Zool.* **1992**, *38*, 277–291.

96. Barros, R.M.S. *Variabilidade Cromossômica em Proechimys e Oryzomys (Rodentia) do Amazonas*; Universidade de São Paulo: São Paulo, Brazil, 1978.

97. Thomson, R.L. B chromosomes in *Rattus fuscipes* II. The transmission of B chromosomes to offspring and population studies: Support for the "parasitic" model. *Heredity* **1984**, *52*, 363–372. [CrossRef]

98. Yosida, T.H. Studies on the karyotype differentiation of the Norway rat. *Proc. Jpn. Acad.* **1986**, *62*, 65–68. [CrossRef]

99. Wahrman, J.; Gourevitz, P. The chromosome biology of the 2n = 38 black rat, *Rattus rattus*. In *Chromosomes Today 4*; Wahrman, J., Lewis, K.R., Eds.; John Wiley and Sons: New York, NY, USA, 1973; pp. 433–434.

100. Yong, H.S.; Dhaliwal, S.S. Supernumerary (B-) chromosomes in the Malayan house rat, *Rattus rattus diardii* (Rodentia, Muridae). *Chromosoma* **1972**, *36*, 256–262. [CrossRef] [PubMed]

101. Ladron de Guevara, R.G.; de la Guardia, R.D. Frequency of chromosome polymorphism for pericentric inversions and B-chromosomes in Spanish populations of *Rattus rattus frugivorus*. *Genetica* **1981**, *57*, 99–103. [CrossRef]

102. Yosida, T.H. Population survey of B-chromosomes in black rat. *Ann. Rep. Nat. Inst. Genet.* **1976**, *26*, 33.

103. Gropp, A.; Marshall, J.; Flatz, G.; Olbrich, M.; Manyanondha, K.; Santadust, A. Chromosomenpolymorphismus durch uberzahuge Autosomen Beobachtungen an der hausratte (*R. rattus*). *Z. Saugetierkd.* **1970**, *35*, 363–371.

104. Baverstock, R.P.; Watts, S.C.H.; Hogarth, T.J.; Robinson, C.A.; Robinson, J.F. Chromosome evolution in Australian rodents. II. The Rattus group. *Chromosoma* **1977**, *61*, 227–241. [CrossRef] [PubMed]

105. Blanks, G.A.; Shellhammer, H.S. Chromosome polymorphism in California populations of harvest mice. *J. Mammal.* **1968**, *49*, 726–731. [CrossRef]

106. Robbins, L.W. Sex chromosome polymorphisms in *Reithrodontomys montanus* (Rodentia: Cricetidae). *Southwest. Nat.* **1981**, *26*, 201–202. [CrossRef]

107. Zimmerman, E.G. Karyology, systematics and chromosomal evolution in the rodent genus *Sigmodon*. *Michigan State Univ. Publ. Mus. Biol. Ser.* **1970**, *4*, 385–394.

108. Silva, M.J.J.; Yonenaga-Yassuda, Y. B chromosomes in Brazilian rodents. *Cytogenet. Genome Res.* **2004**, *106*, 257–263. [CrossRef] [PubMed]

109. Suárez-Villota, E.; Di-Nizo, C.; Neves, C.; de Jesus Silva, M.J. First cytogenetic information for *Drymoreomys albimaculatus* (Rodentia, Cricetidae), a recently described genus from Brazilian Atlantic Forest. *Zookeys* **2013**, *303*, 65–76.

110. Gordon, D.H. Discovery of another species of tree rat. *Transvaal Museum Bull.* **1987**, *22*, 30–32.

111. Gordon, D.H.; Rautenbach, I.L. Species complexes in medically important rodents: Chromosome studies of *Aethomys*, *Tatera* and *Saccostomus* (Rodentia: Muridae, Cricetidae). *S. Afr. J. Sci.* **1980**, *76*, 559–561.

112. Hafner, J.C.; Hafner, D.J.; Patton, J.L.; Smith, M.E. Contact zones and the genetics of differentiation in the pocket gopher *Thomomys bottae* (Rodentia: Geomyidae). *Syst. Zool.* **1983**, *32*, 1–20. [CrossRef]

113. Yonenaga-Yassuda, Y.; DeSouza, M.J.; Kasahara, S.; L'Abbate, M.; Chu, H.T. Supernumerary system in *Proechimys iheringi iheringi* (Rodentia, Echimydae), from the state of Sao Paulo, Brazil. *Caryologia* **1985**, *38*, 179–194. [CrossRef]

114. Fagundes, V.; Camacho, J.P.M.; Yonenaga-Yassuda, Y. Are the dot-like chromosomes in *Trinomys iheringi* (Rodentia, Echimyidae) B chromosomes? *Cytogenet. Genome Res.* **2004**, *106*, 159–164. [CrossRef] [PubMed]

115. Baverstock, P.R.; Gelder, M.; Jahnke, A. Cytogenetic studies of the Australian rodent *Uromys caudimaculatus*, a species showing extensive heterochromatin variation. *Chromosoma* **1982**, *84*, 517–533. [CrossRef] [PubMed]

116. Mammal Diversity Database. Available online: www.mammaldiversity.org (accessed on 10 July 2018).

117. Belayev, D.K.; Volobouev, V.T.; Radjabli, S.I.; Trut, L.N. Polymorphism and mosaicism for additional chromosomes in silver foxes. *Genetika* **1974**, *X*, 58–67.

118. Bekasova, T.S.; Vorontsov, N.N.; Korobitsyina, K.V.; Korablev, V.P. B-chromosomes and comparative karyology of the mice of the genus *Apodemus*. *Genetica* **1980**, *52/53*, 33–43. [CrossRef]

119. Gileva, E. Chromosomal diversity and the aberrant genetic system of sex determination in the artic lemming, *Dicrostonyx torquatus* Pallas, 1779. *Genetica* **1980**, *52/53*, 99–103. [CrossRef]

120. Szczerbal, I.; Kaczmarek, M.; Switonski, M. Compound mosaicism, caused by B chromosome variability, in the Chinese raccoon dog (*Nyctereutes procyonoides procyonoides*). *Folia Biol.* **2005**, *53*, 155–159. [CrossRef]

121. Borisov, Y.M. The Polymorphism and distribution of B chromosomes in germline and somatic cells of *Tscherskia triton* de Winton (Rodentia, Cricetinae). *Russ. J. Genet.* **2012**, *48*, 538–542. [CrossRef]

122. Radzhabli, S.I.; Borisov, Y.M. B-chromosomes system variants among continental forms of *Apodemus peninsulae* (Rodentia, Muridae). *Dokl. Akad Nauk SSSR* **1979**, *248*, 979–981.

123. Kartavtseva, I.V.; Roslik, G.V. A complex B chromosome system in the Korean Field mouse *Apodemus peninsulae*, Cytogenet. *Cytogenet. Genome Res.* **2004**, *106*, 271–278. [CrossRef] [PubMed]

124. Roslik, G.V.; Kartavtseva, I.V. Polymorphism and mosaicism of B chromosome number in Korean field mouse *Apodemus peninsulae* (Rodentia) in the Russian Far East. *Cell Tissue Biol.* **2010**, *4*, 77–89. [CrossRef]

125. Koh, H. Systematic studies of Korean Rodents: II. A chromosome analysis in Korean field mice, *Apodemus peninsulae peninsulae* Thomas (Muridae, Rodentia), from Mungyong, with the comparison of morphometric characters of these Korean Field Mice to sympatric striped. *Korean J. Syst. Zool.* **1986**, *2*, 1–10.

126. Volobujev, V.T.; Timina, N.Z. Unusally high number of B-chromosomes and mosaicism by them in *Apodemus peninsulae* (Rodentia, Muridae). *Tsitologia* **1980**, *14*, 43–45.

127. Abril, V.V.; Carnelossi, E.A.G.; González, S.; Duarte, J.M.B. Elucidating the evolution of the red brocket deer *Mazama americana* Complex (Artiodactyla; Cervidae). *Cytogenet. Genome Res.* **2010**, *128*, 177–187. [CrossRef] [PubMed]

128. Malleret, M.; Labaroni, C.; García, G.V.; Ferro, J.; Marti, D.A.; Lanzone, C. Chromosomal variation in Argentine populations of *Akodon montensis* Thomas, 1913 (Rodentia, Cricetidae, Sigmodontinae). *Comp. Cytogenet.* **2016**, *10*, 129–140. [PubMed]

129. Kral, B.; Herzig-Straschil, B.; Štreba, O. Karyotypes of certain small mammals from Austria. *Folia Zool.* **1979**, *28*, 5–11.

130. Rovatsos, M.T.; Mitsainas, G.P.; Tryfonopoulos, G.A.; Stamatopoulos, C.; Giagia-Athanasopoulou, E.B. A chromosomal study on Greek populations of the genus *Apodemus* (Rodentia, Murinae) reveals new data on B chromosome distribution. *Acta Theriol. (Warsz)* **2008**, *53*, 157–167. [CrossRef]

131. Sablina, O.V.; Radzabli, S.I.; Golenisev, F.N. B-chromosomes in the karyotype of *A. flavicollis* from Leningrad district. *Zool. J.* **1985**, *LXIV*, 1901–1903.

132. Giagia, E.; Soldatović, B.; Savić, I.; Zimonjić, D. Karyotype study of the genus *Apodemus* (Kaup 1829) populations from the Balkan peninsula. *Acta Vet.* **1985**, *35*, 289–298.

133. Vujošević, M.; Blagojević, J.; Radosavljević, J.; Bejaković, D. B chromosome polymorphism in populations of *Apodemus flavicollis* in Yugoslavia. *Genetica* **1991**, *83*. [CrossRef]

134. Nadjafova, R.S.; Bulatova, N.S.; Chasovnikarova, Z.; Gerassimov, S. Karyological differences between two *Apodemus* species in Bulgaria. *Z. Saugetierkd.* **1993**, *38*, 232–239.

135. Wolf, U.; Voiculescu, I.; Zenzes, M.T.; Vogel, W.; Engel, W. Chromosome polymorphism in Apodemus flavicollis, possibly due to creation of a new centromere. In *Modern Aspects of Cytogenetics: Constitutive Heterochromatin in Man*; Pfeiffer, A., Ed.; Schattauer F. K.: Stuttgart, Germany; New York, NY, USA, 1972; pp. 163–168.

136. Ramalhinho, M.G.; Libois, R. First report on the presence in France of a B-chromosome polymorphism in *Apodemus flavicollis*. *Mammalia* **2002**, *66*, 300–303.

137. Castiglia, R. Cytogenetic analysis of *Apodemus flavicollis* in Italy: First report of B chromosomes and X chromosome heteromorphism. *Mammalia* **2003**, *67*, 1–3. [CrossRef]

138. Wójcik, J.M.; Wójcik, A.M.; Macholán, M.; Piálek, J.; Zima, J. The mammalian model for population studies of B chromosomes: The wood mouse (*Apodemus*). *Cytogenet. Genome Res.* **2004**, *106*, 264–270. [CrossRef] [PubMed]

139. Vujošević, M.; Blagojević, J. New localities with B chromosomes in *Apodemus flavicollis* (Rodentia, Mammalia). *Arch. Biol. Sci.* **1994**, *46*, 15.

140. Blagojević, J.; Vujošević, M. Do B chromosomes affect morphometric characters in yellow-necked mice *Apodemus flavicollis* (Rodentia, Mammalia)? *Acta Theriol. (Warsz)* **2000**, *45*, 129–135. [CrossRef]

141. Vujošević, M.; Blagojević, J. Does environment affect polymorphism of B chromosomes in the yellow-necked mouse *Apodemus flavicollis*? *Z. Saugetierkd.* **2000**, *65*, 313–317.

142. Vujošević, M.; Blagojević, J.; Jojić-Šipetić, V.; Bugarski-Stanojević, V.; Adnađević, T.; Stamenković, G. Distribution of B chromosomes in age categories of the yellow-necked mouse *Apodemus flavicollis* (Mammalia, Rodentia). *Arch. Biol. Sci.* **2009**, *61*, 653–658. [CrossRef]

143. Vujošević, M.; Jojić, V.; Bugarski-Stanojević, V.; Blagojević, J. Habitat quality and B chromosomes in the yellow-necked mouse *Apodemus flavicollis*. *Ital. J. Zool.* **2007**, *74*, 313–316. [CrossRef]

144. Shellhammer, H.S. Supernumerary chromosomes of the harvest mouse, *Reithrodontomys megalotis*. *Chromosoma* **1969**, *27*, 102–108. [CrossRef] [PubMed]

145. Boyeskorov, G.; Zagorodnyuk, I.; Belyanin, A.N.; Lyapunova, E.A. B-chromosomes in *Apodemus flavicollis* from eastern Europe. *Pol. Ecol. Stud.* **1994**, *20*, 523–526.

146. Kartavtseva, I.V.; Roslik, G.V.; Pavlenko, M.V.; Amachaeva, E.Y.; Sawaguchi, S.; Obara, Y. The B-chromosome system of the Korean field mouse *Apodemus peninsulae* in the Russian Far East. *Chromosom. Sci.* **2000**, *4*, 21–29.

147. Ostromyshenskii, D.I.; Kuznetsova, I.S.; Podgornaya, O.I.; Kartavtseva, I.V. Appearance of B chromosomes like structures in *Apodemus peninsulae* primary cell culture. *Res. J. Zool.* **2018**, *1*, 1.

148. Volobuev, V.T. Karyological analysis of three Siberian populations of *Apodemus peninsulae* (Rodentia, Muridae). *Dokl. AN SSSR* **1979**, *248*, 1452–1454.

149. Borisov, Y.M. Variation of the cytogenetic structure of the population of *Apodemus peninsulae* (Rodentia, Muridae) in western Sayan Mountains. *Genetika* **1990**, *26*, 1484–1491. [PubMed]

150. Borisov, Y.M. Cytogenetic differentiation of the population of *Apodemus peninsulae* (Rodentia, Muridae) in eastern Siberia. *Genetika* **1990**, *26*, 1828–1839. [PubMed]

151. Borisov, Y.M. Cytogenetic Structure of the Population of *Apodemus peninsulae* (Rodentia, Muridae) on the bank of lake Teletskoe (Altai). *Genetika* **1990**, *26*, 1212–1220.

152. Borisov, Y.M.; Malygin, M.V. Cline variability of B-chromosome system in *Apodemus peninsulae* (Rodentia, Muridae) from the Buryatia and Mongolia. *Citologija* **1991**, *33*, 106–111.

153. Roslik, G.V.; Kartavtseva, I.V. Geographic differentiation of B chromosomes in *Apodemus peninsulae* (Rodentia) from the east Asia. In Proceedings of the International Symposium: Modern Achievements in Population, Evolutionary, and Ecological Genetics, Vladivostok, Russia, 3–9 September 2017.

154. Ruban, A.; Schmutzer, T.; Scholz, U.; Houben, A. How next-generation sequencing has aided our understanding of the sequence composition and origin of B chromosomes. *Genes* **2017**, *8*, 294. [CrossRef] [PubMed]

155. Page, B.T.; Wanous, M.K.; Birchler, J.; Ames, A. Characterization of a maize chromosome 4 centromeric sequence: Evidence for an evolutionary relationship with the B chromosome centromere. *Genetics* **2001**, *159*, 291–302. [PubMed]

156. Cheng, Y.-M.; Lin, B.-Y. Cloning and characterization of maize B chromosome sequences derived from microdissection. *Genetics* **2003**, *164*, 299–310. [PubMed]

157. Bugrov, A.G.; Karamysheva, T.V.; Perepelov, E.A.; Elisaphenko, E.A.; Rubtsov, D.N.; Warchałowska-Śliwa, E.; Tatsuta, H.; Rubtsov, N.B. DNA content of the B chromosomes in grasshopper *Podisma kanoi* Storozh. (Orthoptera, Acrididae). *Chromosom. Res.* **2007**. [CrossRef] [PubMed]

158. Valente, G.T.; Nakajima, R.T.; Fantinatti, B.E.A.; Marques, D.F.; Almeida, R.O.; Simões, R.P.; Martins, C. B chromosomes: From cytogenetics to systems biology. *Chromosoma* **2017**, *126*, 73–81. [CrossRef] [PubMed]

159. Lamb, J.C.; Meyer, J.M.; Corcoran, B.; Kato, A.; Han, F.; Birchler, J.A. Distinct chromosomal distributions of highly repetitive sequences in maize. *Chromosom. Res.* **2007**, *15*, 33–49. [CrossRef] [PubMed]

160. Tanić, N.; Dedovic, N.; Vujošević, M.; Dimitrijević, B. DNA profiling of B chromosomes from the yellow-necked mouse *Apodemus flavicollis* (Rodentia, Mammalia). *Genome Res.* **2000**, *10*, 55–61. [PubMed]

161. Schmid, M.; Ziegler, C.G.; Steinlein, C.; Nanda, I.; Schartl, M. Cytogenetics of the bleak (*Alburnus alburnus*), with special emphasis on the B chromosomes. *Chromosom. Res.* **2006**, *14*, 231–242. [CrossRef] [PubMed]

162. Wang, J.; Zhao, X.; Qi, H.; Koh, H.S.; Zhang, L.; Guan, Z.; Wang, C. Karyotypes and B chromosomes of *Apodemus peninsulae* (Rodenita, Mammalia). *Acta Theriol. Sin.* **2000**, *20*, 289–296.

163. Trifonov, V.A.; Perelman, P.L.; Kawada, S.I.; Iwasa, M.A.; Oda, S.I.; Graphodatsky, A.S. Complex structure of B-chromosomes in two mammalian species: *Apodemus peninsulae* (Rodentia) and *Nyctereutes procyonoides* (Carnivora). *Chromosom. Res.* **2002**, *10*, 109–116. [CrossRef]

164. Karamysheva, T.V.; Andreenkova, O.V.; Bochkarev, M.N.; Borisov, Y.M.; Bogdanchikova, N.; Borodin, P.M.; Rubtsov, N.B. B chromosomes of Korean field mouse *Apodemus peninsulae* (Rodentia, Murinae) analysed by microdissection and FISH. *Cytogenet. Genome Res.* **2002**, *96*, 154–160. [CrossRef] [PubMed]

165. Rubtsov, N.B.; Karamysheva, T.V.; Andreenkova, O.V.; Bochkaerev, M.N.; Kartavtseva, I.V.; Roslik, G.V.; Borissov, Y.M. Comparative analysis of micro and macro B chromosomes in the Korean field mouse *Apodemus peninsulae* (Rodentia, Murinae) performed by chromosome microdissection and FISH. *Cytogenet. Genome Res.* **2004**, *106*, 289–294. [CrossRef] [PubMed]

166. Wurster-Hill, D.H.; Ward, O.G.; Davis, B.H.; Park, J.P.; Moyzis, R.K.; Meyne, J. Fragile sites, telomeric DNA sequences, B chromosomes, and DNA content in raccoon dogs, *Nyctereutes procyonoides*, with comparative notes on foxes, coyote, wolf, and raccoon. *Cytogenet. Genome Res.* **1988**, *49*, 278–281. [CrossRef] [PubMed]

167. Szczerbal, I.; Switonski, M. B chromosomes of the Chinese raccoon dog (*Nyctereutes procyonoides procyonoides* Gray) contain inactive NOR-like sequences. *Caryologia* **2003**, *56*, 213–216. [CrossRef]

168. Graphodatsky, A.S.; Kukekova, A.V.; Yudkin, D.V.; Trifonov, V.A.; Vorobieva, N.V.; Beklemisheva, V.R.; Perelman, P.L.; Graphodatskaya, D.A.; Trut, L.N.; Yang, F.; et al. The proto-oncogene C-KIT maps to canid B-chromosomes. *Chromosom. Res.* **2005**, *13*, 113–122. [CrossRef] [PubMed]

169. Yudkin, D.V.; Trifonov, V.A.; Kukekova, A.V.; Vorobieva, N.V.; Rubtsova, N.V.; Yang, F.; Acland, G.M.; Ferguson-Smith, M.A.; Graphodatsky, A.S. Mapping of KIT adjacent sequences on canid autosomes and B chromosomes. *Cytogenet. Genome Res.* **2007**, *116*, 100–103. [CrossRef] [PubMed]

170. Wnuk, M.; Oklejewicz, B.; Lewinska, A.; Zabek, T.; Bartosz, G.; Slota, E.; Bugno-Poniewierska, M. PRINS detection of 18S rDNA in pig, red fox and Chinese raccoon dog, and centromere DNA in horse. *Hereditas* **2010**, *147*, 320–324. [CrossRef] [PubMed]

171. Becker, S.E.D.; Thomas, R.; Trifonov, V.A.; Wayne, R.K.; Graphodatsky, A.S.; Breen, M. Anchoring the dog to its relatives reveals new evolutionary breakpoints across 11 species of the Canidae and provides new clues for the role of B chromosomes. *Chromosom. Res.* **2011**, *19*, 685–708. [CrossRef] [PubMed]

172. Li, Y.M.; Zhang, Y.; Zhu, W.J.; Yan, S.Q.; Sun, J.H. Identification of polymorphisms and transcriptional activity of the proto-oncogene KIT located on both autosomal and B chromosomes of the Chinese raccoon dog. *Genet. Mol. Res.* **2016**, *15*, 1–6. [CrossRef] [PubMed]

173. Makunin, A.; Romanenko, S.; Beklemisheva, V.; Perelman, P.; Druzhkova, A.; Petrova, K.; Prokopov, D.; Chernyaeva, E.; Johnson, J.; Kukekova, A.; et al. Sequencing of supernumerary chromosomes of red fox and raccoon dog confirms a non-random gene acquisition by B chromosomes. *Genes* **2018**, *9*, 405. [CrossRef] [PubMed]

174. Trifonov, V.A.; Dementyeva, P.V.; Larkin, D.M.; O'Brien, P.C.; Perelman, P.L.; Yang, F.; Ferguson-Smith, M.A.; Graphodatsky, A.S. Transcription of a protein-coding gene on B chromosomes of the Siberian roe deer (*Capreolus pygargus*). *BMC Biol.* **2013**, *11*, 90. [CrossRef] [PubMed]

175. Makunin, A.I.; Kichigin, I.G.; Larkin, D.M.; O'Brien, P.C.M.; Ferguson-Smith, M.A.; Yang, F.; Proskuryakova, A.A.; Vorobieva, N.V.; Chernyaeva, E.N.; O'Brien, S.J.; et al. Contrasting origin of B chromosomes in two cervids (Siberian roe deer and grey brocket deer) unravelled by chromosome-specific DNA sequencing. *BMC Genom.* **2016**, *17*, 618. [CrossRef] [PubMed]

176. Soares, A.A.; Castro, J.P.; Balieiro, P.; Dornelles, S.; Degrandi, T.M.; Sbalqueiro, I.J.; Ferreira Artoni, R.; Hass, I. B Chromosome diversity and repetitive sequence distribution in an isolated population of *Akodon montensis* (Rodentia, Sigmodontinae). *Cytogenet. Genome Res.* **2018**, *154*, 79–85. [CrossRef] [PubMed]

177. Boyeskorov, G.G.; Kartavseva, I.V.; Belyanin, A.N.; Liapunova, E.A. Nucleolus organizer regions and B-chromosomes of wood mice (Mammalia, Rodentia, *Apodemus*). *Russ. J. Genet.* **1995**, *31*, 156–163.

178. Tanić, N.; Vujošević, M.; Dedović-Tanić, N.; Dimitrijević, B. Differential gene expression in yellow-necked mice *Apodemus flavicollis* (Rodentia, Mammalia) with and without B. chromosomes. *Chromosoma* **2005**, *113*, 418–427. [CrossRef] [PubMed]

179. Rajičić, M.; Adnađević, T.; Stamenković, G.; Blagojević, J.; Vujošević, M. Screening of B chromosomes for presence of two genes in yellow-necked mice, *Apodemus flavicollis* (Mammalia, Rodentia). *Genetika* **2015**, *47*, 311–321. [CrossRef]

180. Bugarski-Stanojević, V.; Stamenković, G.; Blagojević, J.; Liehr, T.; Kosyakova, N.; Rajičić, M.; Vujošević, M. Exploring supernumeraries—A new marker for screening of B-chromosomes presence in the yellow necked mouse *Apodemus flavicollis*. *PLoS ONE* **2016**, *11*, e0160946. [CrossRef] [PubMed]

181. Makunin, A.I.; Rajičić, M.; Karamysheva, T.V.; Romanenko, S.A.; Druzhkova, A.; Blagojević, J.; Vujošević, M.; Rubtsov, N.B.; Graphodatsky, A.S.; Trifonov, V.A. Low-pass single-chromosome sequencing of human small supernumerary marker chromosomes (sSMCs) and *Apodemus* B chromosomes. *Chromosoma* **2018**, *127*, 301–311. [CrossRef] [PubMed]

182. Matsubara, K.; Yamada, K.; Umemoto, S.; Tsuchiya, K.; Ikeda, N.; Nishida, C.; Chijiwa, T.; Moriwaki, K.; Matsuda, Y. Molecular cloning and characterization of the repetitive DNA sequences that comprise the constitutive heterochromatin of the A and B chromosomes of the Korean field mouse (*Apodemus peninsulae*, Muridae, Rodentia). *Chromosom. Res.* **2008**, *16*, 1013–1026. [CrossRef] [PubMed]

183. Ventura, K.; O'Brien, P.C.M.; do Nascimento Moreira, C.; Yonenaga-Yassuda, Y.; Ferguson-Smith, M.A. On the Origin and evolution of the extant system of B chromosomes in Oryzomyini Radiation (Rodentia; Sigmodontinae). *PLoS ONE* **2015**, *10*, e0136663. [CrossRef] [PubMed]

184. De Jesus Silva, M.J.; Yonenaga-Yassuda, Y. Heterogeneity and meiotic behaviour of B and sex chromosomes, banding patterns and localization of (TTAGGG)n sequences by fluorescence in situ hybridization in the neotropical water rat *Nectomys* (Rodentia, Cricetidae). *Chromosome Res.* **1998**, *6*, 455–462. [CrossRef] [PubMed]

185. Stitou, S.; Díaz de La Guardia, R.; Jiménez, R.; Burgos, M. Inactive ribosomal cistrons are spread throughout the B chromosomes of *Rattus rattus* (Rodentia, Muridae). Implications for their origin and evolution. *Chromosome Res.* **2000**, *8*, 305–311. [CrossRef] [PubMed]

186. Cavagna, P.; Stone, G.; Stanyon, R. Black rat (*Rattus rattus*) genomic variability characterized by chromosome painting. *Mamm. Genome* **2002**, *13*, 157–163. [CrossRef] [PubMed]

187. Peppers, J.A.; Wiggins, L.E.; Baker, R.J. Nature of B chromosomes in the harvest mouse *Reithrodontomys megalotis* by fluorescence in situ hybridization (FISH). *Chromosome Res.* **1997**, *5*, 475–479. [CrossRef] [PubMed]

188. Rajičić, M.; Romanenko, S.A.; Karamysheva, T.V.; Blagojević, J.; Adnađević, T.; Budinski, I.; Bogdanov, A.S.; Trifonov, V.A.; Rubtsov, N.B.; Vujošević, M. The origin of B chromosomes in yellow-necked mice (*Apodemus flavicollis*)—Break rules but keep playing the game. *PLoS ONE* **2017**, *12*, e0172704. [CrossRef] [PubMed]

189. Bugno-Poniewierska, M.; Solek, P.; Wronski, M.; Potocki, L.; Jezewska-Witkowska, G.; Wnuk, M. Genome organization and DNA methylation patterns of B chromosomes in the red fox and Chinese raccoon dogs. *Hereditas* **2014**, *151*, 169–176. [CrossRef] [PubMed]

190. Jones, R.N.; Houben, A. B chromosomes in plants: Escapees from the A chromosome genome? *Trends Plant Sci.* **2003**, *8*, 417–423. [CrossRef]

191. Sapre, A.B.; Deshpande, D.S. Origin of B chromosomes in Coix L. through spontaneous interspecific hybridization. *J. Hered.* **1987**, *78*, 191–196. [CrossRef]

192. McAllister, B.F. Isolation and characterization of a retroelement from B chromosome (PSR) in the parasitic wasp *Nasonia vitripennis*. *Insect Mol. Biol.* **1995**, *4*, 253–262. [CrossRef] [PubMed]

193. Guioli, S.; Lovell-Badge, R.; Turner, J.M.A. Error-Prone ZW Pairing and no evidence for meiotic sex chromosomeinactivation in the chicken germ line. *PLoS Genet.* **2012**, *8*, e1002560. [CrossRef] [PubMed]

194. Gu, L.; Walters, J.R. Evolution of sex shromosome dosage compensation in animals: A beautiful theory, undermined by facts and bedeviled by details. *Genome Biol. Evol.* **2017**, *9*, 2461–2476. [CrossRef] [PubMed]

195. Daish, T.J.; Casey, A.E.; Grutzner, F. Lack of sex chromosome specific meiotic silencing in platypus reveals origin of MSCI in therian mammals. *BMC Biol.* **2015**, *13*, 106. [CrossRef] [PubMed]

196. Berend, S.A.; Hale, D.W.; Engstrom, M.D.; Greenbaum, I.F. Cytogenetics of collared lemmings (*Dicrostonyx groenlandicus*). II. Meiotic behavior of B chromosomes suggests a Y-chromosome origin of supernumerary chromosomes. *Cytogenet. Genome Res.* **2001**, *95*, 85–91. [CrossRef] [PubMed]

197. Kolomeits, O.L.; Borbiev, T.E.; Safronova, L.D.; Borisov, Y.M. Synaptonemal analysis of B-chromosome behaviour in meiotic prophase I in the East-Asiatic mouse *Apodemus peninsulae* (Muridae, Rodentia). *Cytogenet. Cell Genet.* **1988**, *48*, 183–187. [CrossRef] [PubMed]

198. Matsubara, K.; Nishida-Umehara, C.; Tsuchiya, K.; Nukaya, D.; Matsuda, Y. Karyotypic evolution of *Apodemus* (Muridae, Rodentia) inferred from comparative FISH analyses. *Chromosom. Res.* **2004**, *12*, 383–395. [CrossRef] [PubMed]

199. Thomson, R.L.; Westerman, M.; Murray, N.D. B chromosomes in *Rattus fuscipes* I. Mitotic and meiotic chromosomes and the effects of B on chiasma frequency. *Heredity (Edinb)* **1984**, *52*, 355–362. [CrossRef]

200. Rubtsov, N.B.; Borisov, Y.M.; Karamysheva, T.V.; Bochkarev, M.N. The mechanisms of formation and evolution of B chromosomes in Korean field mice *Apodemus peninsulae* (Mammalia, Rodentia). *Russ. J. Genet.* **2009**, *45*, 389–396. [CrossRef]

201. Ishak, B.; Maetz, J.L.; Rumpler, Y. Absence of transcriptional activity in the B-chromosomes of *Apodemus flavicollis* during pachytene. *Chromosoma* **1991**, *100*, 278–281. [CrossRef]

202. Karamysheva, T.V.; Torgasheva, A.A.; Yefremov, Y.R.; Bogomolov, A.G.; Liehr, T.; Borodin, P.M.; Rubtsov, N.B. Spatial organization of fibroblast and spermatocyte nuclei with different B-chromosome content in Korean field mouse, *Apodemus peninsulae* (Rodentia, Muridae). *Genome* **2017**, *60*, 815–824. [CrossRef] [PubMed]

203. Vujošević, M.; Radosavljević, J.; Živković, S. Meiotic behavior of B chromosomes in yellow necked mouse *Apodemus flavicollis*. *Arch. Biol. Sci.* **1989**, *41*, 39–42.

204. Banaszek, A.; Jadwiszczak, K.A. B-chromosomes behaviour during meiosis of yellow-necked mouse, *Apodemus flavicollis*. *Folia Zool.* **2006**, *55*, 113–122.

205. Świtoński, M.; Gustavsson, I.; Höjer, K.; Plöen, L. Synaptonemal complex analysis of the B-chromosomes in spermatocytes of the silver fox (*Vulpes fulvus* Desm.). *Cytogenet. Genome Res.* **1987**, *45*, 84–92. [CrossRef]

206. Sosnowski, J.; Łukasiewicz, A.; Migalska, L.; Wojnowska, M.; Polański, Z. Different levels of a lack of X-Y chromosome pairing in pachytene spermatocytes of red fox (*Vulpes vulpes*) and Chinese raccoon dog (*Nyctereutes procyonoides procyonoides*). *Ann. Anim. Sci.* **2011**, *11*, 71–81.

207. Patton, J.L. B-chromosome system in the pocket mouse, *Perognathus baileyi*: Meiosis and C-band studies. *Chromosoma* **1977**, *60*, 1–14. [CrossRef] [PubMed]

208. Radjabli, S.I.; Isaenko, A.A.; Volobuev, V.T. Investigation of the nature and role of additional chromosomes in silver fox. *Genetika* **1978**, *XIV*, 438–443.

209. Borisov, Y.M. Increase in the number of the B-chromosomes and variants of their system in mouse *Apodemus peninsulae* in Mountain Altai population over 26 years. *Genetika* **2008**, *44*, 1227–1237. [CrossRef] [PubMed]

210. Gileva, E.A.; Chebotar, N.A. Fertile X0 males and females in the varying lemming, *Dicrostonyx torquatus* Pall. (1779). *Heredity (Edinb)* **1979**, *42*, 67–77. [CrossRef]

211. Gileva, E. The B chromosome system in the varying lemming *Dicrostonyx torquatus* Pall., 1779 from natural and laboratory populations. *Russ. J. Genet.* **2004**, *40*, 1399–1406. [CrossRef]

212. Yosida, T.H. Some genetic analysis of supernumerary chromosomes in the black rat in laboratory matings. *Proc. Jpn. Acad.* **1978**, *50*, 440–445. [CrossRef]

213. Stitou, S.; Zurita, F.; Díaz de la Guardia, R.; Jiménez, R.; Burgos, M. Transmission analysis of B chromosomes in *Rattus rattus* from Northern Africa. *Cytogenet. Genome Res.* **2004**, *106*, 344–346. [CrossRef] [PubMed]

214. Palestis, B.G.; Burt, A.; Jones, R.N.; Trivers, R. B chromosomes are more frequent in mammals with acrocentric karyotypes: Support for the theory of centromeric drive. *Proc. R. Soc. B Biol. Sci.* **2004**, *271*, S22–S24. [CrossRef] [PubMed]

215. Pardo-Manuel de Villena, F.; Sapienza, C. Female meiosis drives karyotypic evolution in mammals. *Genetics* **2001**, *159*, 1179–1189. [PubMed]

216. Kociucka, B.; Sosnowski, J.; Kubiak, A.; Nowak, A.; Pawlak, P.; Szczerbal, I. Three-dimensional positioning of B chromosomes in fibroblast nuclei of the red fox and the Chinese raccoon dog. *Cytogenet. Genome Res.* **2013**, *139*, 243–249. [CrossRef] [PubMed]

217. Wurster-Hill, D.H.; Ward, O.G.; Kada, B.H.; Whittmore, S. Banded chromosome studies and B chromosomes in wild-caught raccoon dogs, *Nyctereutes procyonides viverinus. Cytogenet. Cell Genet.* **1986**, *42*, 85–93. [CrossRef]

218. Zima, J.; Piálek, J.; Macholán, M. Possible heterotic effects of B chromosomes on body mass in a population of *Apodemus flavicollis. Can. J. Zool.* **2003**, *81*, 1312–1317. [CrossRef]

219. Jojić, V.; Blagojević, J.; Ivanović, A.; Bugarski-Stanojević, V.; Vujošević, M. Morphological integration of the mandible in yellow-necked field mice: The effects of B chromosomes. *J. Mammal.* **2007**, *88*, 689–695. [CrossRef]

220. Jojić, V.; Blagojević, J.; Vujošević, M. B chromosomes and cranial variability in yellow-necked field mice (*Apodemus flavicollis*). *J. Mammal.* **2011**, *92*, 396–406. [CrossRef]

221. Blagojević, J.; Vujošević, M. B chromosomes and developmental homeostasis in the yellow-necked mouse, *Apodemus flavicollis* (Rodentia, Mammalia): Effects on nonmetric traits. *Heredity (Edinb)* **2004**, *93*, 249–254. [CrossRef] [PubMed]

222. Adnađević, T.; Jovanović, V.M.; Blagojević, J.; Budinski, I.; Čabrilo, B.; Bjelić-Čabrilo, O.; Vujošević, M. Possible influence of B chromosomes on genes included in immune response and parasite burden in *Apodemus flavicollis. PLoS ONE* **2014**, *9*, e112260. [CrossRef] [PubMed]

223. Jovanović, V.M.; Čabrilo, B.; Budinski, I.; Bjelić-Čabrilo, O.; Adnađević, T.; Blagojević, J.; Vujošević, M. Host B chromosomes as potential sex ratio distorters of intestinal nematode infrapopulations in the yellow-necked mouse (*Apodemus flavicollis*). *J. Helminthol.* **2018**. [CrossRef] [PubMed]

224. Belayev, D.K.; Volobouev, V.T.; Radjabli, S.I.; Trut, L.N. Investigation of the nature and role of additional chromosomes in silver fox II. Additional chromosomes and breeding of animals for behaviour. *Genetika* **1974**, *X*, 83–91.

225. Volobuev, V.T.; Radjabli, S.I. Investigation of the nature and the role of additional chromosomes in silver fox. I. Comparative analysis of additional chromosomes in different tissues and types of preparations and at different seasons. *Genetika* **1974**, *X*, 77–82.

226. Volobujev, V.T.; Radzabli, S.I.; Belajeva, E.S. Investigation of the nature and the role of additional chromosomes in silver foxes. III. Replication pattern in additional chromosomes. *Genetika* **1976**, *XII*, 30–34.

227. Östergren, G. Parasitic nature of extra fragment chromosomes. *Bot. Not.* **1945**, *2*, 157–163.

228. White, M.J.D. *Animal Cytology and Evolution*, 3rd ed.; Cambridge University Press: Cambridge, UK, 1973; ISBN 0-521-07071-6.

229. Camacho, J.P.M.; Shaw, M.W.; López–León, M.D.; Pardo, M.C.; Cabrero, J. Population dynamics of a selfish B chromosome neutralized by the standard genome in the grasshopper *Eyprepocnemis plorans. Am. Nat.* **1997**, *149*, 1030–1050. [CrossRef] [PubMed]

230. Vujošević, M. B-chromosome polymorphism in *Apodemus flavicollis* (Rodentia, Mammalia) during five years. *Caryologia* **1992**, *45*, 347–352. [CrossRef]

231. Blagojević, J.; Vujošević, M. The role of B chromosomes in the population dynamics of yellow-necked wood mice *Apodemus flavicollis* (Rodentia, Mammalia). *Genome* **1995**, *38*, 472–478. [CrossRef] [PubMed]

232. Vujošević, M.; Blagojević, J. Seasonal changes of B-chromosome frequencies within the population of *Apodemus flavicollis* (Rodentia) on Cer mountain in Yugoslavia. *Acta Theriol. (Warsz)* **1995**, *40*, 131–137. [CrossRef]

233. Borisov, Y.M.; Sheftel, B.I.; Safronova, L.D.; Aleksandrov, D.Y. Stability of the Population Variants of the B-Chromosome System in the East-Asian Mouse *Apodemus peninsulae* from the Baikal Region and Northern Mongolia. *Russ. J. Genet.* **2012**, *48*, 1020–1028. [CrossRef]

234. Blagojević, J.; Stamenković, G.; Jojić Šipetić, V.; Bugarski-Stanojević, V.; Adnađević, T.; Vujošević, M. B chromosomes in populations of yellow-necked mice—Stowaways or contributing genetic elements? *Ital. J. Zool.* **2009**, *76*, 250–257. [CrossRef]

235. Adnađević, T.; Bugarski-Stanojević, V.; Blagojević, J.; Stamenković, G.; Vujošević, M. Genetic differentiation in populations of the yellow-necked mouse, *Apodemus flavicollis*, harbouring B chromosomes in different frequencies. *Popul. Ecol.* **2012**, *54*, 537–548. [CrossRef]

© 2018 by the authors. Licensee MDPI, Basel, Switzerland. This article is an open access article distributed under the terms and conditions of the Creative Commons Attribution (CC BY) license (http://creativecommons.org/licenses/by/4.0/).

![genes logo] *genes*

MDPI

Review

The Behavior of the Maize B Chromosome and Centromere

Handong Su [1,2], Yalin Liu [1], Yang Liu [1,2], James A. Birchler [3,*] and Fangpu Han [1,*]

1   State Key Laboratory of Plant Cell and Chromosome Engineering, Institute of Genetics and Developmental Biology, Chinese Academy of Sciences, Beijing 100101, China; shdong@genetics.ac.cn (H.S.); ylliu@genetics.ac.cn (Y.L.); yangliu@genetics.ac.cn (Y.L.)
2   University of Chinese Academy of Sciences, Beijing 100049, China
3   Division of Biological Sciences, University of Missouri, Columbia, MO 65211, USA
*   Correspondence: birchlerj@missouri.edu (J.A.B.); fphan@genetics.ac.cn (F.H.)

Received: 23 August 2018; Accepted: 25 September 2018; Published: 1 October 2018

Abstract: The maize B chromosome is a non-essential chromosome with an accumulation mechanism. The dispensable nature of the B chromosome facilitates many types of genetic studies in maize. Maize lines with B chromosomes have been widely used in studies of centromere functions. Here, we discuss the maize B chromosome alongside the latest progress of B centromere activities, including centromere misdivision, inactivation, reactivation, and de novo centromere formation. The meiotic features of the B centromere, related to mini-chromosomes and the control of the size of the maize centromere, are also discussed.

Keywords: maize B chromosome; centromere; inactivation; reactivation; de novo centromere formation; epigenetics

1. Introduction

The B chromosome is a non-essential chromosome that does not pair with the normal (A) chromosomes during meiosis. B chromosomes are widespread in fungi, plants and animals species [1]. The maize B chromosome is the first B chromosome to be observed [2]. Subsequent cytological observations in the pachytene stage of meiosis [3–5] defined the chromosome structure as including a centromeric chromatin, four heterochromatic blocks (DH1–DH4) in the long arm, and two euchromatic regions that are proximal and distal (diagram of the maize B chromosome in Figure 1a). The maize B chromosome has been studied for several decades for cytology, genetics and sequence [6]. Maize lines containing B chromosomes have been constructed and are widely used in the studies of maize genetics, engineered mini-chromosomes, gene dosage, and centromere functions [7,8]. However, the key issues on the origin, evolution and molecular mechanism of its accumulation in maize populations were still largely unknown. In this review, we will focus on the recent progress of studies on the maize B chromosome, with new insight into B centromere activities, including centromere misdivision, inactivation, reactivation, and de novo centromere formation.

Genes **2018**, *9*, 476

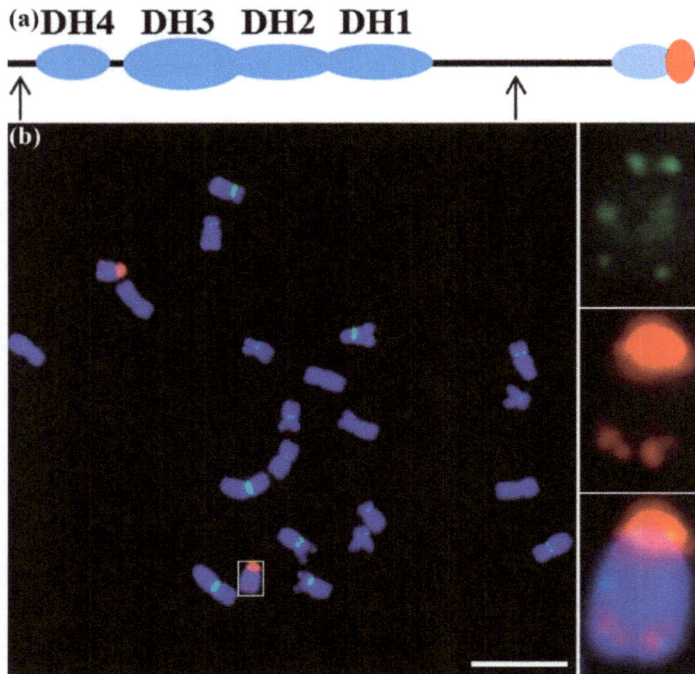

Figure 1. The structure and organization of the maize B chromosome and its centromere. (**a**) Diagram of the maize B chromosome with centric heterochromatin, four heterochromatic blocks (DH1–DH4) in the long arm, and two euchromatic regions—one proximal and one distal. The red oval indicates the centric heterochromatin. The light-blue oval indicates the knob signals. The two arrows indicate the two potential regions that are essential for nondisjunction. (**b**) The distribution of CentC (green) and maize B centromere repeat sequence (B-repeat) (red) signals along the maize B chromosome. The white box indicates the B chromosome, and the insets show a higher magnification view of the B chromosome. 4′,6-diamidino-2-phenylindole (DAPI)-stained chromosomes are blue. Bar = 10 μm.

2. The Structure and Organization of the Maize B Chromosome

The maize B chromosome is not detrimental to plant development unless it is present at approximately 15 copies and more [9]. The study of the maize B chromosome's sequence organization is fundamental in order to understand the origin and evolution of the B chromosome. All known retrotransposon DNA sequences from maize A chromosomes were found to be distributed on the B chromosome [10]. A collection of repeat DNA sequences from A centromeres including CentC, centromere retrotransposon sequence CRM, and CentA were also found to be widely distributed on the B chromosome and, interestingly, not only concentrated in the B centromere [11]. However, several maize B chromosomal specific DNA sequences were isolated. The first one was the maize B centromere repeat sequence (B-repeats or ZmBs) with a unit size over 1.0 kb [12]. The B-repeat has similarity with the Cent4 sequence, the pericentromeric sequence on maize chromosome 4 [13]. Recent work, using de novo assembly from RNA sequencing (RNA-seq) data sets of maize B73 and B chromosomes, also identified two B-chromosome-specific long terminal repeat (LTR) retrotransposons with only partial sequences homologous to the A genome [14]. The transcriptome analysis revealed the transcription of B chromosome genes, including one of the two B-chromosome-specific LTRs. This analysis also showed that the transcription from A chromosome genes was affected by the presence of the B chromosome, as an increase of B chromosome numbers correlated with an increase of the effect [14], indicating that the B chromosome is not generally inert. The transcription of the B chromosome may occur in a condition-specific manner, like the nondisjunction property discussed below. Many sequence variations, including single nucleotide polymorphisms (SNPs) and insertions/deletions (Indels), were accumulated in the B-located gene fragments, compared with their A chromosome counterparts [14].

3. The Composition of the Maize B Centromere and Its Misdivision

The centromere mediates the assembly of the kinetochore, which is essential for faithful chromosome segregation during cell division [15]. Centromere identity, in most species, is determined by the presence of the histone H3 variant CENH3 [16,17]. The epigenetic mechanism for centromere specification also includes the phosphorylation of histone H2AThr133 in plants [18]. The DNA components of maize A chromosome centromeres mainly consist of two types of DNA sequences, as do most species, being the 156-bp tandem repeat CentC and CRM [8,19]. The isolated B-specific repeat sequences were confirmed to be scattered throughout, and around, the B centromere, and numerous copies had high variation. CentC and CRM sequences were embedded within the multimegabases of B-repeat in the B centromere (Figure 1b). The colocalization of the sequences with CENH3 labeling indicated they confer for centromere function, but only small portions of B-repeats were occupied by CENH3 [11,20]. In addition, the B-repeat, CentC, and CRM sequences were also located in many other distinct regions of the B chromosome (Figure 1b). Both of these results indicate that the DNA sequence alone cannot determine centromere function. What distinguishes the binding, or not, with CENH3 for the same DNA sequence is still unknown.

Misdivision is the process of improper division of a univalent chromosome centromere during meiosis, resulting in isochromosome, telocentric and/or ring chromosomes [21]. A B–A translocation line TB-9Sb that involved a B chromosome and the short arm of chromosome 9 with several phenotypic markers was extensively used [22,23]. Misdivisions of the centromere of this translocation chromosome [24,25] produced a series of B centromere variant lines, which is a good model for the study of centromere function (Figure 2a). A series of telocentric chromosomes with different sized B centromeres were used to study the B centromere. The mapping of the B centromere, using Fiber-fluorescent in situ hybridization (FISH) with five B centromere misdivision derivatives, showed a 700 kb core centromere region with B-repeat, CentC, and CRM retrotransposon [20]. The transposon display with CRM2, using these misdivision lines, also generated a map of the B centromere core with 33 markers that directly interacted with CENH3. These efforts to accumulate the complete B centromere map will provide an opportunity to precisely record the process of centromere misdivision events, and to further understand the molecular basis of centromere function.

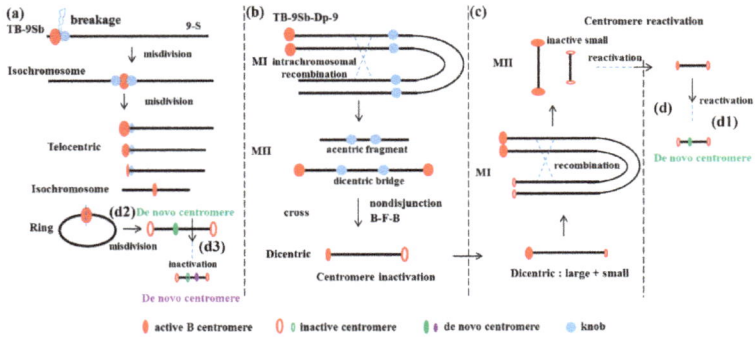

Figure 2. Diagrams of centromere behavior associated with the B chromosome. (**a**) The first misdivision process of a TB-9Sb chromosome produced a pseudoisochromosome; and different types of B chromosome variants formed, including telocentric chromosomes, ring chromosomes, and isochromosomes. (**b**) The TB-9Sb-Dp9 chromosome underwent intrachromosomal recombination during meiosis I, and an acentric fragment and a dicentric bridge formed during meiosis II. Centromere inactivation of the dicentric chromosome occurred from the nondisjunction, breakage-fusion-bridge (B-F-B) cycle, and the misdivision process. (**c**) A large-small dicentric chromosome underwent intrachromosomal recombination. Centromere reactivation occurred on the previously inactive small sister chromatids. (**d**) De novo centromere formation occurred in the ectopic genomic region from the reactivation of a dicentric chromosome (d1) or TB-9Sb centromere misdivision (d2) process. Inactivation occurred at the original active de novo centromere position, and a subsequent de novo formation was formed in another chromosomal region (d3).

4. B Chromosome Nondisjunction

The accumulation mechanism of the B chromosome works to maintain it in populations. It consists of nondisjunction at the second pollen mitosis followed by preferential fertilization of the egg by the B chromosome that contains sperm [26,27]. Genetic studies have proven that the B centromere adjacent heterochromatin, and at least two other regions that are located in the very distal tip and proximal euchromatin, are essential for this process [4,6,28,29], as nondisjunction does not occur in the centromere when these regions are removed. The target site for nondisjunction may be located in B-repeat sequences, as CentC and CRM repeats are also located in A centromeres but nondisjunction does not occur for A chromosomes [5]. When an inactive B centromere was crossed with a whole B chromosome, the translocated chromosome, with the inactive B centromere attached to the short arm of chromosome 9, regained the property of nondisjunction, indicating that centromere function and B chromosome nondisjunction are two independent processes [30]. These two sites may generate transcripts or protein factors that act in trans on the B-repeat for the nondisjunction.

5. Centromere Inactivation, Reactivation and De Novo Centromere Formation

The chromosome type breakage-fusion-bridge (B-F-B) cycle was initiated by the fusion of two terminal deletion chromosomes, and resulted in a dicentric chromosome, which formed a bridge during anaphase that in turn would break to repeat the cycle. Translocation chromosomes between A and B chromosomes can undergo continuous centromere misdivision, nondisjunction, and chromosome type B-F-B cycles during cell divisions, resulting in a large number of centromere variant lines. A high frequency of centromere inactivation, generating stable maize dicentric chromosomes, was reported [31]. To produce the dicentrices, the TB-9Sb-Dp9 chromosome was constructed by recombining a duplicated chromosome 9 to the B–9 chromosome [32]. Intrachromosomal recombination can occur in the duplication of the 9S chromosome. Dicentric chromosomes with two B-repeat regions and one acentric fragment were observed during meiosis II. Only the chromosomes with one active and one inactive centromere could be stably transmitted to the next generation (Figure 2b). An inactive B centromere translocated to the tip of the short arm of chromosome 9 remained in the inactive state, without CENH3 labeling, over several generations [31]. The same DNA sequence in the dicentromeres provides additional evidence that DNA sequence is not the determinant for centromere function.

From a different perspective, a large-small dicentric chromosome was produced by crossing the TB-9Sb-Dp9 to T3-5 (+) with a truncated B centromere from a TB-9Sb misdivision line [33]. The large and small centromeres were in an active state. The foldback structure allowed recombination between the two chromosomes during meiosis and generated structures with two large centromeres and two small centromeres joined together. When the two centromeres were joined together in a stable chromosome that was inherited, the smaller centromere was inactive.

When this chromosome with a large active centromere and a small inactive centromere was studied, the foldback nature allowed the large and small centromeres to be separated from each other by recombination. They could be separated in meiosis I, and the two small-inactive centromere structures could be detected in the progeny, indicating that the centromere was reactivated at the original inactive centromere (Figure 2c). The CENH3 protein was indeed detected binding with one of the formerly inactive centromere regions. The reactivation process of the inactive centromere suggests that the centromere DNA-repeat sequences display a preference for kinetochore assembly, or other CENH3 chaperone factors such as the human holiday junction recognition protein (HJURP) in humans, the chromosome alignment defect 1 protein (CAL1) in *Drosophila*, or the suppressor of chromosome missegregation protein 3 (SCM3) in yeast [34].

A different process can be adopted for the transmission of this chromosome with two small inactive centromeres. De novo centromere formation in an ectopic region can regain centromere function instead of centromere activation, as has been reported in many other systems [35–37]. sDic15 is a dicentric chromosome that is generated from the process mentioned above, which includes intrachromosomal recombination and centromere reactivation (Figure 2 (d1)) [38]. No detectable CentC signals, and strongly reduced CRM and B-repeat signals, were observed in the centromere of the sDic15 mini-chromosome [38]. A 723 kb genomic region from chromosome 9 confers neocentromere function [38]. The misdivision of the chromosomes created breaks in the B centromeric region and produced many new centromere variants. The B centromere sequences were deleted but some B-repeat copies were still remaining in the 3–3 derivative. De novo centromere formation was detected in the 3–3 derivative with a 288 kb genomic region from chromosome 9 (Figure 2 (d2)). In a subsequent misdivision derivative of 3–3, a new centromere variant, without the previously formed neocentromere sequences, was detected; and other genomic regions, including an additional 200 kb of DNA sequences in chromosome 9, became the de novo centromere in derivative 3–3–11 (Figure 2 (d3)). These two neocentromeres were both misdivision derivatives of TB-9Sb, but with different centromere sequences. Centromere breakage and the B-F-B cycle process can both generate de novo centromeres.

The behavior of the maize B centromere, which includes centromere inactivation, reactivation and de novo centromere formation, reveals an epigenetic mechanism for centromere specification [39]. However, the molecular basis for these processes is still largely unknown. We have checked the DNA methylation state of the active and inactive B centromeres, and found that distinct DNA methylation patterns, between them, occur with a hypermethylation state in the inactive centromere [40]. The epigenetic difference is more likely to result after the formation of a stable chromosome state. Distinct DNA methylation patterns were also discovered in the satellite DNA sequences, between the pericentromere and centromere, in maize and *Arabidopsis thaliana* [41]. The comparison of DNA methylation in the ectopic region before and after de novo centromere formation showed a similar methylation level in this region, compared to normal centromeres, whether the original methylation level was high or low [38,42]. This indicates that there was maintenance of the centromere chromatin by an epigenetic mechanism. However, further work is needed on the detailed mechanism that establishes the centromeric chromatin, and on how centromere function is regained, as a de novo centromere or as a reactivation, at the original inactivation sites. Recently, the transcription and/or the transcripts from centromeric regions have been reported in many model systems, including maize [43–47]. They are involved in CENH3 loading, kinetochore assembly, and cell-cycle progression [48–50]. It may provide another aspect of the molecular basis of the transmission of different centromere chromatin states.

6. Meiotic Behavior of the Maize B Derivative Chromosome

Meiosis is a special cell division process that reduces chromosome numbers by half in the daughter cells. The centromere has an important role in meiosis I, homologous chromosome pairing, and segregation [51,52]. The orientation of sister chromatids is a key point for the process, especially the mono-orientation of the sister chromatids toward the same spindle pole in meiosis I [53]. The misdivision derivatives of TB-9Sb produce a series of B centromere deletion lines [54]. They provide a good model to study the meiotic behavior of centromeres with different sizes. Iwata-Otsubo et al. found that the CENP-A chromatin expansion, induced by the amplified centromere repeats, promoted increased transmission through the female germline in a model using mice [55]. We recently reported the observation of the combination of chromosomes with different sized B centromeres, during meiosis, in maize [56]. Only slightly higher frequencies, of a few percent, were observed for the larger centromere in several cases. However, the size of the centromeric DNA was not necessarily correlated with the signals of phH2AThr133 [56], suggesting the regulation of the functional size of the centromere.

A number of mini-chromosomes were derived from the chromosome type B-F-B cycle of the TB-9Sb-Dp9 chromosome [31]. They ranged in both chromosome size and in the orientation of the sister chromatids during meiosis I [57]. The mini-chromosome derivatives 3–3 and sDic15 were generated from the maize B chromosome, through misdivision or intrachromosomal recombination, and de novo centromere formation was reported in these mini-chromosomes [38,58]. The mini-chromosomes #11 and sDic15 displayed mono-orientation in anaphase I, as the normal A chromosomes behaved mono-orientation in this stage. However, the mini-chromosome #3, #9, and derivative 3–3 showed bi-orientation, and their sister chromatids separated in anaphase I [38,57,58]. The distribution of B-repeat signals in two cells in telophase I indicated that the sister chromatids of the mini-chromosome #9 previously separated in anaphase I (Figure 3). What the differences are between the centromeres of the normal A chromosome and the different types of mini-chromosomes, and how they are distinguished for the molecular machinery of chromosome orientation, remain interesting questions.

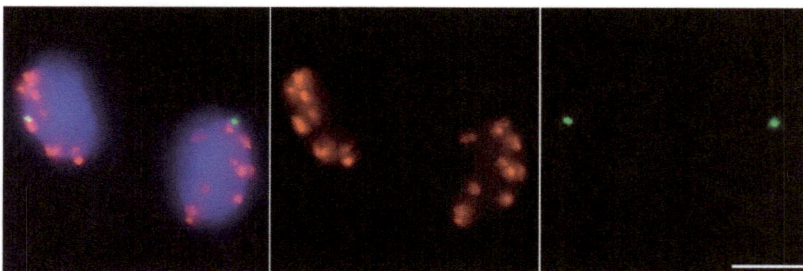

Figure 3. Labeling of phH2AThr133 in maize B mini-chromosome #9. The distribution of phH2AThr133 signals at telophase I stage of meiosis for maize B mini-chromosome #9. DAPI-stained chromosomes are blue, phH2AThr133 are red, and B-repeats are green. Bar = 10 μm. The B-repeat signals distributed in two cells in telophase I indicate that the sister chromatids of this mini-chromosome previously separated in anaphase I.

All these mini-chromosomes contain functional centromeres with CENH3, phH2AThr133, phH3T3, and phH3Ser10 loading in mitotic cells [38,57–59]. The spindle assembly checkpoint kinase, ZmBub1, was detected in the mini-chromosome centromeres as normal A centromeres [60]. We speculate that the phosphorylation of histone H2AThr133 occurs during meiosis for these mini-chromosomes. The signal of phH2AThr133 loaded well in the mini-chromosome #9 (Figure 3), but no differences in phosphorylation levels were found in anaphase I during meiosis between the mini-chromosome #9, with mono-orientation or bi-orientation, and the normal A chromosomes [60]. The mechanical basis of the equational division of these mini-chromosomes may provide clues for the mechanism of chromosome orientation at meiosis I.

7. The Control of the Maize B Centromere Size

The canonical centromere size in any one species seems to assume a certain size, which is strongly correlated with genome size rather than the chromosome size [61]. The mean centromere size in rice is about several hundred kilobases, while the size of the maize centromere is about several megabases [8,62]. The centromere size is determined by the equilibrium between bound and unbound CENH3 proteins, which is regulated by genetic and epigenetic factors [63]. The maize centromere sizes expand to adjust to the oat background, in which the oat genome size is about 4-fold larger than the maize genome size [64]. The transcription of genes, and the corresponding epigenetic modifications, may restrict the centromere size [64,65].

The genome size will certainly be changed as one intact B chromosome is about the size of maize chromosome 10. However, accumulation of B chromosome does not change cell size as does ploidy. The maize B chromosome has been used to produce many lines with different sized centromeres. The CENH3-bound B centromeres may be not with expansion in the maize background even with high copy numbers of B chromosome. The smallest centromere size, resulting from misdivision, was only several hundred kilobase [66]. The small sized centromeres in maize B mini-chromosomes can be used to study the minimal DNA component required to assemble kinetochore. In addition, de novo centromeres from the misdivision lines were detected with centromeres ranging in size, and up to several hundred kilobases [58]. The centromeres of the mini-chromosomes are much smaller than the canonical maize centromere. However, the maize B centromere may expand when the B chromosome is introduced to the oat background as the oat-maize additional line.

8. Concluding Remarks and Future Perspectives

The establishment and maintenance of the centromere chromatin is an interesting topic in chromosome biology. The molecular underpinnings of the transitions among the different chromatin states of the maize B chromosome centromere, including activation, inactivation and de novo centromere formation, is of great interest and is a good model to study. These processes may be involved in the coordination of complex epigenetic machineries that include DNA methylation, histone modification, chaperones, chromatin remodeling, and noncoding RNAs. Further work with the B centromere on these aspects will be carried out with the aid of maize B chromosome genome sequencing and the fruitful B centromere variants.

Funding: Research on this topic was supported by the National Natural Science Foundation of China (31630049 and 31320103912) and the National Science Foundation (NSF) grant IOS-1444514.

Conflicts of Interest: The authors declare no conflicts of interest.

References

1. Jones, N. B chromosomes in plants: Escapees from the a chromosome genome? *Trends Plant Sci.* **2003**, *8*, 417–423. [CrossRef]
2. Longley, A.E. Supernumerary chromosomes in *Zea mays*. *J. Agric. Res.* **1927**, *35*, 769–784.

3.   McClintock, B. The association of non-homologous parts of chromosomes in the mid-prophase of meiosis in *Zea mays*. *Z. Zellforsch. Mikrosk. Anat.* **1933**, *19*, 191–237. [CrossRef]

4.   Ward, E.J. The heterochromatic B chromosome of maize: The segments affecting recombination. *Chromosoma* **1973**, *43*, 177–186. [CrossRef]

5.   Pryor, A.; Faulkner, K.; Rhoades, M.M.; Peacock, W.J. Asynchronous replication of heterochromatin in maize. *Proc. Natl. Acad. Sci. USA* **1980**, *77*, 6705–6709. [CrossRef] [PubMed]

6.   Carlson, W. The B chromosome of maize. In *Handbook of Maize: Genetics and Genomics*; Bennetzen, J.L., Hake, S., Eds.; Springer: New York, NY, USA, 2009; pp. 459–480.

7.   Nannas, N.J.; Dawe, R.K. Genetic and genomic toolbox of *Zea mays*. *Genetics* **2015**, *199*, 655–669. [CrossRef] [PubMed]

8.   Wolfgruber, T.K.; Sharma, A.; Schneider, K.L.; Albert, P.S.; Koo, D.H.; Shi, J.; Gao, Z.; Han, F.; Lee, H.; Xu, R.; et al. Maize centromere structure and evolution: Sequence analysis of centromeres 2 and 5 reveals dynamic loci shaped primarily by retrotransposons. *PLoS Genet.* **2009**, *5*, e1000743. [CrossRef] [PubMed]

9.   Randolph, L.F. Genetic characteristics of the B chromosomes in maize. *Genetics* **1941**, *26*, 608–631. [PubMed]

10.   Theuri, J.; Phelps-Durr, T.; Mathews, S.; Birchler, J. A comparative study of retrotransposons in the centromeric regions of A and B chromosomes of maize. *Cytogenet. Genome Res.* **2005**, *110*, 203–208. [CrossRef] [PubMed]

11.   Lamb, J.C.; Kato, A.; Birchler, J.A. Sequences associated with A chromosome centromeres are present throughout the maize B chromosome. *Chromosoma* **2005**, *113*, 337–349. [CrossRef] [PubMed]

12.   Alfenito, M.R.; Birchler, J.A. Molecular characterization of a maize B chromosome centric sequence. *Genetics* **1993**, *135*, 589–597. [PubMed]

13.   Page, B.T.; Wanous, M.K.; Birchler, J.A. Characterization of a maize chromosome 4 centromeric sequence: Evidence for an evolutionary relationship with the B chromosome centromere. *Genetics* **2001**, *159*, 291–302. [PubMed]

14.   Huang, W.; Du, Y.; Zhao, X.; Jin, W. B chromosome contains active genes and impacts the transcription of A chromosomes in maize (*Zea mays* L.). *BMC Plant Biol.* **2016**, *16*, 88. [CrossRef] [PubMed]

15.   Kursel, L.E.; Malik, H.S. Centromeres. *Curr. Biol.* **2016**, *26*, R487–R490. [CrossRef] [PubMed]

16.   Niikura, Y.; Kitagawa, R.; Kitagawa, K. The inheritance of centromere identity. *Mol. Cell. Oncol.* **2016**, *3*, e1188226. [CrossRef] [PubMed]

17.   Fukagawa, T.; Earnshaw, W.C. The centromere: Chromatin foundation for the kinetochore machinery. *Dev. Cell* **2014**, *30*, 496–508. [CrossRef] [PubMed]

18.   Dong, Q.; Han, F. Phosphorylation of histone H2A is associated with centromere function and maintenance in meiosis. *Plant J.* **2012**, *71*, 800–809. [CrossRef] [PubMed]

19.   Comai, L.; Maheshwari, S.; Marimuthu, M.P.A. Plant centromeres. *Curr. Opin. Plant Biol.* **2017**, *36*, 158–167. [CrossRef] [PubMed]

20.   Jin, W.; Lamb, J.C.; Vega, J.M.; Dawe, R.K.; Birchler, J.A.; Jiang, J. Molecular and functional dissection of the maize B chromosome centromere. *Plant Cell* **2005**, *17*, 1412–1423. [CrossRef] [PubMed]

21.   Birchler, J.A.; Han, F. Barbara McClintock's unsolved chromosomal mysteries: Parallels to common rearrangements and karyotype evolution. *Plant Cell* **2018**, *30*, 771–779. [CrossRef] [PubMed]

22.   Robertson, D.S. Crossing over and chromosomal segregation involving the B^9 element of the a-B translocation B-9b in maize. *Genetics* **1967**, *55*, 433–449. [PubMed]

23.   Carlson, W.R. Nondisjunction and isochromosome formation in the B chromosome of maize. *Chromosoma* **1970**, *30*, 356–365. [CrossRef]

24.   McClintock, B. The behavior in successive nuclear divisions of a chromosome broken at meiosis. *Proc. Natl. Acad. Sci. USA* **1939**, *25*, 405–416. [CrossRef] [PubMed]

25.   McClintock, B. The stability of broken ends of chromosomes in *Zea mays*. *Genetics* **1941**, *26*, 234–282. [PubMed]

26.   Roman, H. Mitotic nondisjunction in the case of interchanges involving the B-type chromosome in maize. *Genetics* **1947**, *32*, 391–409. [PubMed]

27.   Roman, H. Directed fertilization in maize. *Proc. Natl. Acad. Sci. USA* **1948**, *34*, 36–42. [CrossRef] [PubMed]

28.   Carlson, W.R. A procedure for localizing genetic factors controlling mitotic nondisjunction in the B chromosome of maize. *Chromosoma* **1973**, *42*, 127–136. [CrossRef]

29.   Lin, B.Y. Regional control of nondisjunction of the B chromosome in maize. *Genetics* **1978**, *90*, 627.

30. Han, F.; Lamb, J.C.; Yu, W.; Gao, Z.; Birchler, J.A. Centromere function and nondisjunction are independent components of the maize B chromosome accumulation mechanism. *Plant Cell* **2007**, *19*, 524–533. [CrossRef] [PubMed]

31. Han, F.; Lamb, J.C.; Birchler, J.A. High frequency of centromere inactivation resulting in stable dicentric chromosomes of maize. *Proc. Natl. Acad. Sci. USA* **2006**, *103*, 3238–3243. [CrossRef] [PubMed]

32. Zheng, Y.Z.; Roseman, R.R.; Carlson, W.R. Time course study of the chromosome-type breakage-fusion-bridge cycle in maize. *Genetics* **1999**, *153*, 1435–1444. [PubMed]

33. Han, F.; Gao, Z.; Birchler, J.A. Reactivation of an inactive centromere reveals epigenetic and structural components for centromere specification in maize. *Plant Cell* **2009**, *21*, 1929–1939. [CrossRef] [PubMed]

34. Rosin, L.F.; Mellone, B.G. Centromeres drive a hard bargain. *Trends Genet.* **2017**, *33*, 101–117. [CrossRef] [PubMed]

35. Fukagawa, T.; Earnshaw, W.C. Neocentromeres. *Curr. Biol.* **2014**, *24*, R946–R947. [CrossRef] [PubMed]

36. Sullivan, L.L.; Maloney, K.A.; Towers, A.J.; Gregory, S.G.; Sullivan, B.A. Human centromere repositioning within euchromatin after partial chromosome deletion. *Chromosome Res.* **2016**, *24*, 451–466. [CrossRef] [PubMed]

37. Shang, W.H.; Hori, T.; Martins, N.M.; Toyoda, A.; Misu, S.; Monma, N.; Hiratani, I.; Maeshima, K.; Ikeo, K.; Fujiyama, A.; et al. Chromosome engineering allows the efficient isolation of vertebrate neocentromeres. *Dev. Cell* **2013**, *24*, 635–648. [CrossRef] [PubMed]

38. Zhang, B.; Lv, Z.; Pang, J.; Liu, Y.; Guo, X.; Fu, S.; Li, J.; Dong, Q.; Wu, H.J.; Gao, Z.; et al. Formation of a functional maize centromere after loss of centromeric sequences and gain of ectopic sequences. *Plant Cell* **2013**, *25*, 1979–1989. [CrossRef] [PubMed]

39. Liu, Y.; Su, H.; Zhang, J.; Liu, Y.; Han, F.; Birchler, J.A. Dynamic epigenetic states of maize centromeres. *Front. Plant Sci.* **2015**, *6*, 904. [CrossRef] [PubMed]

40. Koo, D.H.; Han, F.; Birchler, J.A.; Jiang, J. Distinct DNA methylation patterns associated with active and inactive centromeres of the maize B chromosome. *Genome Res.* **2011**, *21*, 908–914. [CrossRef] [PubMed]

41. Zhang, W.; Lee, H.R.; Koo, D.H.; Jiang, J. Epigenetic modification of centromeric chromatin: Hypomethylation of DNA sequences in the CENH3-associated chromatin in *Arabidopsis thaliana* and Maize. *Plant Cell* **2008**, *20*, 25–34. [CrossRef] [PubMed]

42. Su, H.; Liu, Y.; Liu, Y.X.; Lv, Z.; Li, H.; Xie, S.; Gao, Z.; Pang, J.; Wang, X.J.; Lai, J.; et al. Dynamic chromatin changes associated with de novo centromere formation in maize euchromatin. *Plant J.* **2016**, *88*, 854–866. [CrossRef] [PubMed]

43. Topp, C.N.; Zhong, C.X.; Dawe, R.K. Centromere-encoded RNAs are integral components of the maize kinetochore. *Proc. Natl. Acad. Sci. USA* **2004**, *101*, 15986–15991. [CrossRef] [PubMed]

44. Wong, L.H.; Brettingham-Moore, K.H.; Chan, L.; Quach, J.M.; Anderson, M.A.; Northrop, E.L.; Hannan, R.; Saffery, R.; Shaw, M.L.; Williams, E.; et al. Centromere RNA is a key component for the assembly of nucleoproteins at the nucleolus and centromere. *Genome Res.* **2007**, *17*, 1146–1160. [CrossRef] [PubMed]

45. Chueh, A.C.; Northrop, E.L.; Brettingham-Moore, K.H.; Choo, K.H.; Wong, L.H. Line retrotransposon RNA is an essential structural and functional epigenetic component of a core neocentromeric chromatin. *PLoS Genet.* **2009**, *5*, e1000354. [CrossRef]

46. Chan, F.L.; Wong, L.H. Transcription in the maintenance of centromere chromatin identity. *Nucleic Acids Res.* **2012**, *40*, 11178–11188. [CrossRef] [PubMed]

47. Bouzinba-Segard, H.; Guais, A.; Francastel, C. Accumulation of small murine minor satellite transcripts leads to impaired centromeric architecture and function. *Proc. Natl. Acad. Sci. USA* **2006**, *103*, 8709–8714. [CrossRef] [PubMed]

48. Rosic, S.; Kohler, F.; Erhardt, S. Repetitive centromeric satellite RNA is essential for kinetochore formation and cell division. *J. Cell Biol.* **2014**, *207*, 335–349. [CrossRef] [PubMed]

49. Blower, M.D. Centromeric transcription regulates Aurora-B localization and activation. *Cell Rep.* **2016**, *15*, 1624–1633. [CrossRef] [PubMed]

50. Grenfell, A.W.; Strzelecka, M.; Heald, R. Transcription brings the complex(ity) to the centromere. *Cell Cycle* **2017**, *16*, 235–236. [CrossRef] [PubMed]

51. Da Ines, O.; White, C.I. Centromere associations in meiotic chromosome pairing. *Ann. Rev. Genet.* **2015**, *49*, 95–114. [CrossRef] [PubMed]

52. Unhavaithaya, Y.; Orr-Weaver, T.L. Centromere proteins CENP-C and CAL1 functionally interact in meiosis for centromere clustering, pairing, and chromosome segregation. *Proc. Natl. Acad. Sci. USA* **2013**, *110*, 19878–19883. [CrossRef] [PubMed]

53. Watanabe, Y. Geometry and force behind kinetochore orientation: Lessons from meiosis. *Nat. Rev. Mol. Cell Biol.* **2012**, *13*, 370–382. [CrossRef] [PubMed]

54. Kaszas, E.; Birchler, J.A. Misdivision analysis of centromere structure in maize. *EMBO J.* **1996**, *15*, 5246–5255. [CrossRef] [PubMed]

55. Iwata-Otsubo, A.; Dawicki-McKenna, J.M.; Akera, T.; Falk, S.J.; Chmatal, L.; Yang, K.; Sullivan, B.A.; Schultz, R.M.; Lampson, M.A.; Black, B.E. Expanded satellite repeats amplify a discrete CENP-A nucleosome assembly site on chromosomes that drive in female meiosis. *Curr. Biol.* **2017**, *27*, 2365–2373e8. [CrossRef] [PubMed]

56. Han, F.; Lamb, J.C.; McCaw, M.E.; Gao, Z.; Zhang, B.; Swyers, N.C.; Birchler, J.A. Meiotic studies on combinations of chromosomes with different sized centromeres in maize. *Front. Plant Sci.* **2018**, *9*, 785. [CrossRef] [PubMed]

57. Han, F.; Gao, Z.; Yu, W.; Birchler, J.A. Minichromosome analysis of chromosome pairing, disjunction, and sister chromatid cohesion in maize. *Plant Cell* **2007**, *19*, 3853–3863. [CrossRef] [PubMed]

58. Liu, Y.; Su, H.; Pang, J.; Gao, Z.; Wang, X.J.; Birchler, J.A.; Han, F. Sequential de novo centromere formation and inactivation on a chromosomal fragment in maize. *Proc. Natl. Acad. Sci. USA* **2015**, *112*, E1263–E1271. [CrossRef] [PubMed]

59. Liu, Y.; Su, H.; Liu, Y.; Zhang, J.; Dong, Q.; Birchler, J.A.; Han, F. Cohesion and centromere activity are required for phosphorylation of histone H3 in maize. *Plant J.* **2017**, *92*, 1121–1131. [CrossRef] [PubMed]

60. Su, H.; Liu, Y.; Dong, Q.; Feng, C.; Zhang, J.; Liu, Y.; Birchler, J.A.; Han, F. Dynamic location changes of Bub1-phosphorylated-H2AThr133 with CENH3 nucleosome in maize centromeric regions. *New Phytol.* **2017**, *214*, 682–694. [CrossRef] [PubMed]

61. Zhang, H.; Dawe, R.K. Total centromere size and genome size are strongly correlated in ten grass species. *Chromosome Res.* **2012**, *20*, 403–412. [CrossRef] [PubMed]

62. Yan, H.; Talbert, P.B.; Lee, H.R.; Jett, J.; Henikoff, S.; Chen, F.; Jiang, J. Intergenic locations of rice centromeric chromatin. *PLoS Biol.* **2008**, *6*, e286. [CrossRef] [PubMed]

63. Wang, N.; Dawe, R.K. Centromere size and its relationship to haploid formation in plants. *Mol. Plant* **2017**, *11*, 398–406. [CrossRef] [PubMed]

64. Wang, K.; Wu, Y.; Zhang, W.; Dawe, R.K.; Jiang, J. Maize centromeres expand and adopt a uniform size in the genetic background of oat. *Genome Res.* **2014**, *24*, 107–116. [CrossRef] [PubMed]

65. Dong, Z.; Yu, J.; Li, H.; Huang, W.; Xu, L.; Zhao, Y.; Zhang, T.; Xu, W.; Jiang, J.; Su, Z.; et al. Transcriptional and epigenetic adaptation of maize chromosomes in oat-maize addition lines. *Nucleic Acids Res.* **2018**, *46*, 5012–5028. [CrossRef] [PubMed]

66. Kaszas, E.; Birchler, J.A. Meiotic transmission rates correlate with physical features of rearranged centromeres in maize. *Genetics* **1998**, *150*, 1683–1692. [PubMed]

© 2018 by the authors. Licensee MDPI, Basel, Switzerland. This article is an open access article distributed under the terms and conditions of the Creative Commons Attribution (CC BY) license (http://creativecommons.org/licenses/by/4.0/).

genes

MDPI

Article

B Chromosome System in the Korean Field Mouse *Apodemus peninsulae* Thomas 1907 (Rodentia, Muridae)

Yuri M. Borisov [1] and **Igor A. Zhigarev** [2,*]

[1] Severtzov Institute of Ecology and Evolution, Russia Academy of Sciences, Moscow 119071, Russia; boriss-spb@yandex.ru

[2] Institute of Biology and Chemistry, Moscow State University of Education (MSPU), Moscow 129164, Russia

* Correspondence: ia.zhigarev@mpgu.edu; Tel.: +7-925-837-3070

Received: 30 August 2018; Accepted: 20 September 2018; Published: 27 September 2018

Abstract: In this paper, we analyzed B chromosome variation in Korean field mouse *Apodemus peninsulae* Thomas 1907 (Rodentia, Muridae) based on a 40-year study of karyotypes collected from geographically distant populations in East Siberia, North Mongolia, China, the Russian Far East, and Japan (Hokkaido). We developed the database of individual variants of B chromosome systems in *A. peninsulae*. In Siberian populations all animals had Bs. The karyotypes of the studied animals contain from 1 to 30 Bs differing in size and morphology. Analysis of B chromosome systems in 598 individuals from different localities of the range shows the presence of 286 variants of Bs combinations in these animals. Unique sets of B morphotypes make up most of these variants ($64.7 \pm 1.3\%$), probably suggesting that individual Bs systems normally result from stochastic processes in the populations. The proportion of animals with a large number of Bs gradually decreases, probably due to increased complexities in the inheritance of large numbers of Bs. *A. peninsulae* is thus proposed as a good model for studying the origin and evolution of extra elements in the karyotype.

Keywords: B chromosomes; dot-like (micro) Bs; karyotypic characteristics; Bs; B morphotypes; *Apodemus peninsulae*

1. Introduction

The Korean field mouse (*Apodemus peninsulae* Thomas, 1907) is widely distributed from East Siberia and North Mongolia, the Russian Far East to China, Korea, and Japan (Hokkaido) (Figure 1). *A. peninsulae* belongs to the genus *Apodemus*, in which six species have been shown to carry B chromosomes [1]. *A. peninsulae* shows one of the widest spectra of Bs variability among animals. Through the wide geographical range mice karyotypes contain from 48 to 78 chromosomes and the vast majority of individuals of this species have supernumerary (B) chromosomes [2–17]. The only population lacking B chromosomes is known from Sakhalin Island, Russia [10]. The Korean field mouse B chromosomes highly vary in morphology. Most mammals with Bs usually have B chromosomes of one type, such as acrocentric in *Apodemus (Sylvaemus) flavicollis* Melchior, 1834 or metacentric in *Rattus rattus* Linnaeus, 1758 [18]. In *A. peninsulae* up to five morphotypes were revealed [19].

Polymorphism of B chromosomes in *A. peninsulae* was found in the 1970s while investigating mice karyotypes from Hokkaido, Japan [20]. A range from small dot-like to large metacentric chromosomes was discovered. However, a relative homogeneity of Bs including only small and medium metacentric chromosomes was found in the mainland part of the species range (from Altai to Primorsky Region) [2,4,11]. Further expanding of catching localities in Central Siberia has demonstrated that mainland populations could also have a high variety of B chromosomes [12,14] leading to onward investigation of geographic variability of B morphotypes in the Korean field mouse [1,5–7,14,17,19,21–26].

The accumulated knowledge of morphological systems of B chromosomes in *A. peninsulae* raises new questions: how to estimate population variability of the species through patterns of its B chromosome variability; what meaning B chromosome morphology could have related to its molecular features; how B chromosomes are originated and inherited?

For over 40 years of the current study nearly 600 individuals of *A. peninsulae* from different parts of its range were karyotyped, making it possible to create an extensive database (http://sev-in.ru/ru/bdhromosomes-apodemus) and to analyze extra chromosome morphotypes in various geographically distant local populations.

The purpose of the study was to identify common statistical patterns in the distribution of *A. peninsulae* B morphotypes in various populations of a wide range.

The Korean field mouse has thus become a good mammalian model for studies of evolutionary dynamics and effects of Bs on the host genome. The aim of this paper is to report new data on B chromosome distribution in local populations of the Korean field mouse *A. peninsulae* (Rodentia, Muridae) that would determine future directions for investigations.

2. Materials and Methods

We examined chromosomal data in 598 *A. peninsulae* individuals collected at 39 local populations in Russia, Mongolia, China, Korea, and Japan. 418 individuals from 30 localities were collected in Siberia (Central Siberia, Altai, Khakassia, Tyva, Baikal Lake region and Buryatia); 94 individuals were collected from Primorsky krai (the Russian Far East); 60 individuals were collected in Mongolia; 8 individuals were collected in Gansu province of China. Data on B chromosomes were partially published earlier [9,12,19,21]. The study protocol was approved by the Ethics Committee of the Severtsov Institute of Ecology and Evolution Russian Academy of Sciences (2017-03-17). All experiments with mice were performed in accordance with the rules approved by the European Convention for the Protection of Vertebrate Animals used for Experimental and other Scientific Purposes. We also used published data on 18 individuals from Hokkaido, Japan [20] (Figure 1, Table 1).

Figure 1. Geographical distribution of *A. peninsulae* and collection numbers. Arabic numbers beside points are location numbers shown in Table 1.

Table 1. B chromosome characteristics in the Korean field mouse *Apodemus peninsulae* Thomas, from different localities. For geographical map of localities see Figure 1. mB Index: mass quantity of B chromosomes.

No. of Locality	Locality	No. of Animals	No. of B Chromosomes			mB Index (M ± SE)
			Total	Macro	Micro	
		Russia				
1	Novosibirsk Region	20	3–14	0–11	0–6	15.8 ± 1.4
2	Kemerovo Region	10	4–12	3–8	0–4	18.7 ± 2.1
3–6	The Altai Republic	146	1–10	1–10	0–3	14.5 ± 0.5
7	Angara mouth (between the left bank of the Angara River and the right bank of the Yenisei River)	15	6–9	6–9	0	18.5 ± 1.1
8	Central Siberia, the north of Krasnoyarsk territory, (the left bank of the Yenisei River)	26	4–30	0	4–30	18.1 ± 1.1
9	Central Siberia, the south of Krasnoyarsk territory, (the right bank of the Yenisei River)	40	3–18	0–6	1–14	11.2 ± 0.7
10–11	Central Siberia, the east of Krasnoyarsk territory, Zelenogorsk	5	8–17	3–4	5–13	17.6 ± 1.5
12–14	Central Siberia, Krasnoyarsk	47	1–10	0–4	0–8	7.7 ± 0.5
15–18	The south of Republic of Khakassia	30	1–10	1–6	0–7	8.3 ± 0.7
19–20	The north of Republic of Khakassia	3	5–9	5–9	0–1	23.0 ± 4.6
21	The Tyva Republic	2	12–15	2	10–13	15.5 ± 1.5
22–24	Western Baikal (Irkutsk Region, the western shore of the lake Baikal)	28	1–3	1–3	0	5.9 ± 0.4
25–30	Southern Baikal (The Republic of Buryatia)	46	2–13	0–5	0–12	11.3 ± 0.5
36–38	Primorye (Primorsky krai)	94	0–5	0–5	0–1	4.1 ± 0.3
		Mongolia				
31–33	Northern Mongolia (West Khentei)	59	2–11	0–6	0–9	7.2 ± 0.4
34	The Great Khingan	1	1	1	0	3.0
		China				
35	The Gansu Province	8	7–14	0–3	6–11	12.4 ± 1.5
		Japan				
39	Island of Hokkaido	18	3–13	0–5	1–9	10.3 ± 0.9
	Total	598	0–30	0–11	0–30	10.7 ± 0.3

Chromosome preparations were obtained from bone marrow and spleen cells after a routine technique with colchicine treatment [27]. At least 20 metaphase plates from each individual were taken for karyotype analysis. In this paper, we used only modal number of chromosomes; for mosaic animals (with one or none B chromosomes) we took those with one B.

Our own developed formula was applied for morphotype numerical coding of B chromosomes [19]. The first number indicates the amount of Bs. The second number indicates B chromosomes of I class: large bi-armed chromosomes equal to 1–8 pairs of A chromosomes (large metacentrics). The third number indicated B chromosomes of II class: medium sized bi-armed chromosomes equal to 9–16 pairs of A chromosomes (medium metacentrics). The forth number indicates B chromosomes of III class: small bi-armed chromosomes equal to 17–23 pairs of A chromosomes (small metacentrics). The fifth number indicates B chromosomes of IV class: small acrocentric chromosomes equal to 17–23 pairs of A chromosomes. Finally, the sixth number indicates B chromosomes of V class or dot-like chromosomes (micro Bs). For example, the formula 5.1.1.1.1.1

means that an individual has five B chromosomes, one in each of five classes. In our opinion, both the ratio of different B morphotypes and the variety of total amount of chromatin in B chromosomes are highly significant to reveal the origin and specific existence of B chromosomes [22,23]. To estimate the amount of chromatin, we used conditional mass quantity of B chromosomes (mB index), developed by G.V. Roslik and I.V. Kartavtseva [13]. In this case, one conditional point is assigned to each size class: one point to dot-like (micro Bs) (V class), two points to larger and relatively similar in size chromosomes of IV and III classes, three points to II class and four points to I class chromosomes. The sum of points demonstrates a certain characteristic of Bs mass quantity.

Localities were aggregated according to distances and presence/absence of geographical, especially water, barriers between them.

The statistical analysis using standard techniques was conducted in Statistica 8.0 Software [28].

3. Results and Discussion

The analysis of 598 *A. peninsulae* karyotypes collected at 39 local populations in Russia, Mongolia, China, and Japan revealed presence of B chromosomes in 97.7% of the species population (Table 1). 12 individuals from the Russian Far East (12.8% of individuals collected in Primorsky krai) are the only exception with no B chromosomes in their karyotypes (localities no. 36–38, Figure 1). Table 1 summarizes the B chromosome data, including their number and morphology, in each examined locality. We also calculated the frequency of animals with macro and micro B chromosomes using our own information (Table 2).

Table 2. Frequency of *A. peninsulae* individuals with different macro and micro B chromosomes in studied local populations.

Chromosome Type	Frequency of Individuals (P)	±SE
Large metacentric	0.28	0.018
Medium metacentric	0.61	0.020
Small metacentric	0.62	0.019
Acrocentric	0.12	0.013
Dot-like micro chromosome	0.53	0.020
No extra chromosomes	0.02	0.006

The standard (A) diploid set of *A. peninsulae* contains 48 acrocentric chromosomes gradually decreasing in size (Figures 2 and 3a). In addition, up to 30 B chromosomes may be found in some individuals. In this paper, we divide B chromosomes into two groups according to their size. The first group includes Bs of visible morphology under light microscopy (macro Bs), which are larger or equal in size to the smallest A chromosome. The second group includes only dot-like chromosomes which are much smaller than A chromosomes and without clear morphology (micro Bs). The macro B chromosomes are divided into classification types according to their morphology and relative size in comparison with the A chromosomes (Figure 3b). Despite high variability of B chromosomes, A chromosome aberrations were not found.

We have previously described four types of macro Bs [12]: (1) large metacentrics or submetacentrics (Lm-sm); (2) medium-to-small metacentrics or submetacentrics (Mm-sm); (3) small metacentric (mm); (4) large-to-small acrocentrics or subtelocentrics (A-St). The most frequent macro Bs in *A. peninsulae* are Mm-sm (Figure 3a), whereas A-St morphological variants of medium or small size are rare. In some cases, we could identify the morphology of micro Bs in good quality metaphase plates, but in most cases micro Bs resembled very small structures without clear morphology.

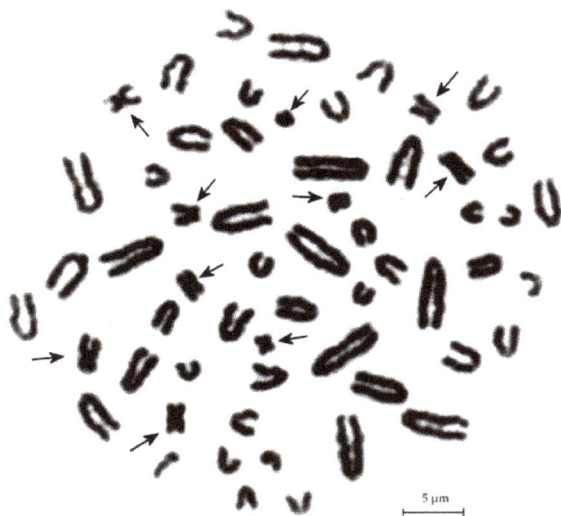

Figure 2. Metaphase plates of *A. peninsulae* with 10 different Bs.

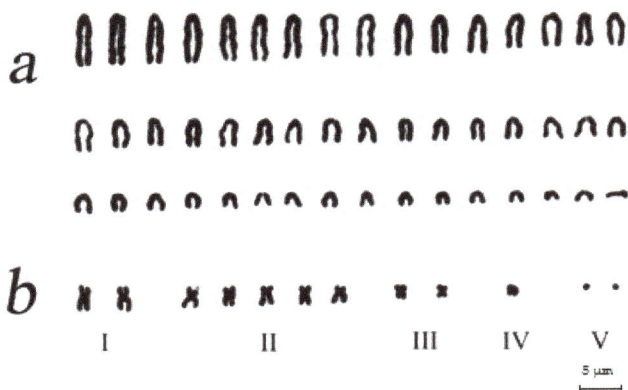

Figure 3. Karyotype of *A. peninsulae*: (**a**) the main set of 48 acrocentric A chromosomes and (**b**) variants of five (I–V) classes of B chromosomes. Formulas of Bs are 12.2.5.2.1.2.

We suppose that dot-like (micro Bs) chromosomes are initial and originated by duplication of centromeric areas of A chromosomes with further reorganization into macro Bs through DNA amplification [22,23].

The frequencies of *A. peninsulae* individuals with one or another extra chromosome type are not equal in nature. Animals carrying even one medium or small metacentric B chromosome are more frequent (Table 2). Dot-like Bs also have relatively high frequency. In contrast, the frequency of large metacentrics across the Korean field mouse range is substantially lower; less than one third of mice caught in nature would have these chromosomes ($p = 0.28 \pm 0.018$). Acrocentric B chromosomes, in their turn, are extremely rare. The similar tendency is shown for the Far East population of *A. peninsulae* based on the analysis of 367 individuals [13].

The number of large metacentric B chromosomes carried by a single karyotype varies from 0 to 5 across local populations. Their maximum number (4–5) occurs extremely rarely: only two such individuals were recorded in Siberia (Altai and Khakassia). Some mice from Altai, Novosibirsk and West Khentei (Mongolia) carried three large chromosomes, though one or two large B chromosome were met much more frequently. Meanwhile, the majority of examined individuals (72.2 ± 1.83%) did not have chromosomes of that type at all (Table 3).

Table 3. Proportion of *A. peninsulae* individuals with different number of B chromosomes (%, ±SE) over the species range.

Number of B Chromosomes	Proportion of Individuals with B Chromosomes				
	Large Metacentric	Medium Metacentric	Small Metacentric	Acrocentric	Dot-Like (micro Bs)
0	72.2 ± 1.83	39.1 ± 2.00	37.8 ± 1.98	88.13 ± 1.32	47.5 ± 2.04
1	16.1 ± 1.50	25.4 ± 1.78	30.8 ± 1.90	8.9 ± 1.16	14.7 ± 1.45
2	8.7 ± 1.15	22.1 ± 1.70	19.4 ± 1.62	2.51 ± 0.64	7.36 ± 1.07
3	2.7 ± 0.70	6.7 ± 1.00	6.52 ± 1.00	0.17 ± 0.17	5.52 ± 0.93
4	0.17 ± 0.17	3.34 ± 0.74	3.5 ± 0.75	0.17 ± 0.17	4.18 ± 0.82
5	0.17 ± 0.17	2.0 ± 0.57	1.0 ± 0.41	0.17 ± 0.17	3.18 ± 0.72
6	-	0.84 ± 0.37	0.84 ± 0.37	-	2.17 ± 0.60
7	-	0.33 ± 0.23	-	-	1.67 ± 0.52
8	-	0.17 ± 0.17	0.17 ± 0.17	-	3.18 ± 0.72
9	-	-	-	-	2.34 ± 0.62
10	-	-	-	-	2.17 ± 0.60
11	-	-	-	-	0.84 ± 0.37
12	-	-	-	-	0.84 ± 0.37
13	-	-	-	-	0.33 ± 0.24
14	-	-	-	-	0.84 ± 0.37
15	-	-	-	-	0.17 ± 0.17
16	-	-	-	-	0.84 ± 0.37
18	-	-	-	-	0.67 ± 0.33
20	-	-	-	-	0.5 ± 0.29
22	-	-	-	-	0.33 ± 0.24
24	-	-	-	-	0.33 ± 0.24
26	-	-	-	-	0.33 ± 0.24
30	-	-	-	-	0.17 ± 0.17

The number of medium and small metacentric B chromosomes varies from 0 to 8 across different populations. Unlike macro metacentric chromosomes, their presence in populations is more significant as over 60% of karyotypes contain Bs of these types (Table 3). A tendency of decreasing proportion of individuals with increasing number of B chromosomes remains (Figure 4). The maximum number of medium (8) and small (8) metacentric chromosomes was recorded only once in karyotypes from left bank of Angara River and Novosibirsk, respectively. Mice with a small number of Bs (1–3) were recorded more often.

A tendency of distribution of acrocentric B chromosomes (IV class) in populations of *A. peninsulae* is similar with that of large metacentric chromosomes (Table 3, Figure 4). The maximum of acrocentric chromosomes (5) was found in a single karyotype from Novosibirsk area.

The number of extra dot-like chromosomes (micro Bs) in different populations varies from 0 to 30 (Table 3). Nearly half of examined Korean field mice (47.5 ± 2.04%) had no dot-like Bs. The proportion of mice with many dot-like Bs gradually decreased as in case of large B chromosomes (Figure 5).

It is necessary to emphasize that the correlation between the number and the frequency of chromosomes is rather high, negative, and significant in all cases ($p < 0.05$) (Table 4). This may indicate an increase in complexity of inheritance with an increase in the number of B chromosomes in the karyotype, particularly in case of large Bs (I, II and III classes).

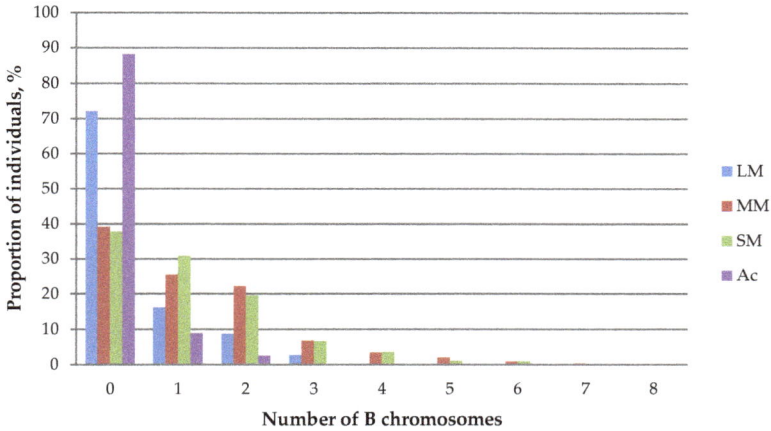

Figure 4. Ratio of *A. peninsulae* individuals with different numbers of B chromosomes across studied populations. LM: large metacentrics; MM: medium metacentric SM: small metacentric Ac: acrocentric.

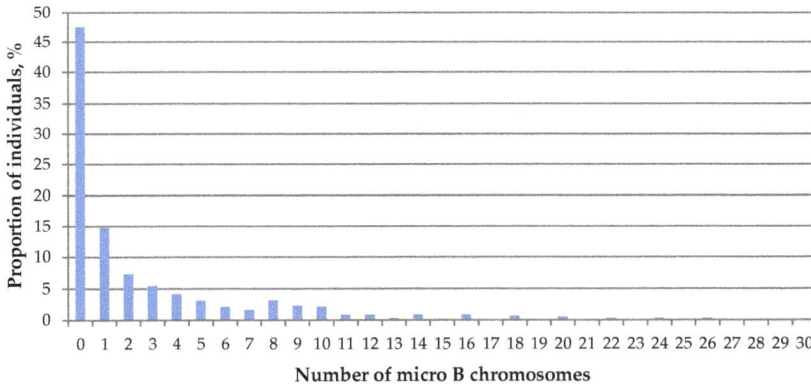

Figure 5. Ratio of *A. peninsulae* karyotypes with a certain number of micro B chromosomes across studied populations.

Table 4. Correlation between the number and frequency of B chromosomes in karyotypes of *A. peninsulae* (Spearman's rank test, $p < 0.05$).

	Type of B Chromosome				
	Large Metacentric	Medium Metacentric	Small Metacentric	Acrocentric	Dot-Like (Micro Bs)
r_s	−0.986	−1.00	−0.98	−0.94	−0.9

The frequency of *A. peninsulae* karyotypes with different numbers of all B chromosomes differs in some way from the frequencies of certain classes (Figure 6). Individuals with two B chromosomes are recorded much more frequently (16.39 ± 2.30%) than those with only one B chromosome type (4.85 ± 0.77%). In general, two thirds of mice in different local populations have from two to seven B chromosomes. Thus, we suppose that not just the presence of extra chromosomes, but the presence of several (not one) Bs in *A. peninsulae* karyotypes is normal for this species.

Overall, even and odd numbers of B chromosomes in a karyotype is not statistically equal (Pearson Chi-square test, $\chi^2 = 6.64$, $p < 0.05$). Individuals with an even number of Bs are recorded significantly more often (57.5%) that indirectly indicates the pattern of inheritance or chromosome division-merging processes.

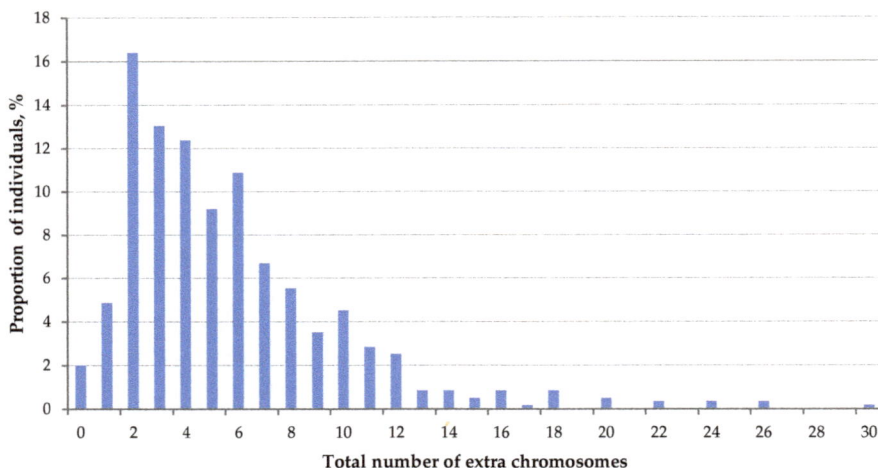

Figure 6. Ratio of *A. peninsulae* karyotypes with total number of extra chromosomes regardless of the type across examined local populations.

The analysis of extra chromosome systems of 598 *A. peninsulae* individuals from different parts of the species range demonstrates the high variety of B combinations (286). Wherein, the unique Bs combinations are dominated (64.7 ± 1.3%, Table 5) that is likely due to the prevalence of stochastic processes in developing of individual Bs systems. Almost the third of examined animals (31%) carries the unique combinations of B chromosomes.

Table 5. The number and ratio of B chromosome systems among *A. peninsulae* karyotypes across the species range.

Frequency of B Chromosome System across the Range	Number of Combinations	%	SE, %
Once (unique)	185	64.7	1.3
Twice	49	17.1	0.7
3 times	23	8.0	0.5
4 times	11	3.8	0.3
5 times	4	1.4	0.2
6 times	2	0.7	0.1
7 times	2	0.7	0.1
8 times	2	0.7	0.1
9 times	1	0.3	0.1
12 times	2	0.7	0.1
13 times	1	0.3	0.1
15 times	1	0.3	0.1
22 times	1	0.3	0.1
26 times	1	0.3	0.1
31 times	1	0.3	0.1
Total number of Bs systems	286		

The unique combinations of B chromosomes in mice karyotypes are met more frequently in local populations in Central Siberia and Japan (Figure 7). On the other hand, mice with repeated variants (non-unique) dominated in some well-studied populations such as local populations from the Far East (localities 36–38, *n* = 94), Western Baikal (localities 22–24, *n* = 28) and the left bank of Yenisei River (the locality 8, *n* = 26) (Figure 7).

Figure 7. The proportion of animals in *A. peninsulae* local populations carrying unique B chromosome combinations (dark shading). Arabic numbers beside points are location numbers shown in Table 1.

Moreover, combinations with few numbers of Bs (0–4) are more frequent (the absence of B chromosomes was considered as one of 286 variants). The number of chromosomes in a combination negatively correlated with its frequency (Spearman's rank test, $r_s = -0.33$, $p < 0.05$). In other words, the more there are B chromosomes in a karyotype, the less these combinations repeat across the species range and more often become unique. The most repeated was the combination of two small metacentrics recorded 31 times in different geographical regions: Khakassia, Western Baikal, Northern Mongolia and Primorye. The combination of one medium and one small metacentric was recorded 26 times and the combination of two medium metacentrics was recorded 15 times in Altai, Khakassia, Western Baikal, Northern Mongolia and Primorye populations. The variant with one small metacentric was found 22 times Khakassia, Western Baikal and Primorye populations. Combinations recorded twice constitute a considerable part ($17.1 \pm 0.7\%$).

The analysis of conditional mass quantity of B chromosomes (mB index) enabled us to compare samples with different amounts of extra chromatin in mice cells from different geographical populations. The lowest values were recorded in Primorsky krai (4.1 ± 0.3, Table 1) and the highest values were recorded in Kemerovo region and the south of Krasnoyarsk region (mB index > 18). Overall, the average mB index value across the western part of the *A. peninsulae* range (Central Siberia; 13.4 ± 0.35) almost twice higher than that across the eastern and southern parts (Baikal region, the Russian Far East, Japan, Mongolia, and China; 7.0 ± 0.26) (Student's *t*-test, $t = 13.94$, $p < 0.05$). Meanwhile, neighboring local populations in Central Siberia could also substantially differ from each other by mB index. For instance, *A. peninsulae* populations from Krasnoyarsk region (localities 12–14, Figure 1) and the right bank of Yenisei River (locality 9) have relatively low mB index values (7.7 ± 0.5 and 11.2 ± 0.7, respectively, Table 1), although they are surrounded by the local populations with very high mB index values (14.5–18.7). A similar situation is found in Khakassia, where the mB index reaches 23.0 ± 4.6 ($n = 3$) in the north and 8.3 ± 0.7 in the south (Student's *t*-test, $t = 5.5$, $p < 0.05$).

The patterns of frequency distribution of mB index values (Figure 8) further reinforce the idea of adaptability of not just presence of B chromosomes but the presence of their certain mass in

A. peninsulae cells. Mice carrying both very low and very high mB index value were rarely met in examined populations. More than half of individuals have mB index between 4 and 11.

In view of ecological prosperity of *A. peninsulae*, having wide species range and high adaptive potential, occupying various habitats, we assume that high Bs variety at least does not deteriorate the species adaptability. It is testified both by the fact that almost all karyotypes carry Bs and that Bs total masses within cells (mB index) have mainly average values. Many issues are likely to be removed after discovering Bs active genes or ascertaining an interaction system between A and B chromosomes. For instance, ribosomal genes of *A. peninsulae* have been already found [23].

From the other point of view, high polymorphism of quantity and composition of B chromosomes does not support the idea of strong control by the selection. The high proportion of unique B combinations evidently confirms it.

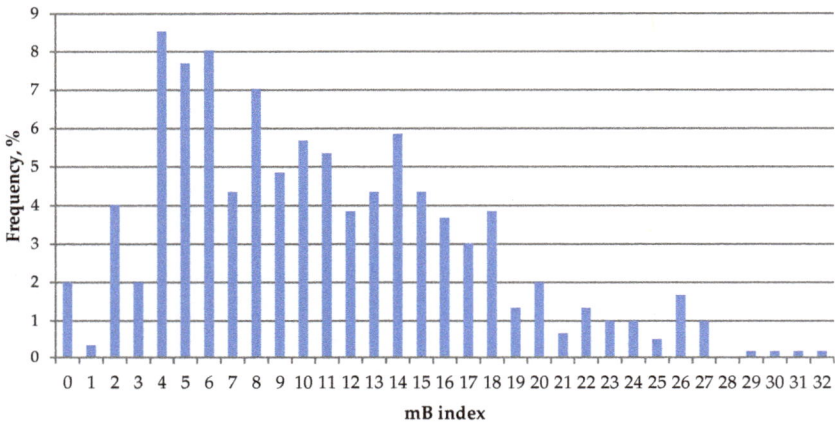

Figure 8. Frequency distribution of mB index values across studied *A. peninsulae* local populations.

Thus, the specific features and the structure of B chromosomes, found in our research, make *A. peninsulae* a promising model for studying mammalian Bs. Further investigation of B chromosome population variability and its relation to other phenomenon of chromosome instability of DNA sequences in A and B chromosomes would enable the clarification of microevolutional processes in mammals.

Author Contributions: Conceptualization, Y.M.B. and I.A.Z.; methodology, Y.M.B.; investigation, I.A.Z.; resources, Y.M.B.; writing—original draft preparation, I.A.Z and Y.M.B.; writing—review and editing, Y.M.B. and I.A.Z.; visualization, I.A.Z., data curation, I.A.Z.

Funding: This research received no external funding.

Conflicts of Interest: The authors declare no conflict of interest.

References

1. Wójcik, J.M.; Wójcik, A.M.; Macholan, M.; Pialek, J.; Zima, J. The mammalian model for population studies of B chromosomes: The wood mouse (*Apodemus*). *Cytogenet. Genome Res.* **2004**, *106*, 264–270. [CrossRef] [PubMed]

2. Bekasova, T.S.; Vorontsov, N.N.; Korobitsyna, K.V.; Korablev, V.P. B-chromosomes and comparative karyology of the mice of the genus *Apodemus. Genetica* **1980**, *52*, 33–43. [CrossRef]

3. Bekasova, T.S.; Vorontsov, N.N. Populational chromosome polymorphism in Asiatic Forest mice *Apodemus peninsulae. Russ. J. Genet.* **1975**, *11*, 89–94.

4. Bekasova, T.S. B-chromosomes of Asiatic wood mice *Apodemus peninsulae* (Rodentia, Muridae). *Voprosy Izmenchivosti i Zoogeographii Mlekopitaushih* **1984**, 14–29.

5. Borbiev, T.E.; Kolomiets, O.L.; Borisov, Y.M.; Safronova, L.D.; Bogdanov, Y.F. Synaptonemal complexes of A and B-chromosomes in spermatocytes of the East Asiatic mouse Apodemus peninsulae. *Tsitologiya* **1990**, *32*, 193–196.

6. Borisov, Y.M.; Afanas'ev, A.G.; Lebedev, T.T.; Bochkarev, M.N. Multiplicity of B microchromosomes in a Siberian population of mice *Apodemus peninsulae* (2n = 48 + 4 − 30 B-chromosomes). *Russ. J. Genet.* **2010**, *46*, 705–711. [CrossRef]

7. Borisov, Y.M.; Malygin, V.M. Cline variability of the B chromosome system in *Apodemus peninsulae* (Rodentia, Muridae) from the Buryatia and Mongolia. *Tsitologiya* **1991**, *33*, 106–111.

8. Borisov, Y.M. Instability of B-chromosomes in somatic and germline cells of *Apodemus peninsulae*. *Russ. J. Genet.* **2008**, *44*, 959–965. [CrossRef]

9. Borisov, Y.M.; Sheftel, B.I.; Safronova, L.D.; Aleksandrov, D.Y. Stability of the population variants of the B-chromosome system in the East-Asian Mouse *Apodemus peninsulae* from the Baikal region and Northern Mongolia. *Russ. J. Genet.* **2012**, *48*, 1020–1028. [CrossRef]

10. Kartavtseva, I.V.; Roslik, G.V. A complex B chromosome system in the Korean field mouse, *Apodemus peninsulae*. *Cytogenet. Genome Res.* **2004**, *106*, 271–278. [CrossRef] [PubMed]

11. Kral, B. Chromosome characteristics of certain murine rodents (Muridae) of the Asiatic part of the USSR. *Zool. Listy* **1971**, *20*, 331–347.

12. Radjabli, C.I.; Borisov, Y.M. Variants of the system of supernumerary chromosomes in continental forms of *Apodemus peninsulae* (Rodentia, Muridae). *Dokl. Acad. Nauk SSSR* **1979**, *248*, 979–981.

13. Roslik, G.V.; Kartavtseva, I.V. B chromosome morphotypes of *Apodemus peninsulae* (Rodentia) from the Russian Far East. *Tsitologiia* **2012**, *54*, 66–77. [PubMed]

14. Volobuev, V.T. Karyological analysis of three Siberian populations of Asiatic wood mouse, *Apodemus peninsulae* (Rodentia, Muridae). *Dokl. AN SSSR* **1979**, *248*, 1452–1454.

15. Vorontsov, N.N.; Bekasova, T.S.; Kral, B.; Korobitsyna, K.V.; Ivanitskaya, E.Y. On specific status of Asian wood mice of genus *Apodemus* (Rodentia, Muridae) from Siberia and Far East. *Zool. Zhurnal* **1977**, *56*, 437–450.

16. Wang, J.; Zhao, X.; Qi, H.; Koh, H.S.; Zhang, L.; Guan, Z.; Wang, C.H. Karyotypes and B chromosomes of *Apodemus peninsulae* (Rodentia, Mammalia). *Acta Theriol. Sin.* **2000**, *20*, 289–296.

17. Zima, J.; Macholan, M. B-chromosomes in the wood mice (genus *Apodemus*). *Acta Theriol.* **1995**, *40*, 75–86. [CrossRef]

18. Vujosevic, M.; Blagojevic, J. B chromosomes in populations of mammals. *Cytogenet. Genome Res.* **2004**, *106*, 247–256. [CrossRef] [PubMed]

19. Borisov, Y.M. Cytogenetic structure of *Apodemus peninsulae* (Rodentia, Muridae) population on the coast of Teletskoye Lake (Altai). *Russ. J. Genet.* **1990**, *26*, 1212–1220.

20. Hayata, J. Chromosomal polymorphism caused by supernumerary chromosomes in field mouse, *Apodemus giliacus*. *Chromosoma* **1973**, *42*, 403–414. [CrossRef] [PubMed]

21. Borisov, Y.M. System of B chromosomes in wood mouse as an integrating and differentiating characteristic of populations. *Dokl. Akad. Nauk SSSR* **1986**, *288*, 720–724.

22. Rubtsov, N.B.; Karamysheva, T.V.; Andreenkova, O.V.; Bochkarev, M.N.; Kartavtseva, I.V.; Roslik, G.V.; Borisov, Y.M. Micro B chromosomes of Korean field mouse *Apodemus peninsulae* (Rodentia, Murinae): Morphology, DNA contents, and evolution. *Cytogenet. Genome Res.* **2004**, *106*, 289–294. [CrossRef] [PubMed]

23. Rubtzov, N.B.; Borissov, Yu.M.; Karamysheva, T.V.; Bochkarev, M.N. The mechanisms of formation and evolution of B chromosomes in Korean field mice *Apodemus peninsulae* (Mammalia, Rodentia). *Russ. J. Genet.* **2009**, *45*, 389–396. [CrossRef]

24. Volobuev, V.T. B-chromosome system of the Asiatic wood mice *Apodemus peninsulae* (Rodentia, Muridae). I. Structure of karyotype, G- and C-bands, and the character of variation of B-chromosomes. *Russ. J. Genet.* **1980**, *16*, 1277–1284.

25. Volobuev, V.T. The B-chromosome system of mammals. *Genetica* **1980**, *52*, 333–337.

26. Volobuev, V.T. B-chromosome system of the mammals. *Caryologia* **1981**, *34*, 1–23. [CrossRef]

27. Ford, C.F.; Hamerton, J.L. A colchicine hypotonic citrate squash preparation for mammalian chromosomes. *Stain Technol.* **1956**, *31*, 247–251. [CrossRef] [PubMed]

28. StatSoft. *Statistica Electronic Manual*, version 8; Statsoft Inc.: Tulsa, OK, USA, 2008; Available online: http://www.statsoft.com/Textbook (accessed on 30 August 2018).

© 2018 by the authors. Licensee MDPI, Basel, Switzerland. This article is an open access article distributed under the terms and conditions of the Creative Commons Attribution (CC BY) license (http://creativecommons.org/licenses/by/4.0/).

![genes logo] *genes*

MDPI

Article

Euchromatic Supernumerary Chromosomal Segments—Remnants of Ongoing Karyotype Restructuring in the *Prospero autumnale* Complex?

Tae-Soo Jang [1,2], John S. Parker [3] and Hanna Weiss-Schneeweiss [1,*]

1   Department of Botany and Biodiversity Research, University of Vienna, A-1030 Vienna, Austria
2   Department of Biology, College of Bioscience and Biotechnology, Chungnam National University, Daejeon 34134, Korea; jangts@cnu.ac.kr
3   Cambridge University Botanic Garden, Cambridge CB2 1JF, UK; jsp25@cam.ac.uk
*   Correspondence: hanna.schneeweiss@univie.ac.at; Tel.: +43-142-775-4159

Received: 31 August 2018; Accepted: 21 September 2018; Published: 27 September 2018

Abstract: Supernumerary chromosomal segments (SCSs) represent additional chromosomal material that, unlike B chromosomes, is attached to the standard chromosome complement. The *Prospero autumnale* complex (Hyacinthaceae) is polymorphic for euchromatic large terminal SCSs located on the short arm of chromosome 1 in diploid cytotypes AA and B^7B^7, and tetraploid AAB^7B^7 and $B^6B^6B^7B^7$, in addition to on the short arm of chromosome 4 in polyploid $B^7B^7B^7B^7$ and $B^7B^7B^7B^7B^7B^7$ cytotypes. The genomic composition and evolutionary relationships among these SCSs have been assessed using fluorescence in situ hybridisation (FISH) with 5S and 35S ribosomal DNAs (rDNAs), satellite DNA *PaB6*, and a vertebrate-type telomeric repeat TTAGGG. Neither of the rDNA repeats were detected in SCSs, but most contained *PaB6* and telomeric repeats, although these never spanned whole SCSs. Genomic in situ hybridisation (GISH) using A, B^6, and B^7 diploid genomic parental DNAs as probes revealed the consistently higher genomic affinity of SCSs in diploid hybrid B^6B^7 and allopolyploids AAB^7B^7 and $B^6B^6B^7B^7$ to genomic DNA of the B^7 diploid cytotype. GISH results suggest a possible early origin of SCSs, especially that on chromosome 1, as by-products of the extensive genome restructuring within a putative ancestral *P. autumnale* B^7 genome, predating the complex diversification at the diploid level and perhaps linked to B-chromosome evolution.

Keywords: FISH (fluorescence in situ hybridisation); GISH (genomic in situ hybridisation); *Prospero autumnale* complex; supernumerary chromosomal segments (SCS) evolution; tandem repeats

1. Introduction

Karyotypes of plants and animals sometimes carry supernumerary genetic material, either in the form of B-chromosomes (Bs) or physically integrated into the standard chromosome complement as supernumerary chromosomal segments (SCSs) [1,2]. SCSs in plants are widespread [3], but are most frequently found in monocotyledons [4–10]. They are usually heterochromatic, but euchromatic segments also occur [11,12]. SCSs are frequently located terminally, and are thus seen at meiosis as heteromorphic bivalents resulting from chiasma formation proximal to SCSs [3,5,7,13]. SCSs usually behave as selfish genetic elements [9], but their inheritance can sometimes be Mendelian [14], and both of these patterns may be found within a single species [3].

A model system for investigating SCSs is provided by the *Prospero autumnale* complex (L.) Speta (Hyacinthaceae) [6,12]. This species complex has four evolutionarily and phylogenetically well-characterised and genomically distinct diploid cytotypes AA, B^5B^5, B^6B^6, and B^7B^7, each with unique combinations of genome size, base chromosome number ($x = 5$, 6 and 7), and repetitive DNA amounts and distribution [15]. Such chromosome number variation results from the relatively high rate of genome rearrangements in *Prospero*, involving translocations and inversions [6,15]. Hybridisation and polyploidisation of three of these diploid cytotypes (AA, B^6B^6, and B^7B^7) have given rise to further polyploid cytotypes, including autopolyploids of genome B^7 ($x = 7$) [16–20] and two classes of allopolyploids, of A/B^7 (both $x = 7$) and of B^6/B^7 genome composition [20–23]. Both B chromosomes (Bs) and SCSs have been reported in the three diploid cytotypes AA, B^6B^6, and B^7B^7 [15], and in various polyploids [6,17,21,24]. SCSs in *P. autumnale* are preferentially located terminally on the short arms of chromosomes 1 and 4 in the A and B^7 genomes and are frequently quite large. Thus, the SCS on chromosome 1 may be nearly double the length of the chromosome. Remarkably, no phenotypic effects have been ascribed to the presence of these massive elements [16,23], although SCSs are geographically widespread and reach polymorphic proportions in many populations [6].

The origins of SCSs, particularly of the euchromatic ones, are obscure [2,3]. They may result from amplification of part of the genome, such as tandem repeats, especially satellite DNAs [25], or may represent inessential chromosome blocks generated by karyotype rearrangements and retained during chromosomal evolution [11,26,27]. In this paper, the structure, repeat composition, and origins of SCSs have been addressed in the *P. autumnale* complex using molecular cytological tools based on FISH (fluorescence in situ hybridisation) and GISH (genomic in situ hybridisation). In particular, the distribution patterns of tandem repeats (rRNA genes, satellite DNA, telomeric sequences) within SCSs have been established in a range of diploid and polyploid cytotypes to assess whether SCSs reflect single or multiple origins and what their relationships to specific genomes or chromosomal locations might be.

2. Materials and Methods

2.1. Plant Materials

Plants for cytogenetic analysis were collected from natural populations across the Mediterranean basin [15,20] and grown in the Botanical Garden of the University of Vienna. Ten diploid and nine polyploid (4x and 6x) plants carrying SCSs were studied and their collection details are listed in Table 1.

For cytological investigations, root meristems were pretreated with a solution of 0.05% colchicine (Sigma Aldrich, Vienna, Austria) for 4.5 h at room temperature, fixed in ethanol: acetic acid (3:1) for at least 3 h at room temperature, and stored at −20 °C until use.

2.2. Karyotyping and FISH (Fluorescence In Situ Hybridisation with 5S & 35S Ribosomal DNAs, Vertebrate-Type Telomeric Repeats, and Satellite DNA PaB6)

Chromosome numbers and karyotypes were analysed as described by Jang et al. [15,20] using standard Feulgen staining. Chromosomal spreads for FISH and GISH were prepared by enzymatic digestion and squashing, as described in Jang et al. [15,20].

Probes used for FISH were: satellite DNA *PaB6* isolated from the B^6 genome in plasmid pGEM-T easy [28], 35S rDNA (18S/25S rDNA) from *Arabidopsis thaliana* in plasmid pSK+, and 5S rDNA from *Melampodium montanum* (Asteraceae) in plasmid pGEM-T easy, labeled with biotin or digoxygenin (Roche, Vienna, Austria). Probes were labeled either directly by PCR (5S rDNA and satellite DNA *PaB6*) or using a nick translation kit (35S rDNA; Roche). A commercially available, directly Cy3-labelled PNA (peptide nucleic acid) probe to vertebrate telomeric sequences (CCCTAA)$_3$ (Dako, Glostrup, Denmark) was used as described in [10]. FISH was performed as described in Jang et al. [15,20].

Table 1. List of plant material of the *Prospero autumnale* complex studied with voucher information and chromosome numbers. The chromosomal location, genomic affinity (as established using GISH), and length of supernumerary chromosomal segments (SCSs) are indicated.

Cytotype (Accession Number)	Locality; Collector	2n	Genomic Location/Genomic Affinity of SCSs	SCS Length (μm)	Proportion of SCS (%) in the Chromosome [†]	Figure No.
Diploids with SCSs						
AA (H541)	Spain, Huelva; J. Parker	14	A 1/genome B^7	2.67	28.8	Figure 1, Figure 2e,f, Figures 3a and S1a
B^{7}B^7 (H1)	Greece, Rhodos; F. Speta	14	B^7 1	2.02	28.8	Figures 1 and 2a,b
B^{7}B^7 (H247)	Greece, Crete; F. Speta	14	B^7 1	2.02	24.5	Figure 1
B^{7}B^7 (H423)	Montenegro; F. Speta	14	B^7 1	2.07	31.3	Figure 1
B^{7}B^7 (H500)	Greece, Crete; F. Speta	14	B^7 1	2.02	30.6	Figure 1
B^{7}B^7 (H502)	Greece, Crete; F. Speta	14	B^7 1	2.07	33.3	Figure 1
B^{7}B^7 (H614)	Israel, HaCarmel Park; J. Parker	14	B^7 3	2.29	36.4	Figures 1 and 2c,d
B^{7}B^7 (H641)	Spain; J. Parker	14	B^7 1	1.98	32.0	Figure 1, Figures 3d and S1d
B^{7}B^7 (H642)	Spain; J. Parker	14	B^7 1	2.07	27.8	Figure 1
B^{6}B^7 (H258)	Greece, Crete; F. Speta	13	B^6 1	2.11	23.4	Figure 1, Figures 2g,h,e and S1e
Polyploids with SCSs						
AAB^{7}B^7 (H110–1)	Portugal, Cheleiros; F. Speta	28	A 1 *	3.16	21.4	Figure 1, Figures 3b and S1b
AAB^{7}B^7 (H110–2)	Portugal, Cheleiros; F. Speta	28	A 1	2.63	33.3	Figure 1, Figure 2i,j, Figures 3c and S1c
AAB^{7}B^7 (H606)	Portugal, Castro Marin; J. Parker	28	A 1	2.89	26.2	Figure 1
B^{6}B^{6}B^{7}B^7 (H574–1)	Greece, Naxos; F. Speta	28	B^6 1	3.85	27.3	Figure 1, Figure 2m,n, Figures 3f and S1f
B^{6}B^{6}B^{7}B^7 (H574–2)	Greece, Naxos; F. Speta	28	B^6 1	3.21	31.3	Figure 1
B^{7}B^{7}B^{7}B^7 (H360)	Greece, Kos; F. Speta	28	B^7 4	1.94	35.3	Figures 1 and 2k,l
B^{7}B^{7}B^{7}B^{7}B^7 (H31)	No information; F. Speta	42	B^7 4 * and (5 or 6) §	1.92 §	33.3 §	Figures 1 and 2o,p
B^{7}B^{7}B^{7}B^{7}B^{7}B^7 (H308)	Croatia, Solta; F. Speta	42	B^7 4	2.88	34.6	Figure 1
B^{7}B^{7}B^{7}B^{7}B^{7}B^7 (H453)	Croatia, Mosor; F. Speta	42 + 1B	B^7 4 *	2.56	33.3	Figure 1

*: Homozygous SCS; †: Proportion of SCSs within the chromosomes carrying them, §: Only SCSs of chromosomes 4 were measured because the other SCSs could not be unambiguously assigned to chromosome 5 or 6. GISH: Genomic in situ hybridization.

2.3. GISH (Genomic In Situ Hybridisation)

GISH has been performed in one diploid hybrid individual B^6B^7 (H258) and four allopolyploid individuals, two of AAB^7B^7 (H110–1 & 2) and two of $B^6B^6B^7B^7$ (H574–1 & 2) composition, using labelled parental diploid genomic DNA as probes [24,29]. Total genomic DNA from diploid cytotypes AA, B^6B^6, and B^7B^7 was isolated using the CTAB (Cetyltrimethylammonium Bromide) method [29] and sheared at 98 °C for 5 min. Approximately 1 µg of genomic DNA of each cytotype was labeled using either digoxigenin or a biotin nick translation kit (Roche). GISH was carried out following the method of [29].

All preparations after FISH and GISH were analysed with an AxioImager M2 epifluorescent microscope (Carl Zeiss, Vienna, Austria), and images were captured with a CCD camera and processed using AxioVision ver. 4.8 (Carl Zeiss) with only those functions that apply to all pixels of the image equally.

3. Results and Discussion

3.1. Karyotype Structure and Localisation of Supernumerary Chromosomal Segments

Nineteen SCS-carrying plants of the *P. autumnale* complex have been analysed (Table 1; Figure 1): ten diploids (one AA, eight B^7B^7, and one diploid hybrid B^6B^7; Figure 1) and nine polyploids (two B^6/B^7 allotetraploids, three A/B^7 allotetraploids, and four B^7 autopolyploids, one tetra- and three hexaploids; Figure 1). All SCSs are located distally, most frequently on chromosomes 1 and 4, and occasionally on chromosome 3 or either 5 or 6 (it is not possible to unambiguously assign this SCS to a specific chromosome; Figure 1). Chromosomes 1, 3, and 4 carry SCSs on their short arms and are thus are more symmetrical than their standard counterparts, while chromosome 5/6 carries an SCS on the long arm and is thus more asymmetric (Figure 1). The SCSs are massive blocks of chromatin and their length varies from 1.92 µm to 3.85 µm, representing a 21.4% to 36.4% increase in chromosome length, as found previously (Table 1) [6,16,21]. The SCS on chromosome 1 in the allotetraploid $B^6B^6B^7B^7$ is reported here for the first time (Figure 1).

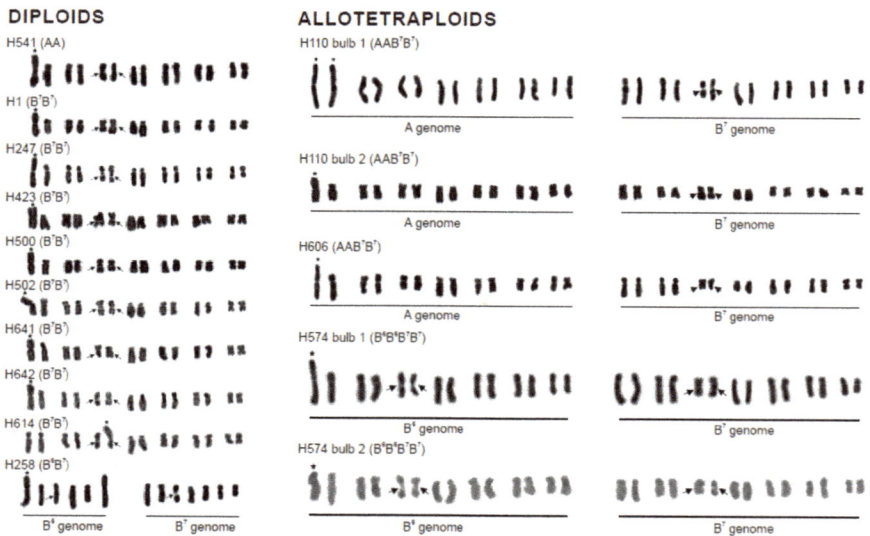

Figure 1. *Cont.*

AUTOPOLYPLOIDS

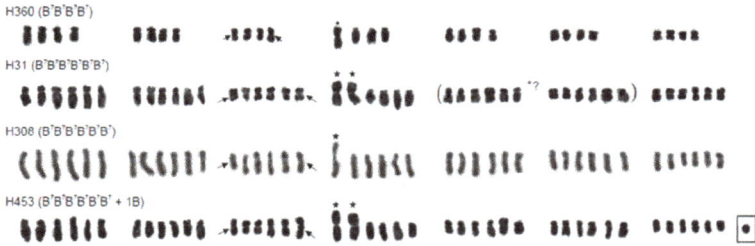

Figure 1. Karyotypes of *P. autumnale* individuals representing diploid and polyploid cytotypes of the *P. autumnale* complex carrying SCSs. Stars and arrows indicate SCSs and nucleolar organizer regions (NORs), respectively. Individual H31: Question mark above the bracket indicates chromosomes that potentially carry the small third SCS, as identified by FISH (fluorescence in situ hybridization) with the *PaB6* satellite DNA (see Figure 2p). Inset shows B-chromosome. Bar = 5 μm.

The SCSs of chromosome 1 are remarkably constant in size across the distribution range of *P. autumnale* and also across ploidy levels. Thus, the SCS in B^7B^7 diploids has a length of about 2 μm in plants from Greece, Montenegro, and Spain, while that in the A genome is about 2.75 μm in length in AA diploids and in A/B^7 allotetraploids (Table 1). The same constancy is not found in SCSs attached to other chromosomes.

3.2. Tandem Repeats in Supernumerary Chromosomal Segments

Four types of tandem repeats–35S rDNA, 5S rDNA, satellite DNA *PaB6*, and telomeric repeats–have been used as FISH probes on standard and SCS-carrying chromosomes of *P. autumnale* cytotypes (Figure 2). No signals for 35 or 5S rDNAs were detected within SCSs (Figure 2).

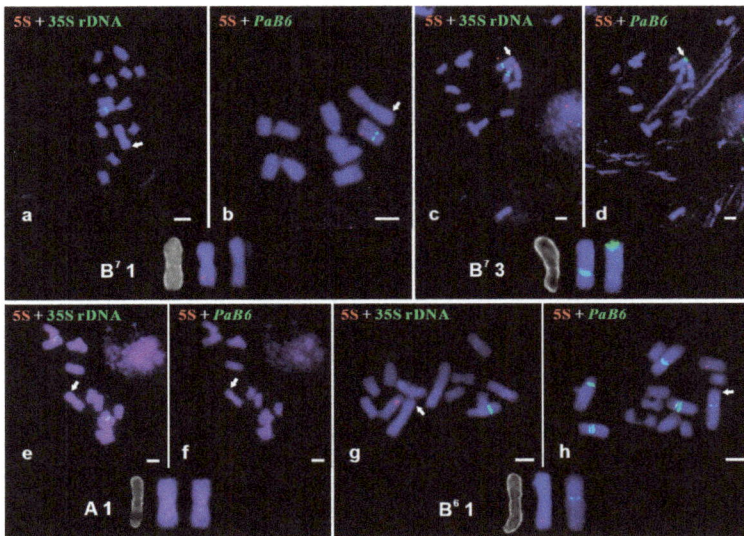

Figure 2. *Cont.*

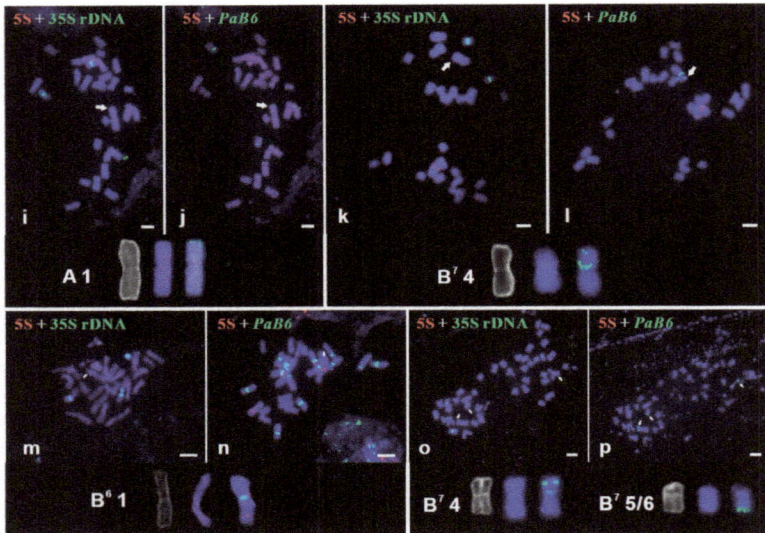

Figure 2. Localisation of 5S and 35S rDNAs and satellite DNA *PaB6* in diploid and polyploid cytotypes of the *P. autumnale* complex carrying SCSs. (**a,b**) B^7B^7 (H1); (**c,d**) B^7B^7 (H614); (**e,f**) AA (H541); (**g,h**) B^6B^7 (H258); (**i,j**) AAB^7B^7 (H110–2); (**k,l**) $B^7B^7B^7B^7$ (H360); (**m,n**) $B^6B^6B^7B^7$ (H574–1); (**o,p**) $B^7B^7B^7B^7B^7B^7$ (H31). Arrows indicate SCSs. Bar = 5 μm.

In standard karyotypes of diploids of *P. autumnale*, satellite DNA *PaB6* is located in pericentromeric regions of at least one chromosome (AA) and up to all chromosomes (B^6B^6, B^7B^7) [28]. By contrast, in SCSs, it is located terminally on chromosome 1 of B^7B^7 (2 of 7 plants), chromosome 1A of AAB^7B^7, chromosome 3 of B^7B^7, chromosomes 4 of $B^7B^7B^7B^7$ and $B^7B^7B^7B^7B^7B^7$, and chromosome 5/6 of $B^7B^7B^7B^7B^7B^7$ individuals (Figures 1 and 2). No *PaB6* signals were detected on the SCSs of the AA diploid, of the B^6B^7 diploid hybrid, and of the five remaining B^7B^7 plants (Figure 2).

In the *P. autumnale* complex, vertebrate-type telomeric signals are found at the termini of both arms of all standard chromosomes (Figure S1) [28], and also in pericentromeric regions, coinciding with satellite DNA *PaB6* as its monomers contain a few full repeats of telomeric sequence TTAGGG (Figure S1b,c,e,f). Telomeric repeat signals in telomeric positions in standard chromosomes (i.e., lacking SCSs) were often very weak. In SCSs, slight amplification of telomeric repeats was detected in SCSs on chromosomes 1 of some B^7 diploids (Figure S1d) and AAB^7B^7 allotetraploids (Figure S1b,c), always subterminally and coincident with amplification of satellite DNA *PaB6*.

3.3. Genomic DNA Affinities of the Supernumerary Chromosomal Segments (Genomic In Situ Hybridisation)

The relationships of SCSs to parental genomes can be established using formamide-free GISH [20,29] in the diploid hybrid B^6B^7 (Figure 3e), and in the allotetraploids $B^6B^6B^7B^7$ (Figure 3f) and AAB^7B^7 (Figure 3b,c). In all cases, the SCSs have a higher affinity for the B^7 genomic probe than for either of the other two parental genomic probes (A or B^6 genomes; Figure 3b,c,e,f).

Remarkably, the SCS of chromosome 1 of the AA diploid shows hybridisation with B^7 genomic DNA (Figure 3a). As expected, no discrimination is shown by the chromosome 1 SCS in B^7B^7 diploid plants used as the control (Figure 3d). A possible explanation for this is hybridisation at the diploid level between AA and B^7B^7 plants. Recombination between the SCSs-carrying B^71 chromosome and an A1 chromosome may have occurred in the diploid background, allowing the transfer of this SCS from B^71 to A1 (A/B chiasma formation has been seen in AB^7B^7 triploids) [30]. Subsequent recurrent backcrossing to AA would restore the AA complement, but with the addition of an SCS derived from B^71.

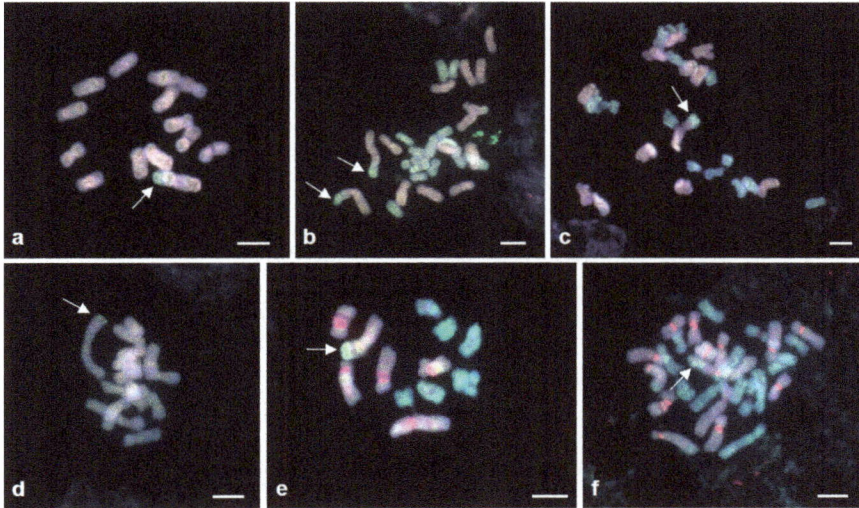

Figure 3. GISH in the *P. autumnale* complex. A–D: Localisation of A (red) and B^7 (green) genomic DNA, (**a**): H541 (cytotype AA, $2n = 2x = 14$, with one SCS), (**b**): H110–1 (cytotype AAB^7B^7, $2n = 4x = 28$, with two SCSs), (**c**): H110–2 (AAB^7B^7, $2n = 4x = 28$, with one SCS), (**d**): H641 (B^7B^7, $2n = 2x = 14$, with one SCS), (**e,f**): Localization of B^6 (red) and B^7 (green) parental genomic DNAs in (**e**): H258 (B^6B^7 hybrid $2n = 2x = 13$ with one SCS), (**f**): H574–1 ($B^6B^6B^7B^7$, $2n = 4x = 28$, with one SCS). Arrows indicate SCSs. Bar = 5 μm.

The current study by means of FISH and GISH indicates that, in the complex evolutionary system of *P. autumnale,* the origin of at least some SCSs can be traced back to the B^7 genome (Figure 4). Previously, it was hypothesized that the ancestral karyotype of the complex closely resembled that of B^7B^7 [15,28]. From this diploid complement, both B^6B^6 and B^5B^5 cytotypes have been derived by one and two independent fusions, respectively [15]. The generation of the SCSs, so remarkably prevalent in the complex, may thus be an outcome of chromosomal rearrangements associated with the generation of new cytotypes with new base chromosome numbers and karyotype structures. At least some of these SCSs might be quite old, their origin preceding the diversification of the extant diploid cytotypes. Extensive chromosomal rearrangements in *P. autumnale* have also been proposed to be the most likely reason for its extraordinary variability (both structural and genomic) and the ongoing origin of B chromosomes [24]. It is possible that the putative ancestral SCS of chromosome 1 has originated from a B chromosome that translocated to chromosome 1 early in the diversification of the *P. autumnale* complex. Other SCSs may be of a different age and their formation might also be ongoing.

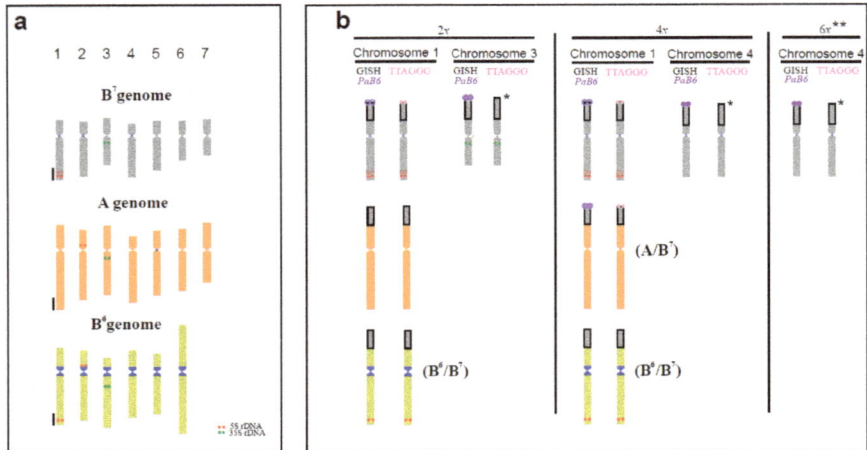

Figure 4. Summary of all SCSs types present in diploid and polyploid individuals of *P. autumnale*: (a) ideograms of standard diploid cytotypes, B^7B^7 (gray), AA (orange), and B^6B^6 (lime) with 5S rDNA, 35S rDNA, and *PaB6* loci indicated (modified from Emadzade et al. 2014); (b) genomic affinity assessed by GISH and distribution of repetitive DNAs (rDNAs, satDNA *PaB6*, and vertebrate-type telomeric sequence TTAGGG) in SCSs in analysed diploids (2*x*), tetraploids (4*x*), and hexaploids (6*x*). *—not analysed; **—chromosome 5/6 is not indicated due to our inability of assigning the SCS to chromosome 5 or 6.

We may postulate, then, that all of the SCSs of chromosomes 1 found in *P. autumnale* derived from a single event, and the presence of a B^7-like SCS on the short arm of chromosome 1 of the A-genome supports this contention. It is clear, however, that the SCSs now differ in their molecular structure between chromosomes and between cytotypes, as we have demonstrated with repetitive DNA probes. This may simply reflect the profound genomic changes that have swept through the complex since its origin [15,28]. This common descent of SCSs, however, should be further explored through GISH studies and molecular analyses of many more plants, making use of the extraordinary levels of chromosomal polymorphism found in natural populations of *P. autumnale* [16,20,21,23].

Supplementary Materials: The following are available online at www.mdpi.com/2073-4425/9/10/468/s1. Figure S1: Localisation of Cy3-labelled telomeric (TTAGGG)n repeats in supernumerary chromosomal segments (SCSs) (arrowed) in diploid and polyploid cytotypes of the *Prospero autumnale* complex. (a) H541 (cytotype AA with one SCS), (b) H110–1 (cytotype AAB7B7 with two SCSs), (c) H110–2 (cytotype AAB7B7 with one SCS), (d) H641 (cytotype B7B7 with one SCS), (e) H258 (B6B7 hybrid with one SCS), (f) H574–1 (cytotype B6B6B7B7 with one SCS).

Author Contributions: Conceptualization, T.-S.J. and H.W.-S.; Methodology, T.-S.J. and H.W.-S.; Formal Analysis, T.-S.J., J.S.P., and H.W.-S.; Writing-Original Draft Preparation, T.-S.J., J.S.P., and H.W.-S.; Writing-Review & Editing, J.S.P. and H.W.-S.; Visualization, T.-S.J. and H.W.-S.; Supervision, H.W.-S.; Project Administration, H.W.-S.; Funding Acquisition, H.W.-S.

Funding: This research was partly funded by Austrian Science Fund (FWF) grant number P21440-B03 to H.W-S.

Acknowledgments: Open access funding provided by University of Vienna.

Conflicts of Interest: The authors declare no conflict of interest. The funders had no role in the design of the study; in the collection, analyses, or interpretation of data; in the writing of the manuscript, and in the decision to publish the results.

References

1. Houben, A.; Banaei-Moghaddam, A.M.; Klemme, S. Biology and evolution of B chromosomes. In *Plant Genome Diversity, Vol 2, Physical Structure, Behavior and Evolution of Plant Genomes*, 1st ed.; Leitch, I.J., Greilhuber, J., Dolezel, J., Wendel, J., Eds.; Springer Verlag: Vienna, Austria, 2013; pp. 149–165.

2. Weiss-Schneeweiss, H.; Schneeweiss, G.M. Karyotype diversity and evolutionary trends in angiosperms. In *Plant Genome Diversity, Vol 2, Physical Structure, Behavior and Evolution of Plant Genomes*, 1st ed.; Leitch, I.J., Greilhuber, J., Dolezel, J., Wendel, J., Eds.; Springer Verlag: Vienna, Austria, 2013; pp. 209–230.

3. Wilby, A.S.; Parker, J.S. The supernumerary segment systems of *Rumex acetosa*. *Heredity* **1988**, *60*, 109–117. [CrossRef]

4. Greilhuber, J.; Speta, F. Quantitative analysis of the C-banded karyotypes and systematic in the cultivated species of the *Scilla siberica* group (*Liliaceae*). *Plant Syst. Evol.* **1978**, *129*, 63–109. [CrossRef]

5. Ruiz Rejón, M.; Oliver, J.L. Genetic variability in *Muscari comosum* (Liliaceae). I. A comparative analysis of chromosome polymorphisms in Spanish and Aegean populations. *Heredity* **1981**, *47*, 403–407. [CrossRef]

6. Parker, J.S.; Lozano, R.; Taylor, S.; Ruiz Rejon, M. Chromosomal structure of populations of *Scilla autumnalis* in the Iberian Peninsula. *Heredity* **1991**, *67*, 287–297. [CrossRef]

7. Stevens, J.P.; Bougourd, S.M. The frequency and meiotic behaviour of structural chromosome variants in natural populations of *Allium schoenoprasum* L. (wild chives) in Europe. *Heredity* **1991**, *66*, 391–401. [CrossRef]

8. Jamilena, M.; Martínez, F.; Garrido-Ramos, M.A.; Ruiz-Rejón, C.; Romero, A.T.; Camacho, J.P.M.; Parker, J.S.; Ruiz-Rejón, M. Inheritance and fitness effects analysis for a euchromatic supernumerary chromosome segment in *Scilla autumnalis* (Liliaceae). *Bot. J. Linn. Soc.* **1995**, *118*, 249–259. [CrossRef]

9. Garrido-Ramos, M.A.; Jamilena, M.; de la Herrán, R.; Ruiz Rejón, C.; Camacho, J.P.M.; Ruiz-Rejón, M. Inheritance and fitness effects of a pericentric inversion and a supernumerary chromosome segment in *Muscari comosum* (Liliaceae). *Heredity* **1998**, *80*, 724–731. [CrossRef]

10. Weiss-Schneeweiss, H.; Riha, K.; Jang, C.G.; Puizina, J.; Scherthan, H.; Schweizer, D. Chromosome termini of the monocot plant *Othocallis siberica* are maintained by telomerase, which specifically synthesizes vertebrate-type telomere sequences. *Plant J.* **2004**, *37*, 484–493. [CrossRef] [PubMed]

11. Ainsworth, C.C.; Parker, J.S.; Horton, D.M. Chromosome variation and evolution in *Scilla autumnalis*. In *Kew Chromosome Conference II*; Brandham, P.E., Bennett, M.D., Eds.; George Allen & Unwin: London, UK, 1983; pp. 261–268.

12. Ebert, I.; Greilhuber, J.; Speta, F. Chromosome banding and genome size differentiation in *Prospero* (Hyacinthaceae): Diploids. *Plant Syst. Evol.* **1996**, *203*, 143–177. [CrossRef]

13. Navas-Castillo, J.; Cabrero, J.; Camacho, J.P.M. Chiasma redistribution in bivalents carrying supernumerary chromosome segments in grasshoppers. *Heredity* **1985**, *55*, 245–248. [CrossRef]

14. Ruiz Rejón, C.; Ruiz Rejón, M. Chromosomal polymorphism for a heterochromatic supernumerary segment in a natural population of *Tulipa australis* Link (Liliaceae). *Can. J. Genet. Cytol.* **1981**, *27*, 633–638. [CrossRef]

15. Jang, T.-S.; Emadzade, K.; Parker, J.; Temsch, E.M.; Leitch, A.R.; Speta, F.; Weiss-Schneeweiss, H. Chromosomal diversification and karyotype evolution of diploids in the cytologically diverse genus *Prospero* (Hyacinthaceae). *BMC Evol. Biol.* **2013**, *13*, 136. [CrossRef] [PubMed]

16. Ainsworth, C.C. The Population Cytology of *Scilla autumnalis*. Ph.D. Thesis, University of London, London, UK, 1980.

17. Ebert, I. Systematische Karyologie und Embryologie von *Prospero* Salisb. und *Barnardia* Lindl. (Hyacinthaceae). Ph.D. Thesis, University of Vienna, Vienna, Austria, 1993.

18. Speta, F. The autumn-flowering squills of the Mediterranean Region. In *Proceedings of the 5th Optima Meeting, Istanbul, Turkey, 18–30 July 1993*; University of Istanbul: Istanbul, Turkey, 1993; pp. 109–124.

19. Speta, F. Beitrag zur Kenntnis der Gattung *Prospero* Salisb (Hyacinthaceae) auf der griechischen Insel Kreta. *Linzer Biol. Beitr.* **2000**, *32*, 1323–1326.

20. Jang, T.-S.; Parker, J.; Emadzade, K.; Temsch, E.M.; Leitch, A.R.; Weiss-Schneeweiss, H. Multiple origins and nested cycles of hybridization result in high tetraploid diversity in the monocot *Prospero*. *Front. Plant Sci.* **2018**, *9*, e433. [CrossRef] [PubMed]

21. Taylor, S. Chromosomal Evolution of *Scilla autumnalis*. Ph.D. Thesis, University of London, London, UK, 1997.

22. Vaughan, H.E.; Taylor, S.; Parker, J.S. The ten cytological races of the *Scilla autumnalis* species complex. *Heredity* **1997**, *79*, 371–379. [CrossRef]

23. Jang, T.-S. Chromosomal Evolution in *Prospero autumnale* Complex. Ph.D. Thesis, University of Vienna, Vienna, Austria, 2013.
24. Jang, T.-S.; Parker, J.; Weiss-Schneeweiss, H. Structural polymorphisms and distinct genomic composition suggest recurrent origin and ongoing evolution of B chromosomes in the *Prospero autumnale* complex (Hyacinthaceae). *New Phytol.* **2016**, *210*, 669–679. [CrossRef] [PubMed]
25. Shibata, F.; Hizume, M.; Kuroki, Y. Molecular cytogenetic analysis of supernumerary heterochromatic segments in *Rumex acetosa*. *Genome* **2000**, *43*, 391–397. [CrossRef] [PubMed]
26. John, B. The cytogenetic systems of grasshoppers and locusts II. The origin and evolution of supernumerary segments. *Chromosoma* **1973**, *44*, 123–146. [CrossRef] [PubMed]
27. Camacho, J.P.M.; Cabrero, J. New hypotheses about the origin of supernumerary chromosome segments in grasshoppers. *Heredity* **1987**, *58*, 341–343. [CrossRef]
28. Emadzade, K.; Jang, T.-S.; Macas, J.; Kovařík, A.; Novák, P.; Parker, J.; Weiss-Schneeweiss, H. Differential amplification of satellite *PaB6* in chromosomally hypervariable *Prospero autumnale* complex (Hyacinthaceae). *Ann. Bot.* **2014**, *114*, 1597–1608. [CrossRef] [PubMed]
29. Jang, T.-S.; Weiss-Schneeweiss, H. Formamide-free genomic in situ hybridization (ff-GISH) allows unambiguous discrimination of highly similar parental genomes in diploid hybrids and allopolyploids. *Cytogenet. Genome Res.* **2015**, *146*, 325–331. [CrossRef] [PubMed]
30. White, J.; Jenkins, G.; Parker, J.S. Elimination of multivalents during meiotic prophase in *Scilla autumnalis*. I. Diploid and triploid. *Genome* **1988**, *30*, 930–939. [CrossRef]

© 2018 by the authors. Licensee MDPI, Basel, Switzerland. This article is an open access article distributed under the terms and conditions of the Creative Commons Attribution (CC BY) license (http://creativecommons.org/licenses/by/4.0/).

GCAT
TACG
GCAT
genes

MDPI

Review

B Chromosomes in the *Drosophila* Genus

Stacey L. Hanlon [1,*] and R. Scott Hawley [1,2]

1   Stowers Institute for Medical Research, Kansas City, MO 64110, USA; rsh@stowers.org
2   Department of Molecular and Integrative Physiology, University of Kansas Medical Center,
    Kansas City, KS 66160, USA
*   Correspondence: slh@stowers.org

Received: 1 September 2018; Accepted: 20 September 2018; Published: 27 September 2018

Abstract: Our current knowledge of B chromosome biology has been augmented by an increase in the number and diversity of species observed to carry B chromosomes as well as the use of next-generation sequencing for B chromosome genomic analysis. Within the genus *Drosophila*, B chromosomes have been observed in a handful of species, but recently they were discovered in a single laboratory stock of *Drosophila melanogaster*. In this paper, we review the B chromosomes that have been identified within the *Drosophila* genus and pay special attention to those recently found in *D. melanogaster*. These newly-discovered B chromosomes have centromeres, telomeres, and a number of simple satellite repeats. They also appear to be entirely heterochromatic since next-generation sequencing of isolated B chromosomes did not detect sequences associated with known genic regions. We also summarize what effects the B chromosomes have been found to have on the A chromosomes. Lastly, we highlight some of the outstanding questions regarding B chromosome biology and discuss how studying B chromosomes in *Drosophila melanogaster*, which is a versatile model system with a wealth of genetic and genomic tools, may advance our understanding of the B chromosome's unique biology.

Keywords: *Drosophila*; supernumerary; satellite DNA; sSMC

1. Introduction

All members of a single species carry a defined set of essential chromosomes or A chromosomes that are required for the normal growth, development, and reproduction of an organism. In many species, however, a subset of individuals can harbor nonessential, supernumerary chromosomes referred to as B chromosomes. First described by Wilson over a century ago, they have since been identified in hundreds of species spanning many different taxa [1–4]. Our understanding of B chromosome biology is currently undergoing rapid change due to the growing accessibility of advanced molecular and genomic techniques [5–7]. These modern methods are illuminating facets of B chromosome structure and biology that would not have been uncovered through classical cytological approaches such as the presence of active genetic material or the landscape of repetitive elements at a high resolution [8–10].

Since more B chromosomes are molecularly analyzed, an important next step will be to connect their genetic composition to known attributes. For example, several B chromosomes are subject to drive mechanisms that can result in their accumulation within a population, but, in some instances, the details of how those systems work have not been elucidated [11]. When the plant or animal systems carrying these B chromosomes are not conducive to experimentation in a laboratory setting, it can be difficult to definitively connect B chromosome genotypes to observed phenotypes.

Recently, B chromosomes were discovered in a laboratory stock of the robust model organism known as *Drosophila melanogaster* (Figure 1) [12]. These B chromosomes have since undergone a molecular analysis that is beginning to provide insight into their origin, composition, and structure [13].

In this review, we summarize the B chromosomes that have been observed in various species within the genus *Drosophila*, which was followed by an analysis of the emerging B chromosome model system in *D. melanogaster*. Lastly, we highlight some outstanding questions in B chromosome biology that may be well-suited for study in this emerging B chromosome system.

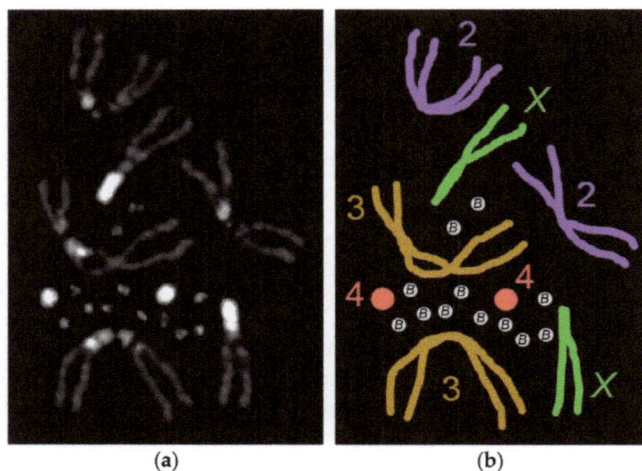

Figure 1. The B chromosomes of *Drosophila melanogaster*. (**a**) A karyotype from a female carrying 12 supernumerary B chromosomes. The DNA is stained with DAPI (photo by authors). (**b**) Illustrated representation of the karyotype in (**a**). Homologous chromosomes are shown in the same color.

2. B Chromosomes within the Genus *Drosophila*

Out of over 1600 documented species within the *Drosophila* genus, more than 600 of these have been sampled to determine the shape and number of their chromosomes [14,15]. To date, supernumerary B chromosomes have been positively identified in just nine *Drosophila* species. The initial report documenting the existence of a B chromosome in each of these species is presented in Table 1. Since B chromosomes are generally small and can be easily overlooked in a karyotype and, because within the same species they can be present in some populations and absent in others, this low frequency of B chromosomes in *Drosophila* is likely an underestimate.

Table 1. Initial reports of B chromosomes in different species of *Drosophila*.

Year Reported	*Drosophila* Species	Location of Sample Collection	Reference
1980	*D. albomicans* [1]	Chiang Mai, Thailand	[16,17]
1983	*D. malerkotliana*	Multiple locations in Southeast Asia	[18]
1994	*D. kikkawai*	Bhubaneswar, India	[19]
1995	*D. subsilvestris*	Tübingen, Germany	[20]
2007	*D. lini*	Malipo county, Yunnan province, China	[21]
2007	*D. pseudoananassae*	Jianfengling, Hainan province, China	[21]
2009	*D. yangana* [2]	Loja province, Ecuador	[22]
2009	*D. huancavilcae* [2]	Manabí province, Ecuador	[22]
2014	*D. melanogaster*	Domesticated laboratory stock	[12]

[1] Also referred to as *D. nasuta albomicana*. [2] B chromosomes reported but not cytologically recorded.

The first reported B chromosome in *Drosophila* was identified by two separate groups who collected independent samples of *D. albomicans* (also referred to as *D. nasuta albomicana*) from wild populations around Chiang Mai, Thailand [16,17]. Clyde collected individual females to create isolines that were each examined cytologically. Three out of eight isolines carried supernumerary "dot" chromosomes in addition to the normal karyotype and individuals within the same isoline could carry up to two copies [16]. The second strain of *D. albomicans* from Thailand was cultured in the laboratory of Osamu Kitagawa [17]. Ramachandra and Ranganath cytologically analyzed this strain and determined that individuals were able to carry one, two, or three copies of the B chromosome [23]. In subsequent studies, they were also able to document the rapid accumulation of B chromosomes within the strain after being kept under laboratory conditions [24]. In just three years, the percentage of individuals carrying at least one B chromosome rose from 67% to 98% and the frequency of individuals carrying three or more B chromosomes jumped from 5% to 40% [25]. The B chromosome of *D. albomicans* likely possesses the ability to promote its rapid propagation within a population since samples collected from diverse locations in Southeast Asia have been found to carry B chromosomes [26]. Though an analysis of heterochromatic composition through various staining techniques revealed that the *D. albomicans* B chromosomes differ from the A chromosomes, no conclusions were drawn regarding how those differences may influence the proliferation of the B chromosome [27]. Perhaps a closer examination of the recent genomic sequence of *D. albomicans* will provide the key [28].

The second *Drosophila* species reported to carry a B chromosome was also from Southeast Asia. Samples of *D. malerkotliana* collected throughout the region—from Mauritius and Seychelles to all along the coast of India, Myanmar (Burma), Thailand, and Malaysia—were shown cytologically to carry B chromosomes [18]. Another species collected from India, *D. kikkawai*, carries two distinct B chromosome variants in which each is considerably larger than the B chromosomes from *D. albomicans* and *D. malerkotliana* [19]. When these stocks were selected for the presence of B chromosomes in a laboratory setting, the maximum copy number of the B chromosomes we were able to reach was four [29]. Even though this is the only report of B chromosomes in *D. kikkawai*, it would be interesting to re-examine early cytological data while keeping the possibility of an unidentified B chromosome in mind. For example, in a different sample of *D. kikkawai*, there appears to be a darkly-staining fragment present in one of the mitotic spreads (see Figure 2B in Reference [30]). Since this is the only image taken of the sample, it is not known if other metaphases from the same sample also had this fragment, which makes it impossible to determine if the fragment is a B chromosome or simply an artifact of the experiment. A similar oversight in *D. flavopilosa* may have occurred in 1962: Bencic briefly mentions "an extra dot-like chromosome" that is not present in all samples, which very well may have been a B chromosome [31].

B chromosomes have also been found in *Drosophila* specimens collected outside of Southeast Asia. Samples of *D. subsilvestris* collected on two separate occasions from Southwest Germany each carried B chromosomes [20]. The first of the established lines had up to five dot-like chromosomes shortly after its collection, but after ~14 years in the lab, the dot-like chromosomes were all but lost. The second line was collected from the wild and established when the first line died out. This new stock also carried dot-like chromosomes and they were soon determined to be B chromosomes since they did not appear to undergo polytenization, which indicates that they were heterochromatic. Additionally, the B chromosomes carry the heterochromatic satellite DNA repeat *pSsP216*, which indicates that they may have arisen from the dot chromosome that also carries this repeat [20].

A large cytological survey of 34 species from the *melanogaster* species group revealed that *D. lini* and *D. pseudoananassae* collected from China can have extra chromosomes that are potential B chromosomes [21]. The size of the *D. lini* supernumerary chromosomes—as well as the consistency of the A chromosome karyotype—suggests that these are likely to be B chromosomes. The extreme range of karyotypic variations observed in *D. pseudoananassae*, however, confounds the assessment of whether some of the chromosomes are truly supernumerary and nonessential and a closer examination of this species is required. More cytological studies are also necessary in *D. yangana* and *D. huancavilcae*, which reportedly have supernumerary chromosomes but have not been cytologically documented in the literature [22].

Recently, B chromosomes were detected in *D. melanogaster*, which is the best-studied species of *Drosophila* [12]. As a widely used model organism with an extensive history of laboratory experimentation, *D. melanogaster* has a wealth of genetic and molecular tools that can be utilized to understand important aspects of B chromosome biology. Some of these aspects are discussed below.

3. B Chromosomes in *Drosophila melanogaster*

Two studies documenting and molecularly characterizing the B chromosomes in *D. melanogaster* were recently conducted in the Hawley laboratory [12,13]. The findings of the initial study [12] and emerging results from a second study [13] are summarized below.

3.1. Characterization of the Original Drosophila melanogaster B chromosome

The first B chromosomes identified in *D. melanogaster* were discovered in a Hawley laboratory stock that carries a null mutation in *matrimony*, which is a Polo-kinase inhibitor that acts only during female meiosis in *Drosophila* [32–34]. While examining female oocytes during meiotic prometaphase I in this mutant, DAPI-staining fragments that were much smaller than chromosome 4 were observed along the meiotic spindle. Initially, these were assumed to be the common *Drosophila* intercellular bacterial parasite, *Wolbachia*. This presumption was challenged when immunofluorescence (IF) using antibodies recognizing the *Drosophila* centromeric histone Centromeric Identifier (CID) revealed these fragments incorporated CID [12]. IF also revealed that these fragments were coated with H3K9 methylation, which is an epigenetic mark of silenced chromatin. This indicated that they were mostly heterochromatic [12]. The presence of centromeres signified these fragments were chromosomes and, since they were small, heterochromatic and, when they carried in addition to the full complement of A chromosomes, they were classified as supernumerary B chromosomes [12]. Additionally, it appears that these fragments may also have telomeres as indicated by the presence of HOAP (encoded by *cav*), which is a telomere-specific capping protein in *Drosophila* (Figure 2) [13].

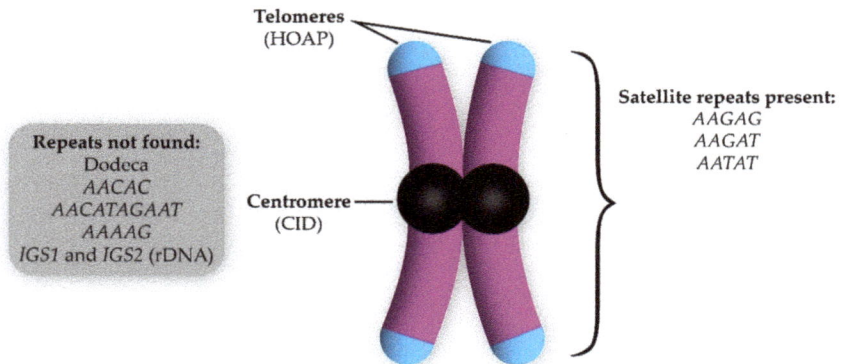

Figure 2. Composition of the *D. melanogaster* B chromosome (initially described in Reference [12]). The B chromosome has telomeres (indicated by the presence of HOAP, which is a telomere capping protein), centromeres (due to the incorporation of the centromeric histone CID), and a variety of satellite repeats revealed by fluorescent in situ hybridization. The repetitive sequences not found on the B chromosomes are listed in the gray box.

The first step toward finding the origin of the *D. melanogaster* B chromosomes was to determine their composition. Fluorescent in situ hybridization (FISH) using probes for a variety of repetitive sequences has revealed that the B chromosomes carry satellite DNA sequences that are also found on the A chromosomes such as the *AATAT* (chromosomes 4, X, and Y) and *AAGAG* (all chromosomes) repeats [12,13]. The repeat sequences from the A chromosomes that were not found on the B chromosomes were Dodeca (chromosome 3), *AACAC* (chromosomes 2 and Y), *AACATAGAAT* (chromosomes 2 and 3), *AAAAG* (chromosomes 2 and Y), and the intergenic spacer (IGS) that is between ribosomal DNA repeats (chromosomes *X* and *Y*) [12,13]. Of the repeats tested, the most prominent satellite repeat on the B chromosomes as assayed by FISH appears to be the *AAGAT* repeat (Figure 2). The only A chromosome that also carries this repeat is chromosome 4, which is consistent with the hypothesis that the B chromosomes originated from chromosome 4 [13].

Chromosome 4 in *D. melanogaster* has euchromatin near the tip of its right arm and the remainder of the chromosome is heterochromatin. Even though the B chromosomes appear to be mostly heterochromatic (based on H3K9 methylation), they may carry part of the euchromatic tip from chromosome 4. Thus, they were initially assayed for the presence of several genes present in this euchromatic region both genetically via complementation tests as well as molecularly through a qPCR-based analysis. Both approaches indicated the B chromosomes do not carry chromosome 4 genes [12].

To conduct a deeper molecular examination of their composition, the B chromosomes were separated from the A chromosomes via pulsed-field gel electrophoresis (PFGE) [13]. Using this method, the B chromosomes appear to be homogeneous in size at approximately 1.8 Mbp. The DNA band corresponding to the B chromosomes was extracted from the gel and subjected to next-generation sequence analysis in the absence of the A chromosomes. Even though no known genic regions were detected, a number of genetic elements most often associated with heterochromatin (e.g., transposons and satellite repeat sequences) were found in which a large portion was full of repetitive sequences also found on chromosome 4 [13].

3.2. Effects of the Drosophila melanogaster B Chromosomes on the A Chromosomes

Stocks that carry an average of 10 to 12 B chromosomes do not have an obvious phenotype. However, the presence of B chromosomes can have a measurable effect on the A chromosomes. During female meiosis, the segregation of chromosome 4 is disrupted when B chromosomes are present, which increases their frequency of mis-segregation by nearly two orders of magnitude [12]. The mechanism of this disruption is presently unknown. The achiasmate (non-crossover) system that segregates chromosome 4 may be encumbered in the presence of B chromosomes or possibly due to the similarity of heterochromatic sequence between the B chromosomes and chromosome 4 is affecting the establishment of heterochromatin associations earlier in meiosis [35–37].

The B chromosomes can also influence position effect variegation (PEV), which occurs when a gene in a euchromatic region is intermittently silenced due to its proximity to a heterochromatic region [38]. Placing the visual genetic marker *white* (*w*), which controls the degree of eye pigmentation, into one of these regions will lead to eyes with a variegated color pattern. Though the mechanism of PEV suppression is not entirely understood, it is thought that the presence of excess heterochromatin (such as an extra *Y* chromosome or an extra chromosome 4) has the ability to act as a sink for silencing proteins. This would reduce the overall spread of heterochromatin into nearby euchromatin, which increases the expression from *w* to the result in eyes with more pigment and less variegation. The B chromosomes were shown to affect PEV at multiple regions in the genome, which indicates that their presence can alter the heterochromatic silencing boundaries of the A chromosomes [12].

4. The *D. melanogaster* B Chromosome as an Emerging Model for B Chromosome Biology

Every B chromosome has its own unique biology and there is no shortage of interesting questions that can be asked. Two recurring questions—how are B chromosomes formed and what is the mechanism of B chromosome maintenance over time—are challenging to answer in non-model systems that lack genetic and molecular tools. The B chromosomes of *D. melanogaster* are poised to be a powerful system to begin studying how B chromosomes arise as well as what is required for them to be maintained and proliferate through a stock.

Currently, cytological and molecular analysis of the *D. melanogaster* B chromosomes indicates they arose from chromosome 4, but the underlying mechanism of their creation is unknown. Examination of the *AAGAT* FISH signal shows that it is more abundant on the B chromosomes, which leads to the hypothesis that the B chromosomes formed after a centromere mis-division of chromosome 4 during meiosis [13]. To test this hypothesis, a chromosome with markers on either end would need to be created and monitored after hundreds—or possibly thousands—of meiotic events. With the genetic toolbox available in *D. melanogaster*, marking a specific site within the genome with a visible genetic marker is a routine task. Additionally, the husbandry of *D. melanogaster* in the laboratory is easily amenable to large-volume cultures, which enables the researcher to screen as many flies as necessary. It is also unclear whether the appearance of the B chromosomes was fortuitous or if the genotype of the stock played a role in their creation. Once a basal frequency of B chromosome formation can be established, this question can be addressed by testing whether the *matrimony* genetic background influences the frequency of B chromosome formation. From here, one can test a variety of other existing mutant *D. melanogaster* stocks that are curated in the Bloomington *Drosophila* Stock Center and available to the research community.

How the original *D. melanogaster* B chromosomes have been maintained in such high copy number for several years is also puzzling. B chromosomes in other systems have been shown to promote their own propagation to the next generation through a variety of drive mechanisms (for a review, see "Transmission and Drive Involving Parasitic B Chromosomes" that is included in this special issue [11]). It is clear that the *D. melanogaster* B chromosomes do not supply their own mechanism of drive [12] and no known genes were detected after their genomic analysis [13]. Therefore, one hypothesis is that the genetic background of the stock may be responsible for their maintenance. It is known that *matrimony* plays an essential role in *Drosophila* female meiosis and is required for the accurate segregation of chromosomes that do not form crossovers [32–34]. Since *D. melanogaster* B chromosomes do not appear to form crossovers (as demonstrated by their dynamics on the meiotic spindle [12]), the disruption of *matrimony* may be influencing their segregation. Observing how the B chromosomes segregate is attainable in *D. melanogaster* either in fixed specimens using established protocols or in live egg chambers with a menagerie of fluorescent tools [39,40]. Ultimately, studying B chromosome formation and behavior in such an established model system will accelerate our understanding of general B chromosome biology, which may be applicable to other systems that are less amenable to laboratory research.

5. Conclusions

The first *Drosophila* species reported to carry B chromosomes was *D. albomicans* from Thailand over 30 years ago. Since then, very few *Drosophila* species have been found to carry B chromosomes, but it is fortunate that one of them is *D. melanogaster*. With its powerful set of genetic and genomic tools, there is much to learn about the *D. melanogaster* B chromosomes. Dissecting the biology of B chromosomes in such a robust and versatile model system will provide insight into other B chromosome systems as well as small supernumerary chromosomes (sSMC) found in humans that can be detrimental to our health. These sSMC are similar to B chromosomes in many ways and have been associated with various syndromes, intellectual disabilities, and infertility [41–43]. Thus, the B chromosomes in *D. melanogaster* stand to be a promising model system with discoveries that will potentially be of value to both biology and medicine.

Author Contributions: S.L.H. and R.S.H. contributed equally to the preparation and editing of this review.

Funding: S.L.H. and R.S.H. are funded by the Stowers Institute for Medical Research. R.S.H. is an American Cancer Society professor.

Conflicts of Interest: The authors declare no conflicts of interest.

References

1. Wilson, E.B. The supernumerary chromosomes of Hemiptera. *Science* **1907**, *26*, 870–871. [CrossRef]
2. Jones, R.N.; Rees, H. *B Chromosomes*; Academic Press: London, UK, 1982; ISBN 9780123900609.

3. Camacho, J. B chromosomes. In *The Evolution of the Genome*; Gregory, T.R., Ed.; Elsevier: San Diego, CA, USA, 2005; pp. 223–286. ISBN 9780123014634.

4. D'Ambrosio, U.; Alonso-Lifante, M.P.; Barros, K.; Kovařík, A.; Mas de Xaxars, G.; Garcia, S. B-chrom: A database on B chromosomes of plants, animals and fungi. *New Phytol.* **2017**, *216*, 635–642. [CrossRef] [PubMed]

5. Banaei-Moghaddam, A.M.; Martis, M.M.; Macas, J.; Gundlach, H.; Himmelbach, A.; Altschmied, L.; Mayer, K.F.X.; Houben, A. Genes on B chromosomes: old questions revisited with new tools. *Biochim. Biophys. Acta* **2015**, *1849*, 64–70. [CrossRef] [PubMed]

6. Ruban, A.; Schmutzer, T.; Scholz, U.; Houben, A. How next-generation sequencing has aided our understanding of the sequence composition and origin of B chromosomes. *Genes (Basel)* **2017**, *8*, 294. [CrossRef] [PubMed]

7. Valente, G.T.; Nakajima, R.T.; Fantinatti, B.E.A.; Marques, D.F.; Almeida, R.O.; Simões, R.P.; Martins, C. B chromosomes: from cytogenetics to systems biology. *Chromosoma* **2017**, *126*, 73–81. [CrossRef] [PubMed]

8. Navarro-Domínguez, B.; Ruiz-Ruano, F.J.; Cabrero, J.; Corral, J.M.; López-León, M.D.; Sharbel, T.F.; Camacho, J.P.M. Protein-coding genes in B chromosomes of the grasshopper *Eyprepocnemis plorans*. *Sci. Rep.* **2017**, *7*, 45200. [CrossRef] [PubMed]

9. Coan, R.L.B.; Martins, C. Landscape of transposable elements focusing on the B chromosome of the cichlid fish *Astatotilapia latifasciata*. *Genes (Basel)* **2018**, *9*, 269. [CrossRef] [PubMed]

10. Makunin, A.I.; Dementyeva, P.V.; Graphodatsky, A.S.; Volobouev, V.T.; Kukekova, A.V.; Trifonov, V.A. Genes on B chromosomes of vertebrates. *Mol. Cytogenet.* **2014**, *7*, 99. [CrossRef] [PubMed]

11. Jones, R.N. Transmission and Drive Involving Parasitic B Chromosomes. *Genes (Basel)* **2018**, *9*, 388. [CrossRef] [PubMed]

12. Bauerly, E.; Hughes, S.E.; Vietti, D.R.; Miller, D.E.; McDowell, W.; Hawley, R.S. Discovery of supernumerary B chromosomes in *Drosophila melanogaster*. *Genetics* **2014**, *196*, 1007–1016. [CrossRef] [PubMed]

13. Hanlon, S.L.; Miller, D.E.; Eche, S.; Hawley, R.S. Origin, Composition, and Structure of the Supernumerary B Chromosome of *Drosophila melanogaster*. *Genetics* **2018**. [CrossRef]

14. O'Grady, P.M.; DeSalle, R. Phylogeny of the Genus *Drosophila*. *Genetics* **2018**, *209*, 1–25. [CrossRef] [PubMed]

15. Clayton, F.E. Published karyotypes of the Drosophilidae. *Dros. Inf. Serv.* **1998**, *81*, 5–125.

16. Clyde, M. Chromosome IV variation in *D. albomicans* Duda. *Dros. Inf. Serv.* **1980**, *55*, 25–26.

17. Hatsumi, M.; Kitagawa, O. Supernumerary chromosomes in *Drosophila albomicans* collected in Thailand. *Abstr. Int. Congr. Entomol.* **1980**, *16*, 124.

18. Tonomura, Y.; Tobari, Y.N. Y-chromosome variation and B-chromosomes in *Drosophila malerkotliana*. *Sci. Rep. Tokyo Women's Christ. Univ.* **1983**, *32*, 705–711.

19. Sundaran, A.K.; Gupta, J.P. A new finding of B chromosomes in *Drosophila kikkawai* Burla. *Rev. Bras. Genet.* **1994**, *17*, 223–224.

20. Gutknecht, J.; Sperlich, D.; Bachmann, L. A species specific satellite DNA family of *Drosophila subsilvestris* appearing predominantly in B chromosomes. *Chromosoma* **1995**, *103*, 539–544. [CrossRef] [PubMed]

21. Deng, Q.; Zeng, Q.; Qian, Y.; Li, C.; Yang, Y. Research on the karyotype and evolution of *Drosophila melanogaster* species group. *J. Genet. Genom.* **2007**, *34*, 196–213. [CrossRef]

22. Mafla-Mantilla, A.B.; Romero, E.G. The heterochromatin of *Drosophila inca*, *Drosophila yangana* and *Drosophila huancavilcae* of the *inca* subgroup, *repleta* group. *Dros. Inf. Serv.* **2009**, *92*, 10–15.

23. Ramachandra, N.B.; Ranganath, H.A. Supernumerary chromosomes in *Drosophila nasuta albomicana*. *Experientia* **1985**, *41*, 680–681. [CrossRef] [PubMed]

24. Ramachandra, N.B.; Ranganath, H.A. Further studies on B-chromosomes in *D. nasuta albomicana*. *Dros. Inf. Serv.* **1985**, *61*, 139–140.

25. Ramachandra, N.B.; Ranganath, H.A. Accumulation of B-chromosomes in *Drosophila nasuta albomicana*. *Curr. Sci.* **1987**, *56*, 850–852.

26. Hatsumi, M. Karyotype polymorphism in *Drosophila albomicans*. *Genome* **1987**, *29*, 395–400. [CrossRef]

27. Ramachandra, N.B.; Ranganath, H.A. Characterization of heterochromatin in the B chromosomes of *Drosophila nasuta albomicana*. *Chromosoma* **1987**, *95*, 223–226. [CrossRef]

28. Zhou, Q.; Zhu, H.; Huang, Q.; Zhao, L.; Zhang, G.; Roy, S.W.; Vicoso, B.; Xuan, Z.; Ruan, J.; Zhang, Y.; et al. Deciphering neo-sex and B chromosome evolution by the draft genome of *Drosophila albomicans*. *BMC Genom.* **2012**, *13*, 109. [CrossRef] [PubMed]

29. Sundaran, A.K.; Gupta, J.P. Accumulation of B chromosomes in *Drosophila kikkawai* Burla. *Cytobios* **1994**, *80*, 211–215. [PubMed]

30. Baimai, V.; Chumchong, C. Karyotype variation and geographic distribution of the three sibling species of the *Drosophila kikkawai* complex. *Genetica* **1980**, *54*, 113–120. [CrossRef]

31. Brncic, D. Chromosomal structure of populations of *Drosophila flavopilosa* studied in larvae collected in their natural breeding sites. *Chromosoma* **1962**, *13*, 183–195. [CrossRef] [PubMed]

32. Bonner, A.M.; Hughes, S.E.; Chisholm, J.A.; Smith, S.K.; Slaughter, B.D.; Unruh, J.R.; Collins, K.A.; Friederichs, J.M.; Florens, L.; Swanson, S.K.; et al. Binding of *Drosophila* Polo kinase to its regulator Matrimony is noncanonical and involves two separate functional domains. *Proc. Natl. Acad. Sci. USA* **2013**, *110*, e1222–e1231. [CrossRef] [PubMed]

33. Harris, D.; Orme, C.; Kramer, J.; Namba, L.; Champion, M.; Palladino, M.J.; Natzle, J.; Hawley, R.S. A deficiency screen of the major autosomes identifies a gene (*matrimony*) that is haplo-insufficient for achiasmate segregation in *Drosophila* oocytes. *Genetics* **2003**, *165*, 637–652. [PubMed]

34. Xiang, Y.; Takeo, S.; Florens, L.; Hughes, S.E.; Huo, L.-J.; Gilliland, W.D.; Swanson, S.K.; Teeter, K.; Schwartz, J.W.; Washburn, M.P.; et al. The inhibition of polo kinase by matrimony maintains G2 arrest in the meiotic cell cycle. *PLoS Biol.* **2007**, *5*, e323. [CrossRef] [PubMed]

35. Dernburg, A.F.; Sedat, J.W.; Hawley, R.S. Direct evidence of a role for heterochromatin in meiotic chromosome segregation. *Cell* **1996**, *86*, 135–146. [CrossRef]

36. Karpen, G.H.; Le, M.H.; Le, H. Centric heterochromatin and the efficiency of achiasmate disjunction in *Drosophila* female meiosis. *Science* **1996**, *273*, 118–122. [CrossRef] [PubMed]

37. Hawley, R.S.; Theurkauf, W.E. Requiem for distributive segregation: A chiasmate segregation in *Drosophila* females. *Trends Genet.* **1993**, *9*, 310–317. [CrossRef]

38. Elgin, S.C.R.; Reuter, G. Position-effect variegation, heterochromatin formation, and gene silencing in *Drosophila*. *CSH Perspect. Biol.* **2013**, *5*, a017780. [CrossRef] [PubMed]

39. Page, A.W.; Orr-Weaver, T.L. Activation of the meiotic divisions in *Drosophila* oocytes. *Dev. Biol.* **1997**, *183*, 195–207. [CrossRef] [PubMed]

40. Endow, S.A.; Komma, D.J. Spindle dynamics during meiosis in *Drosophila* oocytes. *J. Cell Biol.* **1997**, *137*, 1321–1336. [CrossRef] [PubMed]

41. Fuster, C.; Rigola, M.A.; Egozcue, J. Human supernumeraries: Are they B chromosomes? *Cytogenet. Genome Res.* **2004**, *106*, 165–172. [CrossRef] [PubMed]

42. Liehr, T.; Mrasek, K.; Kosyakova, N.; Ogilvie, C.M.; Vermeesch, J.; Trifonov, V.; Rubtsov, N. Small supernumerary marker chromosomes (sSMC) in humans; are there B chromosomes hidden among them. *Mol. Cytogenet.* **2008**, *1*, 12. [CrossRef] [PubMed]

43. Liehr, T. *Small Supernumerary Marker Chromosomes (SMCs), A Guide for Human Geneticists and Clinicians*; Springer: Berlin/Heidelberg, Germany, 2012; ISBN 9783642207655.

© 2018 by the authors. Licensee MDPI, Basel, Switzerland. This article is an open access article distributed under the terms and conditions of the Creative Commons Attribution (CC BY) license (http://creativecommons.org/licenses/by/4.0/).

genes

MDPI

Article

B Chromosomes of the Asian Seabass (*Lates calcarifer*) Contribute to Genome Variations at the Level of Individuals and Populations

Aleksey Komissarov [1,*], Shubha Vij [2,3], Andrey Yurchenko [1,4], Vladimir Trifonov [5,6], Natascha Thevasagayam [2], Jolly Saju [2], Prakki Sai Rama Sridatta [2], Kathiresan Purushothaman [2,7], Alexander Graphodatsky [5,6], László Orbán [2,8,9,*] and Inna Kuznetsova [2,*]

1   Theodosius Dobzhansky Center for Genome Bioinformatics, Saint Petersburg State University, St. Petersburg 199004, Russia; andreyurch@gmail.com
2   Reproductive Genomics Group, Temasek Life Sciences Laboratory, Singapore 117604, Singapore; shubha_vij@rp.edu.sg (S.V.); natmay@gmail.com (N.T.); jolly@tll.org.sg (J.S.); prakki_sr_sridatta@ttsh.com.sg (P.S.R.S.); kathiresan.purushothaman@nord.no (K.P.)
3   School of Applied Science, Republic Polytechnic 9 Woodlands Avenue 9, Singapore 738964, Singapore
4   Institute of Biodiversity, Animal Health & Comparative Medicine, College of Medical, Veterinary & Life Sciences, University of Glasgow, Glasgow G12 8QQ, UK
5   Institute of Molecular and Cellular Biology, Siberian Branch of the Russian Academy of Sciences, Novosibirsk 630090, Russia; vlad@mcb.nsc.ru (V.T.); graf@mcb.nsc.ru (A.G.)
6   Department of Natural Science, Novosibirsk State University, Novosibirsk 630090, Russia
7   Faculty of Biosciences and Aquaculture, Nord University, 8049 Bodø, Norway
8   Department of Animal Sciences, Georgikon Faculty, University of Pannonia, H-8360 Keszthely, Hungary
9   Center for Comparative Genomics, Murdoch University, 6150 Murdoch, Australia
*   Correspondence: ad3002@gmail.com (A.K.); orban@georgikon.hu (L.O.); inna.kuznetcova@gmail.com (I.K.)

Received: 4 August 2018; Accepted: 12 September 2018; Published: 20 September 2018

Abstract: The Asian seabass (*Lates calcarifer*) is a bony fish from the Latidae family, which is widely distributed in the tropical Indo-West Pacific region. The karyotype of the Asian seabass contains 24 pairs of A chromosomes and a variable number of AT- and GC-rich B chromosomes (Bchrs or Bs). Dot-like shaped and nucleolus-associated AT-rich Bs were microdissected and sequenced earlier. Here we analyzed DNA fragments from Bs to determine their repeat and gene contents using the Asian seabass genome as a reference. Fragments of 75 genes, including an 18S rRNA gene, were found in the Bs; repeats represented 2% of the Bchr assembly. The 18S rDNA of the standard genome and Bs were similar and enriched with fragments of transposable elements. A higher nuclei DNA content in the male gonad and somatic tissue, compared to the female gonad, was demonstrated by flow cytometry. This variation in DNA content could be associated with the intra-individual variation in the number of Bs. A comparison between the copy number variation among the B-related fragments from whole genome resequencing data of Asian seabass individuals identified similar profiles between those from the South-East Asian/Philippines and Indian region but not the Australian ones. Our results suggest that Bs might cause variations in the genome among the individuals and populations of Asian seabass. A personalized copy number approach for segmental duplication detection offers a suitable tool for population-level analysis across specimens with low coverage genome sequencing.

Keywords: teleost; population analysis; whole genome resequencing; DNA copy number variation; ribosomal DNA; B chromosomes

1. Introduction

The Asian seabass is an economically important food fish species, the aquaculture production of which is rapidly spreading to a large number of countries extending beyond its native range [1,2].

It is a euryhaline and catadromous species with a wide geographic distribution stretching from Northern Australia to at least the Western part of India. A comprehensive analysis of the identity of the Asian seabass across its geographical range of distribution was described earlier using morphometric data and/or key mitochondrial sequences [3–5] and later based on whole genome resequencing information [6]. Sequence data of three mitochondrial markers (cytochrome c oxidase subunit 1 COI, 16S rDNA and D-loop) pointed to the existence of at least two distinct species, the first representing the Indian subcontinent, and the second South-East Asia (Singapore, Malaysia, Thailand, and Indonesia) together with Australia/Papua New Guinea, with the latter showing signs of splitting from the South-East Asian region [3–6].

The Asian seabass in Australia has a stronger migration capability than in South-East Asia and is adapted to both freshwater and marine environments [7]. The Asian seabass in South-East Asia shows an earlier maturation time than in Australia/Papua New Guinea [8] and was found only in marine water [9]. All Asian seabass populations were examined using 32,433 single-nucleotide polymorphisms (SNPs). [6]. It has revealed a genetic difference between South-East Asian and Australian/Papua New Guinean populations. There has been no recent gene flow between Asian seabass from South-East Asia and Australian/Papua New Guinea; however, the current patterns of genetic variability were likely caused by the co-effects of founder events, random genetic drift, mutations, and local selection [10]. The Asian seabass is a protandrous hermaphrodite with most individuals typically maturing as males at about a year of age and subsequently changing their sex to females [7,8]. Sex reversal in hermaphrodites may happen during a period of a few weeks, or up to several months [11–13]. The reproductive cycle of hermaphroditic teleosts can be affected by various environmental factors and is also known to be associated with the age of fish [11,14]. Comparative analysis of the expression of genes with a sex-associated function (e.g., germ cell markers, *Wnt* and retinoic acid signaling genes, and apoptotic genes) between the two gonad types showed differences in the patterns observed in mammals and other teleosts [13].

The Asian seabass genome was sequenced recently using long-read sequencing technology, followed by a multi-step assembly involving optical mapping and a comparative analysis of syntenic regions among three fish genomes, resulting in an assembly size of 670 Mb [6]. The diploid Asian seabass karyotype contains 24 pairs of A chromosomes and a variable number of either AT- or GC-rich B chromosomes (Bchrs or Bs in short) [6]. Bs, also known as supernumerary or accessory chromosomes, sometimes occur in eukaryotic genomes, typically constitute 1–5% of genome size and represent a non-essential and dynamic component of the genome. Accumulation of Bs may take place in either males or females and copy number polymorphism is often population-specific [15–17]. Bs are found in ~15% of karyotyped eukaryotic species and exhibit significant similarity in their morphology and structure in many species [15,18]. Certain common features have been demonstrated for Bs of various species, including plants. For instance, they have been shown to be comprised of an arrangement of A chromosome fragments [17,19,20]. The presence of Bs in the grass species *Aegilops speltoides* increases the frequency of heterologous synapses and recombination, causing new chromosomal abnormalities [21]. In addition, DNA fragments like rDNAs, tandem repeats, and clustered mobile elements have been documented in Bs of different plants [17,20,22,23], insects [24,25], fishes [7,23,24,26], and mammals [18,27,28]. These impact on the tissue-specific dynamics of mobile elements and tandem repeats [17,21,26]. The presence of genes on Bs was also demonstrated [17–19], however, the functional role and precise structure of Bs in most species are still unclear.

Previously, we isolated and sequenced fragments of 4′,6-diamidino-2-phenylindole (DAPI)-stained supernumerary chromosomes and presented their preliminary analysis, demonstrating homology of Asian seabass B-related fragments to four genomic scaffolds located on four specific linkage groups (LG5, LG9, LG17 and LG19) and several genomic regions that have not been assigned to the LGs [6]. In this study, we performed a more detailed cytogenetic and sequencing analysis of AT-rich Asian seabass Bs. In addition, we analyzed the content contribution of Bs to population-specific genomic changes among Asian seabass populations collected from India, South-East Asia/Philippines and Australia/Papua New Guinea.

2. Materials and Methods

2.1. Fish Material and DNA Extraction

Wild Asian seabass were procured from different locations. The freshly caught fishes were then obtained from the landing centers followed by sample collection. Farmed Asian seabass (Lates calcarifer) were obtained from the Marine Aquaculture Centre (Singapore). All experiments were approved by Agri-food and Veterinary Authority (AVA) Institutional Animal Care and Use Committee (IACUC) (approval ID: AVA-MAC-2012-02) and performed according to guidelines set by the National Advisory Committee on Laboratory Animal Research (NACLAR) for the care and use of animals for scientific research in Singapore.

To study the potential variation in the number of B-associated regions at a population scale, 50 wild-caught Asian seabass samples, collected from 13 geographic regions across the native range: India Western coast (four specimens) and Indian Eastern coast (five), South-East Asia (20), Philippines (four), Australia (12) and Papua New Guinea (five), were re-sequenced to ~7× average sequencing depth using the HiSeq 1500 Illumina platform as reported previously [6].

Asian seabass larvae at 1–2 days post-hatching (dph) were euthanized by placing them on ice, dissected and used for the culturing of primary fibroblasts and chromosome preparation as described previously [6,29]. In addition, we used small pieces of fins from three males and three females (nine months of age) for fibroblast culturing. Liver, skin, ovary and testis of ten male and female (nine-month-old) fish were used for DNA/RNA extractions as well as cell isolation for fluorescence-activated cell sorting (FACS) analysis. DNA was extracted using Qiagen kit (Qiagen, Hilden, Dermany) according to the manufacturer's instructions.

2.2. Estimation of DNA Content in Different Tissues by Flow Cytometry

Nuclei were extracted from male and female gonads, skin and livers as described previously [30]. The nuclei suspension was stained with propidium iodide (PI) for estimation of total nuclear DNA content [31]. The fluorescence peak of G0/G1 nuclei isolated from the liver of chicken (standard) was used as reference standard. For each sample, when possible, at least three independent replicate measurements were performed. Each probe contained 5000–10,000 nuclei, the average coefficient of variation for the probes and dyes ranged between 2–5%. The genome size was calculated by multiplying the size of the standard genome (chicken) with the ratio of their relative fluorescence intensities. The average 2C-value (in pg) was converted to bp size considering that 1 pg of DNA corresponds to 0.978×10^9 bp [32]. A comparison between the individual data was performed using the Student t-test, $p < 0.01$.

2.3. Isolation and Sequencing of B Chromosomes

B chromosome microdissection, amplification, library construction, sequencing and assembly have been described previously [6]. B chromosome-specific libraries (Bchr5 and Bchr6) were sequenced on a MiSeq genome sequencer (Illumina) in the Genomics Core Facility (Institute of Chemical Biology and Fundamental Medicine, Siberian Branch of the Russian Academy of Sciences, Novosibirsk, Russia). After trimming, MiSeq reads were then mapped to the Asian seabass reference genome assembly [6]. The consensus sequence motifs, obtained from the mapping of reads, were further chained together across increasing lengths of a spacer to form pseudo-scaffolds [6].

2.4. Fluorescence In Situ Hybridization (FISH) and Immunohistochemical Analysis

Isolation of Asian seabass *Cot-1* DNA was described in our earlier study [29]. The whole genome amplification (WGA1) PCR-amplified DNA of the micro-dissected B chromosomes and *Cot-1* were re-amplified in the presence of 16-dUTP-biotin and digoxigenin-11-dUTP, respectively (the final concentration of each was 2 µM, Roche (Basel, Switzerland) under the following conditions: (1×) 94 °C for 5 min; (35×) 90 °C for 30 s, 54 °C for 30 s, 72 °C for 30 s using a WGA3 reamplification kit (Sigma-Aldrich, St. Louis, MO, USA). Preparation of the 18S rDNA probe was performed by amplification from genomic DNA using specific primers 5′-GCGAAGGGTAGACACACGCTGA-3′, 5′-CCTCTAGCGGCACAATACGAATG-3′ [33].

The DNA mixture of ~400 ng of the Bchr-derived painting probe and 10 µg of the *Cot-1* fraction of the Asian seabass DNA was prepared. A fluorescent in situ hybridization (FISH) analysis was performed as described earlier [6,28]. Signal detection was accomplished using streptavidin conjugated with AlexaFluor-594 (ThermoFisher Scientific, Waltham, MA, USA) and anti-digoxigenin-fluorescein (Roche) to detect digoxigenin.

After post-hybridization washes, an anti-5-MeC antibody (AbCam, Cambridge, UK) was applied. An anti-mouse AlexaFluor-488 antibody was used to detect the signal. Finally, the slides were counterstained with DAPI and mounted in an antifade solution (Vectashield from Vector laboratories, Burlingame, CA, USA).

Chromosome spreads were incubated with rabbit a centromere-specific protein A (CENP-A) antibody (Aviva Systems Biology, San Diego, CA, USA) in blocking solution overnight at 4 °C, followed by a 1 h room temperature incubation with a secondary biotinylated anti-rabbit antibody (Vector Lab, Peterborough, UK).

Chromosomal images were captured on a Zeiss/MetaMorph epifluorescence microscope equipped with a Nikon (CCD) camera. Fibroblast sections were examined using a Zeiss LSM 510 META inverted microscope. Images were analyzed using Image-Pro Express software V5.0 (Media Cybernetics, CA, USA). The path of the chromosomes was computationally traced and straightened according to the manual provided by the Image J software V1.41 (http://rsb.info.nih.gov/ij).

2.5. Annotation of B Chromosome-Derived Pseudo-Scaffolds

Gene descriptions were assigned to the B-derived pseudo-scaffolds based on gene prediction and annotation of the corresponding regions in the Asian seabass genome [6] and Zebrafish Model Organism Database (ZFIN) [34]. Repeat analysis was performed using the Asian seabass repeats database [6] and the Repbase dataset for vertebrate genomes (http://www.girinst.org/repbase/).

2.6. Data Availability

The scaffolded genome assembly has been submitted to DDBJ/EMBL/NCBI GenBank under the accession number of LLXD00000000.1. Alternatively, it is also available for download at http://seabass.sanbi.ac.za/. The Illumina and PacBio reads utilized for the genome assembly, as well as the whole-genome resequencing reads, have been submitted to NCBI SRA under BioProject accession numbers SRP069219 and SRP069848, respectively. The raw reads obtained from the sequencing of Bchr5 and Bchr6 libraries have been deposited in the NCBI SRA database under the BioProject accession number SRP082620.

2.7. Population-Level Analyses of Asian Seabass B Chromosomes from Re-Sequencing Data

Population scale variation in the number of copies of B-associated regions was estimated using the read-depth based algorithm [35]. Trimmed reads from the 51 sequenced samples, including the reference genome, were aligned to the repeat-masked Asian seabass reference genome (version 2.0) using Bowtie2 [36] with -a mode (search for and report all alignments). Read alignment and SNP-calling in repeats are unreliable because of the high misalignment rate and their problematic assembly; thus, repeat-rich regions were excluded from this analysis. The resulting sequences alignment map (SAM) files were processed using mrCaNaVaR [35] to produce a genome-wide GC-corrected copy number of the genome fragments in 1000 bp windows, excluding repetitive and gapped regions of the genome. A copy number variation analysis was performed using the samples with the most uniform coverage to achieve comparativeness between the studied individuals from the previous study [6].

Two single-exon genes were selected as controls as described earlier [37]: ZNFX1-type zinc finger-containing protein 1-like *(znfx-1)* and galactose-3-*O*-sulfotransferase 3 *(gal3st3)*. The control genome fragment with a size similar to that of Bchr6 (~400 kb) was taken from LG20 (unitig_2 from 6205808 to 6605807). This fragment did not have any sequences similar to Bchr6.

2.8. Estimation of the Density of Single Nucleotide Variants and Polymorphisms

Estimation of the density of single nucleotide variants and polymorphisms (SNVs/SNPs) in the B-associated regions was performed based on SNP-calling results from the Asian seabass resequencing data published earlier [6]. SNP-calling was carried out using the Samtools pipeline [6]. The corresponding coordinates for the 10 kb spacer B-derived pseudo-scaffolds were extracted from the whole genome SNP-calling results. Possible functional effects of SNVs in the B regions were evaluated using SNPeff [38].

2.9. Expression of Predicted B-Chromosomal Genes

Information for the expression of genes identified from B chromosomal pseudo-scaffolds was obtained from microarray-based data that was previously produced for the analysis of the Asian seabass gonadal transcriptome [39].

3. Results

3.1. Asian Seabass B Chromosomes are Repeat-Rich and Associated with Nucleolus

Dot-like Bs were estimated to constitute 0.05–0.5% of the DAPI-stained area of all metaphase chromosomes (Figure 1A). The number of DAPI-stained Bs in the primary fibroblast cell culture did not show any specific association with sex and varied from zero to four across individual metaphase spreads, suggesting an intra-individual mosaicism.

To examine the contribution of Bs to DNA content differences between certain tissues, we estimated the DNA content in liver or skin as well as male and female gonads using a flow cytometric approach. The data showed that the diploid DNA content in the ovary of the mature Asian seabass was significantly lower than that of the testis (705 ± 42 Mb vs. 773 ± 50 Mb; $n = 10$ per sex; $p < 0.01$; Student's *t*-test). On the other hand, there was no statistical difference in genome size between male and female hepatocyte or skin nuclei (734 ± 66 Mb vs. 747 ± 54 Mb; $p < 0.01$; $n = 10$ per sex; Student's *t*-test).

In situ hybridization of a labelled Asian seabass *Cot-1* DNA probe to its own chromosomes revealed that repetitive sequences were abundant in the genome. Their distribution was dispersed along the length of all A chromosomes with an insignificant accumulation on the centromeric and telomeric regions whereas some of the Bs appeared to be largely free from *Cot-1* DNA hybridization signals (Figure 1B).

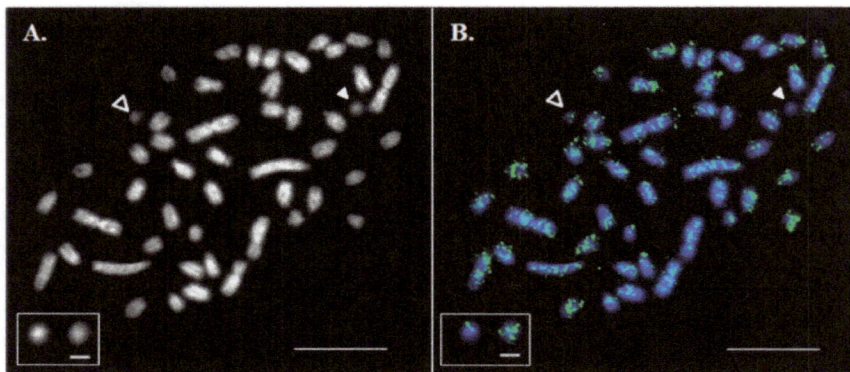

Figure 1. Hybridization of mixed *Cot-1* DNA from liver, ovary and testis on metaphase chromosomes. (**A**) A DAPI-stained metaphase spread. (**B**) Asian seabass *Cot-1* DNA fraction hybridization (green). Chromosomes were contrasted by 4′,6-diamidino-2-phenylindole (DAPI; blue). Empty arrowheads indicate B chromosomes (Bchrs). White arrowheads indicate Bchrs with a low level of *Cot-1* signal. The bar is 5 μm. The inset shows magnified examples of Bchrs with a low (left) and high (right) intensity of *Cot-1* signal; the bar for the framed B chromosomes is 1 μm.

We have shown earlier that fluorescent in situ hybridization signals of Bchr5- and Bchr6-derived painting probes result in an overlapping pattern [6]. In the interphase nuclei, Bchr-derived, labeled probes occupied a highly methylated region, indicating that the chromatin of Bchrs was not active at this stage and was in contact with the nucleolus (Figure 2A). Two pairs of autosomes with an 18S rDNA signal were identified on the metaphase spread. Bchrs with and without 18S rDNA are shown in Figure 2B. The presence of the centromere/kinetochore protein A (CENP-A) could not be detected on Bchrs (Figure 2C).

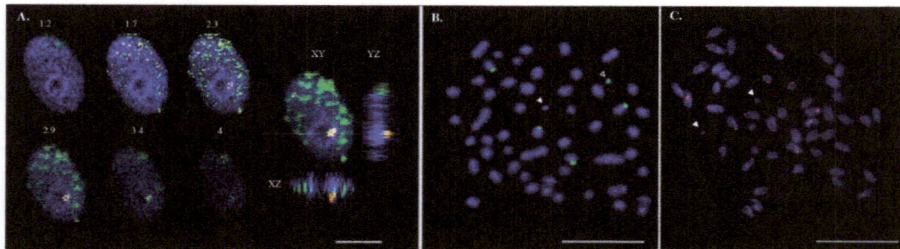

Figure 2. Micro-dissected B chromosome-derived probes associated with the nucleolus and 18S rDNA were detected on both A and B chromosomes of the Asian seabass. (**A**) Hybridization of Bchr6 painting probes on primary fibroblast cell nuclei. Confocal section with a 0.5 μm step and three nuclear projections (XY, YZ, XZ). Micro-dissected probes of Bchr6 (red), anti-5-methylcytosine AB (green). Nuclei were counterstained with DAPI (blue). The bar is 5 μm. (**B**) B chromosomes with and without 18S rDNA (green signal). White arrowheads indicate a Bchr without 18S rDNA; empty arrowheads indicate a Bchr with an 18S rDNA signal. The bar is 5 μm. (**C**) Histone H3-like centromere protein A (CENP-A) stained chromosomes (red signal). White arrowheads indicate a Bchr without CENPA. The bar is 5 μm.

3.2. B Chromosomal Repeats

The assembly size of Bchr5 and Bchr6, together with the 10 kb spacer, was 386,725 bp, representing ~0.067% of the Asian seabass reference genome assembly [6]. This is similar to the cytogenetic estimation of the total size of Bs. Therefore, subsequent downstream analyses were performed on the 10 kb-joined pseudo-scaffolds.

Here we demonstrated that the Bchr6 assembly contained just ~2% of characterized repeats compared to 16.75% present in the whole genome. Short fragments of transposons were identified in both assemblies of Bs (Table S1). In spite of FISH signal accumulation for the Bchr6 and Bchr5 probes in the centromeric region of metaphase A chromosomes. We did not find any sign of the centromeric/pericentromeric satellite DNA of A chromosomes within the Bchrs libraries (Table S1).

Bchr5-specific sequences contained a fragment homologous to LG5 as well as fragments of the Asian seabass genome without any linkage group assignment [6]. In the Asian seabass genome assembly, one of the sequences coding for rDNAs was located as a repeat unit on scaffold_93 assigned to LG5 (Figure 3A). The Bchr5 assembly contained short fragments of unknown DNA and a 24,841 bp contig (contig 1), which aligned with a major ribosomal rRNA region. Three copies of 18S rDNA were found in contig 1 of Bchr5 (Figure 3B, Table S1). The 18S rRNA of Bchr5, as well as scaffold 93, of the Asian seabass genome, was compared with the *Cyprinus carpio* 18S rRNA fragment (JN628435, 9426 bp, RefSeq). The results showed that the 18S ribosomal RNA genes and internal transcribed spacer 1 for both genomes and Bchr5 were similar but not identical (Figure 3B). The 18S rDNA region from the Asian seabass genome assembly (scaffold 93) and Bchr5 assembly showed a complex arrangement of 18S rDNA and various fragments of transposable elements, mainly from the piggyBAC and Gypsy families (Figure 3, Table S1).

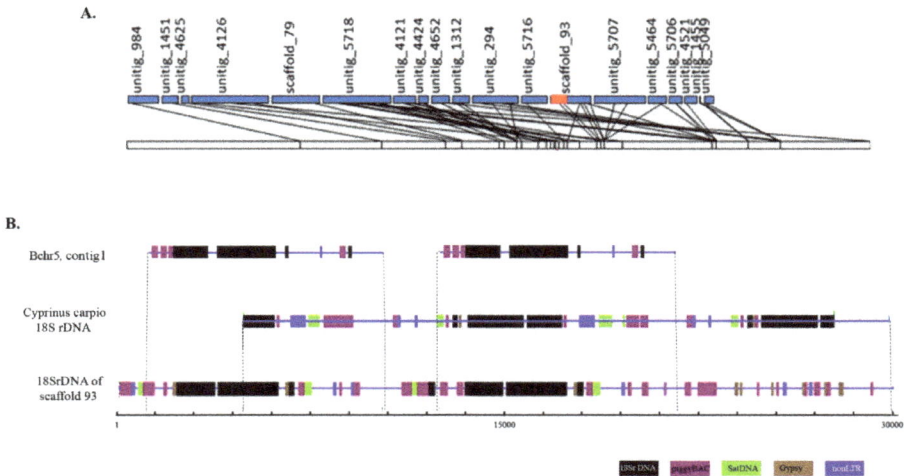

Figure 3. The 18S rDNA region of B chromosome 5 (Bchr5) shows a complex arrangement of various fragments of transposable elements with ribosomal genes and is different from the 18S rDNA region of the Asian seabass autosomal chromosomes. (**A**) Diagrammatic representation of genetic linkage group 5 (LG5). The ribosomal region located on genomic scaffold_93 assigned to LG5 is labeled in red. (**B**) Alignment of ribosomal genes containing the region of scaffold_93 of the Asian seabass genome, contig 1 of micro-dissected Bchr5 and rRNA sequence of *Cyprinus carpio* (9426 bp; JN628435). Various fragments of repetitive elements are indicated as follows: Satellite DNA (green), piggy BAC and R2 (violet), Gypsy (brown), and 18S rDNA (black).

3.3. B chromosomal Genes

In total, 75 genes were identified in the Bchr6 assembly (listed in Table S2), whereas just 18S rDNA was identified from Bchr5. The average gene densities for the assembled genome and the partial Bchrs assembly were 30.2 per kb and 5.16 per kb respectively. Thus, six times more genes per kilobase were present in the assembled genome compared to the Bchr6 assembly. On the other hand, the density of SNVs in the B chromosomal regions was approximately eight per kb (2901 SNVs per 361,036 kb), a value identical to that of the Asian seabass genome (5,642,327 SNVs per 670 Mb; [6]). Genomic fragments representing Bchr6 showed different genomic structural variations including insertions and deletions, however, genomic structure variations such as splice-site disruption and a gain of stop codons could not be identified. No specific function could be attributed to ~30 of the annotated B-specific genes (Table S2), while another ~30 genes were found to be expressed in most of the organs (Table S2, ZFIN [6]). The remaining ten genes were found to have a specific expression associated with the gonads and the brain (*astn2, bre, dpf3, fbxo33, fkbp, gabrb2, myt1l, rab14, rxrab,* and *pacrg,* see Table S2).

3.4. Population-Level Analysis of B Chromosome-Associated Regions Supports the Existence of an Asian Seabass Species Complex

We have performed a detailed analysis of the copy number variation of Bchr6, whereas Bchr5 was excluded from this analysis due to the enrichment of repeats (Figure 3, Table S1). Analyses of sequences representing Bchr6 from the 51 individuals across the different geographical regions (Figure 4A) revealed copy number variations in 438 unique genomic fragments (Table S3). Of these, 198 were annotated while 240 were not described earlier as genes or repeats. DNA fragments representing Bchr6 were observed to have a tendency to multiply (>3 copies per haploid genome) in the genomes (Figure 4C; Table S3). The copy number variation profile amongst the Bchr6 fragments was more variable in the Indian seabass individuals compared to the other two populations. Individuals from Australia and Papua New Guinea had minimal copy number variation (Figure 4C; Table S3).

Samples from the Indian and South-East Asia/Philippines regions had a wider range of copy number variation between specimens compared to those from Australia and Papua New Guinea. A wide range of copy numbers from Indian and South-East Asia/Philippines specimens was detected for gene fragments coding for brain and reproductive organ-expressed proteins (*bre*; copy number range 1.1–30.6 and 1.2–14.6 for India and South-East Asia respectively), phosphatidic acid phosphatase type 2B (*ppap2b*; 1.1–11.2 vs. 1.5–8.25), and Ras protein-specific guanine nucleotide-releasing factor 1 (*rasgrf*; 1.6–25 vs.1.6–15.5) (Table S3, Sheet 1). Fragments corresponding to control genes (*gal3st3* and *znfx-1*, see scaffold 14 and unitig_4699 on Sheet 1 of Table S3) and the control genome fragment corresponding to fragment of LG20 (unitig_2, 6205808 to 6605807; see Sheet 2 of Table S3) had copy numbers typical for normal diploid genomes (1.5–3) for each of the 51 specimens from all three geographical regions (Figure 4C, Table S3, Sheet 2).

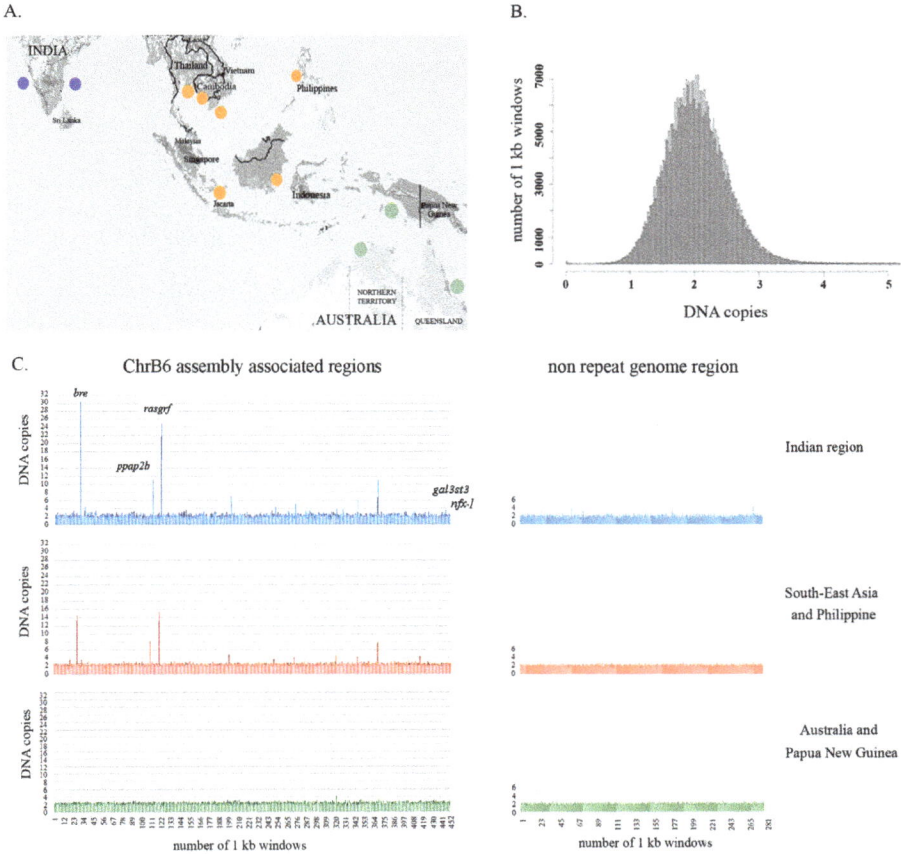

Figure 4. Population-scale variation in the copy number of B chromosome-associated regions in the Asian seabass species complex. (**A**) Geographic regions across the native range of Asian seabass: Western and Eastern Coast of India; South-East Asia and the Philippines, Australia/Papua New Guinea. (**B**) Distribution of inferred genome fragment copy number in 1 kb windows across the Asian seabass genome: Homozygous deletion, 0.1–1.4; heterozygous deletion, 1.5–2.5; normal diploid copy number, 2.6–4; heterozygous duplication, > 4; multiple duplications. X-axis: DNA copies; Y-axis: Number of 1 kb windows. (**C**) Population-scale variations in the copy number of the Bchr6-associated region and control genome fragment were demonstrated for the three geographical regions. X-axis: Number of genomic fragments which correspond to a particular 1 kb window of the reference genome for each individual. Y-axis: Genomic fragment copy number in 1 kb windows across the Bchr6-associated region and non-repeat genomic fragment. The order is indicated in Table S3: Indian region (blue), South-East Asia and Philippine region (orange), Australia/Papua New Guinea region (green). The copy number of control genes (*gal3st3*, *znfx-1*) is presented in 442–452 regions. The estimation of copy number variation was performed after repeat masking, so the number of DNA fragments and size of the analyzed sequences were less than the source. The Asian seabass genome assembly version 2.0 was used as a reference genome.

Thus, the data demonstrated a high level of polymorphism of the Bchr6 content providing further evidence and additional insight into the structuring of the three major geographical groups, which were shown earlier based on mitochondrial and nuclear markers, morphometric data and whole genome re-sequencing [5].

4. Discussion

At the time of its publication, the genome of the Asian seabass had the best assembly metrics among the de novo assembled fish genomes available thus far [6], which made it a good tool for detailed genomic and cytogenetic studies. The species is a protandrous hermaphrodite: Individuals typically mature as males and later reverse their sex to become females. The standard diploid chromosome number for male and female Asian seabass is identical ($2n$ = 48), as demonstrated by the high-resolution linkage map constructed using 790 microsatellites and SNPs [40] and confirmed by cytological analysis [29].

Variable numbers of both AT- and GC-rich Bs were detected in metaphase plates of cultured primary fibroblast cells from adult male and female skin, as well as larvae of the Asian seabass [29]. Some metaphase plates of the Asian seabass fibroblast culture did not have any Bs, which points to an intra-individual mosaicism related to their unstable transmission during mitosis. The variability in the number of the Bs is associated with their typical non-Mendelian inheritance pattern and could be explained by their elimination during mitosis, possibly under the influence of environmental factors [15].

The presence of Bs usually increases the individual's genome size but does not substantially affect the variability within the complex or population, regardless of whether individuals with accessory chromosomes are included [41]. Comparative sequence analyses of the genomes of B-carrying and B-lacking individuals from closely related populations are informative [20,24]. The inter-individual difference between the Asian seabass from various geographical regions was obvious (Figure 4). Investigation of the Asian seabass B-related fragments identified 438 DNA fragments with a tendency for amplifications or deletions (Table S3) in two of the three geographical regions of Asian seabass distribution, namely in India and South-East Asia/Philippines (Figure 4). Control genomic regions were slightly more variable in the Indian specimens than in those from South-East Asia/Philippines and Australia. Therefore, a more stable genome with less population fluctuation was demonstrated for the Australia/Papua New Guinea population. Thus, the copy number variation of the B-related fragments was in agreement with the deep phylogenetic analysis of the Asian seabass populations based on mtDNA markers, re-sequenced genomes [5,6] and SNP analyses [10]. Here, we have demonstrated the genetic diversity and patterns of population differentiation across all of the populations and we have provided significant evidence for genetic differentiation between South-East Asian and Australian/Papua New Guinean populations; in agreement with the results of Wang [10], who assumed that the Asian seabass of these regions had been evolving towards allopatric speciation since the split from the ancestral population during the mid-Pleistocene.

To summarize, the population analysis of the widely distributed Asian seabass has given an indication that supernumerary chromosomes could be involved in genome size variations between populations. The presence or absence of Bs in different populations of the Asian seabass can be linked to environmental factors in different geographical regions.

As described earlier, the sequenced heterochromatin from Asian seabass Bs (Bchr5 and Bchr6) did not contain any pericentromeric or centromeric sequences from autosomes [5,6,29]. In addition, the Bchr-derived repeats also had short fragments of transposable elements (TE) from the host genome (Table S1). This agrees with the cytogenetic examination of the repetitive DNAs (*Cot-1*) distribution on Bs, which was extremely weakly stained by the total repetitive DNA probe as well as with the absence of histone H3-like centromere protein A (CENPA) signals on Bs (Figure 2C). The centromeres of the rye Bs evolved from a standard centromere and lack sequences connected with centromeric, i.e., CENH3-histone-containing nucleosomes or histone H3-like centromeric protein A. Additional repeats accumulated in the centromere of the newly formed B chromosome [17,42]. A similar structure of a centromere may be characteristic for the Asian seabass Bs.

These results suggest that the repetitive DNA of the Asian seabass Bs had a specific set of repeats different from those of autosomal origin and might possess a specific centromere and sub-telomeric blocks (Figure 1B). Full-size transposons were absent from the Bchr5 and Bchr6 assemblies, however, transposon-derived fragments can maintain their functional activity in the genome. The role of particular repetitive elements and their fragments associated with the Bs of teleosts has also been actively discussed [24,41]. Multiplication of transposable element (TE) fragments in Bs of cichlid [20,42,43] and cyprinid [26] fish led to the proposal that TEs are responsible for the insertion of sequences into Bs [17,41]. This mechanism could explain the presence of fragments of different genes with a variable copy number. For example, Bs of the grasshopper *Eyprepocnemis plorans* contained large amounts of rDNA [42], whereas the arrangement of the ribosomal genes and surrounding rDNA-specific TEs, such as R2, Gypsy, RTE and mariner transposable elements is different in Bs and host genomes [18,20,24,25]. It was shown that R2 elements are transcribed in the ovaries and eggs but not in male tissues or embryos and different R2 5′ truncation patterns are found in natural populations, providing evidence for recent retro-transposition activity [24]. A similar organization was demonstrated for rRNA genes in Bs of *Drosophila virilis* [44], grasshopper *E. plorans* [24,25], and teleosts [45,46]. We observed that the Asian seabass Bs are located adjacent to the nucleolus within the nucleus, suggesting their potential contribution to nuclear formation (Figures 2 and 3). The Asian seabass Bchr5 assembly contained fragments of 18S rRNA genes and short R2 fragments in addition to other elements (Table S1). Therefore, the repeat units of the rRNA genes are usually located very close to blocks of constitutive heterochromatin and they are flanked by transposon elements (Figure 3B).

A comparison of Asian seabass Bs with those from different species revealed common features of these genomic elements. Bs typically contain thousands of sequences duplicated from the A chromosomes [18,42]. Most of the genes that originate from Bs are fragmented but some, including those influencing cell metabolism [47] and rRNA genes [48], may still maintain their transcriptional activity. Recently, a novel male-specific sequence, called *setdm* located on extra micro-chromosomes, was described in Gibel carp [49]. Fragments of genes, which are potentially located in Bs, did not contain any nonsense or missense mutations (see Material and Methods) and could be involved in different processes, including sexual development, cell division, immune response, and DNA repair (Table S2).

Cytogenetic analysis of four Japanese hagfish species from the order Myxinida (*Eptatretus okinoseanus*, *Eptatretus burgeri*, *Paramyxine atami*, and *Myxine garmani*) revealed differences in the B numbers between germ cells (spermatocytes and spermatogonia) and somatic cells (liver, blood, gill, and kidney) [50]. The C-banding of metaphase chromosome preparations of germline and somatic cells from each hagfish species revealed that the C-band-positive chromatin in the ancestral somatic cells had been almost completely eliminated [51].

Microarray analysis of the expression of 60,080 Asian seabass genes was used earlier to identify differentially expressed genes in transforming gonad types that are in a sequential order of development [39]. From 75 genes identified on the Bchr6-derived pseudo-scaffolds, 42 were inventoried for differential expression in transforming gonads from the above-mentioned microarray experiment (Table S2). Therefore, we can assume that these 42 genes are presumably active in Bchr6 but final confirmation of their activity would require further experiments.

A number of different sex determination systems have been described for fishes [14,16,52,53]. They can even be different between populations of the same species. For example, a comparative analysis of the *Eigenmannia* genus suggests that their sex chromosomes may have arisen independently in the different populations [52,53]. A connection between Bs and sex determination was described for Amazon molly [54], in Lake Victoria cichlids [55] and some cyprinids [49]. A cytogenetic analysis of four Japanese hagfish species from the order Myxinida revealed differences in the B numbers between male germ cells (spermatocytes and spermatogonia) and somatic cells (liver, blood, gill, and kidney) [49].

In conclusion, Bs are a major source of intraspecific variation in the nuclear DNA amounts in numerous species of plants and create polymorphisms for DNA variation in natural populations [22]. Different populations of the Asian seabass may differ in the number and composition of B chromosomes. A personalized copy number approach for segmental duplication detection offers a suitable tool for a population-level analysis across specimens with low coverage genome sequencing.

Supplementary Materials: The following are available online at ttp://www.mdpi.com/2073-4425/9/10/464/s1, Table S1: Repeats of Bchr5 assembly (Sheet 1), Bchr6 assembly (Sheet 2), and Asian seabass genome scaffold_93 for control (Sheet 3).; Table S2: The list of 75 genes predicted from the B chromosome 6 assembly and two single exon genes, which were used as controls; Table S3: Population-scale variation in the copy number of B chromosome 6-associated regions (Sheet 1) and control fragment of the Asian seabass reference genome (Sheet 2, unitig_2).

Author Contributions: Conceived and designed the experiments: I.K., S.V., L.O., V.T.; performed the experiments: J.S., K.P., I.K.; performed the bioinformatics analysis: P.S.R.S., N.T., A.K., A.Y.; wrote the MS: I.K., A.K., V.T., A.G. All authors revised the manuscript.

Funding: This research was supported by the National Research Foundation, Prime Minister's Office, Singapore under its Competitive Research Program (CRP Award No. NRF-CRP7-2010-01); interdisciplinary grant of the SB RAS No. 137; supported by RFBR grants 17-00-00144 and 17-00-00146 as part of 17-00-00148; VAT was supported by RSF (No. 18-44-04007).

Conflicts of Interest: The authors declare that there are no conflicts of interest regarding the publication of this paper.

References

1. Chou, R.; Lee, H.B. Commercial marine fish farming in Singapore. *Aquacult. Res.* **1997**, *28*, 767–776. [CrossRef]

2. Frost, L.A.; Evans, B.S.; Jerry, D.R. Loss of genetic diversity due to hatchery culture practices in barramundi (*Lates calcarifer*). *Aquaculture* **2006**, *261*, 1056–1064. [CrossRef]

3. Pethiyagoda, R.; Gill, A.C. Description of two new species of sea bass (Teleostei: Latidae: Lates) from Myanmar and Sri Lanka. *Zootaxa* **2012**, *3314*, 1–16. [CrossRef]

4. Ward, R.D.; Zemlak, T.S.; Innes, B.H.; Last, P.R.; Hebert, P.D.N. DNA barcoding Australia's fish species. *Philos. Trans. R. Soc. Lond. B Biol. Sci.* **2005**, *360*, 1847–1857. [CrossRef] [PubMed]

5. Vij, S.; Purushothaman, K.; Gopikrishna, G.; Lau, D.; Saju, J.M.; Shamsudheen, K.V.; Kumar, K.V.; Basheer, V.S.; Gopalakrishnan, A.; Hossain, M.S.; et al. Barcoding of Asian seabass across its geographic range provides evidence for its bifurcation into two distinct species. *Front. Mar. Sci.* **2014**, *1*, 30. [CrossRef]

6. Vij, S.; Kuhl, H.; Kuznetsova, I.S.; Komissarov, A.; Yurchenko, A.A.; Van Heusden, P.; Singh, S.; Thevasagayam, N.M.; Prakki, S.R.S.; Purushothaman, K.; et al. Chromosomal-level assembly of the Asian seabass genome using long sequence reads and multi-layered scaffolding. *PLoS Genet.* **2016**, *12*, e1005954. [CrossRef]

7. Russell, D.; Garrett, R. Early life history of barramundi, *Lates calcarifer* (Bloch), in north-eastern Queensland. *Mar. Freshw. Res.* **1985**, *36*, 191–201. [CrossRef]

8. Moore, R. Spawning and early life history of barramundi, *Lates calcarifer* (Bloch), in Papua New Guinea. *Mar. Freshw. Res.* **1982**, *33*, 647–661. [CrossRef]

9. Jerry, D.R. (Ed.) *Biology and Culture of Asian Seabass*; CRC Press, Taylor & Francis Group: Boca Raton, FL, USA, 2014; ISBN 978-1-4822-0807-8.

10. Wang, L.; Wan, Z.Y.; Lim, H.S.; Yue, G.H. Genetic variability, local selection and demographic history: Genomic evidence of evolving towards allopatric speciation in Asian seabass. *Mol. Ecol.* **2016**, *25*, 3605–3621. [CrossRef] [PubMed]

11. Zohar, Y.; Abraham, M.; Gordin, H. The gonadal cycle of the captivity-reared hermaphroditic teleost *Sparus aurata* (L.) during the first two years of life. *Ann. Biol. Anim. Biochim. Biophys.* **1978**, *18*, 877–882. [CrossRef]

12. Sadovy, Y.; Shapiro, D.Y. Criteria for the diagnosis of hermaphroditism in fishes. *Copeia* **1987**, *1987*, 136. [CrossRef]

13. Ravi, P.; Jiang, J.; Liew, W.C.; Orbán, L. Small-scale transcriptomics reveals differences among gonadal stages in Asian seabass (*Lates calcarifer*). *Reprod. Biol. Endocrinol.* **2014**, *12*, 5. [CrossRef] [PubMed]

14. Guiguen, Y.; Cauty, C.; Fostier, A.; Fuchs, J.; Jalabert, B. Reproductive cycle and sex inversion of the seabass, *Lates calcarifer*, reared in sea cages in French Polynesia: Histological and morphometric description. *Environ. Biol. Fishes* **1994**, *39*, 231–247. [CrossRef]

15. Camacho, J.P.M.; Schmid, M.; Cabrero, J. B chromosomes and sex in animals. *Sex. Dev.* **2011**, *5*, 155–166. [CrossRef] [PubMed]

16. Devlin, R.H.; Nagahama, Y. Sex determination and sex differentiation in fish: An overview of genetic, physiological, and environmental influences. *Aquaculture* **2002**, *208*, 191–364. [CrossRef]

17. Houben, A. B Chromosomes—A matter of chromosome drive. *Front. Plant Sci.* **2017**, *8*, 210. [CrossRef] [PubMed]

18. Makunin, A.I.; Dementyeva, P.V.; Graphodatsky, A.S.; Volobouev, V.T.; Kukekova, A.V.; Trifonov, V.A. Genes on B chromosomes of vertebrates. *Mol. Cytogenet.* **2014**, *7*, 99. [CrossRef] [PubMed]

19. Puertas, M.J. Nature and evolution of B chromosomes in plants: A non-coding but information-rich part of plant genomes. *Cytogenet. Genome Res.* **2002**, *96*, 198–205. [CrossRef] [PubMed]

20. Valente, G.T.; Conte, M.A.; Fantinatti, B.E.A.; Cabral-de-Mello, D.C.; Carvalho, R.F.; Vicari, M.R.; Kocher, T.D.; Martins, C. Origin and evolution of B chromosomes in the cichlid fish *Astatotilapia latifasciata* based on integrated genomic analyses. *Mol. Biol. Evol.* **2014**, *31*, 2061–2072. [CrossRef] [PubMed]

21. Ruban, A.; Fuchs, J.; Marques, A.; Schubert, V.; Soloviev, A.; Raskina, O.; Badaeva, E.; Houben, A. B Chromosomes of *Aegilops speltoides* are enriched in organelle genome-derived sequences. *PLoS ONE* **2014**, *9*, e90214. [CrossRef] [PubMed]

22. Jones, R.N.; Viegas, W.; Houben, A. A Century of B chromosomes in plants: So what? *Ann. Bot.* **2008**, *101*, 767–775. [CrossRef] [PubMed]

23. Lamb, J.C.; Riddle, N.C.; Cheng, Y.-M.; Theuri, J.; Birchler, J.A. Localization and transcription of a retrotransposon-derived element on the maize B chromosome. *Chromosom. Res.* **2007**, *15*, 383–398. [CrossRef] [PubMed]

24. Montiel, E.E.; Cabrero, J.; Ruiz-Estévez, M.; Burke, W.D.; Eickbush, T.H.; Camacho, J.P.M.; López-León, M.D. Preferential occupancy of R2 retroelements on the B chromosomes of the grasshopper *Eyprepocnemis plorans*. *PLoS ONE* **2014**, *9*, e91820. [CrossRef] [PubMed]

25. Montiel, E.E.; Cabrero, J.; Camacho, J.P.M.; López-León, M.D. Gypsy, RTE and Mariner transposable elements populate *Eyprepocnemis plorans* genome. *Genetica* **2012**, *140*, 365–374. [CrossRef] [PubMed]

26. Ziegler, C.G.; Lamatsch, D.K.; Steinlein, C.; Engel, W.; Schartl, M.; Schmid, M. The giant B chromosome of the cyprinid fish *Alburnus alburnus* harbours a retrotransposon-derived repetitive DNA sequence. *Chromosom. Res.* **2003**, *11*, 23–35. [CrossRef]

27. Makunin, A.I.; Kichigin, I.G.; Larkin, D.M.; O'Brien, P.C.M.; Ferguson-Smith, M.A.; Yang, F.; Trifonov, V.A. Contrasting origin of B chromosomes in two cervids (Siberian roe deer and grey brocket deer) unravelled by chromosome-specific DNA sequencing. *BMC Genome* **2016**, *17*, 618. [CrossRef] [PubMed]

28. Makunin, A.I.; Rajičić, M.; Karamysheva, T.V.; Romanenko, S.A.; Druzhkova, A.S.; Blagojević, J.; Vujošević, M.; Rubtsov, N.B.; Graphodatsky, A.S.; Trifonov, V.A. Low-pass single-chromosome sequencing of human small supernumerary marker chromosomes (sSMCs) and *Apodemus* B chromosomes. *Chromosoma* **2018**, *127*, 301–311. [CrossRef] [PubMed]

29. Kuznetsova, I.S.; Thevasagayam, N.M.; Sridatta, P.S.R.; Komissarov, A.S.; Saju, J.M.; Ngoh, S.Y.; Jiang, J.; Shen, X.; Orbán, L. Primary analysis of repeat elements of the Asian seabass (*Lates calcarifer*) transcriptome and genome. *Front. Genet.* **2014**, *5*, 223. [CrossRef] [PubMed]

30. Heinlein, C.; Speidel, D. High-resolution cell cycle and DNA ploidy analysis in tissue samples. *Curr. Protoc. Cytom.* **2011**. [CrossRef] [PubMed]

31. Carvalho, C.R.; Clarindo, W.R.; Praça, M.M.; Araújo, F.S.; Carels, N. Genome size, base composition and karyotype of *Jatropha curcas* L., an important biofuel plant. *Plant Sci.* **2008**, *174*, 613–617. [CrossRef]

32. Dolezel, J.; Bartos, J.; Voglmayr, H.; Greilhuber, J. Nuclear DNA content and genome size of trout and human. *Cytom. A* **2003**, *51*, 127–128. [CrossRef]

33. Mantovani, M.; Dos Santos Abel, L.D.; Moreira-Filho, O. Conserved 5S and variable 45S rDNA chromosomal localisation revealed by FISH in *Astyanax scabripinnis* (Pisces, Characidae). *Genetica* **2005**, *123*, 211–216. [CrossRef] [PubMed]

34. Howe, D.G.; Bradford, Y.M.; Conlin, T.; Eagle, A.E.; Fashena, D.; Frazer, K.; Knight, J.; Mani, P.; Martin, R.; Moxon, S.A.T.; et al. ZFIN, the zebrafish model organism database: Increased support for mutants and transgenics. *Nucleic Acids Res.* **2013**, *41*, D854–D860. [CrossRef] [PubMed]

35. Alkan, C.; Kidd, J.M.; Marques-Bonet, T.; Aksay, G.; Antonacci, F.; Hormozdiari, F.; Kitzman, J.O.; Baker, C.; Malig, M.; Mutlu, O.; et al. Personalized copy number and segmental duplication maps using next-generation sequencing. *Nat. Genet.* **2009**, *41*, 1061–1067. [CrossRef] [PubMed]

36. Langmead, B.; Salzberg, S.L. Fast gapped-read alignment with Bowtie 2. *Nat. Methods* **2012**, *9*, 357–359. [CrossRef] [PubMed]

37. Li, C.; Ortí, G.; Zhang, G.; Lu, G. A practical approach to phylogenomics: The phylogeny of ray-finned fish (Actinopterygii) as a case study. *BMC Evol. Biol.* **2007**, *7*, 44. [CrossRef] [PubMed]

38. Platt, A.; Gugger, P.F.; Pellegrini, M.; Sork, V.L. Genome-wide signature of local adaptation linked to variable CpG methylation in oak populations. *Mol. Ecol.* **2015**, *24*, 3823–3830. [CrossRef] [PubMed]

39. Jiang, J. Functional Genomic Analysis of Gonad Development in the Protandrous Asian Seabass. Ph.D. Thesis, National University of Singapore, Singapore, 2014.

40. Wang, C.M.; Bai, Z.Y.; He, X.P.; Lin, G.; Xia, J.H.; Sun, F.; Lo, L.C.; Feng, F.; Zhu, Z.Y.; Yue, G.H. A high-resolution linkage map for comparative genome analysis and QTL fine mapping in Asian seabass, *Lates calcarifer*. *BMC Genom.* **2011**, *12*, 174. [CrossRef] [PubMed]

41. Chumová, Z.; Mandáková, T.; Trávníček, P. Are B-chromosomes responsible for the extraordinary genome size variation in selected Anthoxanthum annuals? *Plant Syst. Evol.* **2016**, *302*, 731–738. [CrossRef]

42. Fantinatti, B.E.A.; Mazzuchelli, J.; Valente, G.T.; Cabral-de-Mello, D.C.; Martins, C. Genomic content and new insights on the origin of the B chromosome of the cichlid fish *Astatotilapia latifasciata*. *Genetica* **2011**, *139*, 1273–1282. [CrossRef] [PubMed]

43. Shirak, A.; Grabherr, M.; Di Palma, F.; Lindblad-Toh, K.; Hulata, G.; Ron, M.; Kocher, T.D.; Seroussi, E. Identification of repetitive elements in the genome of *Oreochromis niloticus*: Tilapia repeat masker. *Mar. Biotechnol.* **2010**, *12*, 121–125. [CrossRef] [PubMed]

44. Abdurashitov, M.A.; Gonchar, D.A.; Chernukhin, V.A.; Tomilov, V.N.; Tomilova, J.E.; Schostak, N.G.; Zatsepina, O.G.; Zelentsova, E.S.; Evgen'ev, M.B.; Degtyarev, S. Medium-sized tandem repeats represent an abundant component of the *Drosophila virilis* genome. *BMC Genom.* **2013**, *14*, 771. [CrossRef] [PubMed]

45. López-Flores, I.; Garrido-Ramos, M.A. The Repetitive DNA Content of Eukaryotic Genomes. *Genome Dyn.* **2012**, *7*, 1–28. [CrossRef] [PubMed]

46. Gornung, E. Twenty years of physical mapping of major ribosomal RNA genes across the teleosts: A review of research. *Cytogenet. Genome Res.* **2013**, *141*, 90–102. [CrossRef] [PubMed]

47. Trifonov, V.A.; Dementyeva, P.V.; Larkin, D.M.; O'Brien, P.C.M.; Perelman, P.L.; Yang, F.; Ferguson-Smith, M.A.; Graphodatsky, A.S. Transcription of a protein-coding gene on B chromosomes of the Siberian roe deer (*Capreolus pygargus*). *BMC Biol.* **2013**, *11*, 90. [CrossRef] [PubMed]

48. Leach, C.R.; Houben, A.; Field, B.; Pistrick, K.; Demidov, D.; Timmis, J.N. Molecular evidence for transcription of genes on a B chromosome in *Crepis capillaris*. *Genetics* **2005**, *171*, 269–278. [CrossRef] [PubMed]

49. Li, X.-Y.; Liu, X.-L.; Ding, M.; Li, Z.; Zhou, L.; Zhang, X.-J.; Gui, J.-F. A novel male-specific SET domain-containing gene *setdm* identified from extra microchromosomes of gibel carp males. *Sci. Bull.* **2017**, *62*, 528–536. [CrossRef]

50. Nakai, Y.; Kubota, S.; Kohno, S. Chromatin diminution and chromosome elimination in four Japanese hagfish species. *Cytogenet. Cell Genet.* **1991**, *56*, 196–198. [CrossRef] [PubMed]

51. Kloc, M.; Zagrodzinska, B. Chromatin elimination—An oddity or a common mechanism in differentiation and development? *Differentiation* **2001**, *68*, 84–91. [CrossRef] [PubMed]

52. Henning, F.; Trifonov, V.; Ferguson-Smith, M.A.; de Almeida-Toledo, L.F. Non-homologous sex chromosomes in two species of the genus Eigenmannia (Teleostei: Gymnotiformes). *Cytogenet. Genome Res.* **2008**, *121*, 55–58. [CrossRef] [PubMed]

53. Silva, D.S.; Milhomem, S.S.; Pieczarka, J.C.; Nagamachi, C.Y. Cytogenetic studies in *Eigenmannia virescens* (Sternopygidae, Gymnotiformes) and new inferences on the origin of sex chromosomes in the Eigenmannia genus. *BMC Genet.* **2009**, *10*, 74. [CrossRef] [PubMed]

54. Lamatsch, D.K.; Nanda, I.; Epplen, J.T.; Schmid, M.; Schartl, M. Unusual triploid males in a microchromosome-carrying clone of the Amazon molly, *Poecilia formosa*. *Cytogenet. Cell Genet.* **2000**, *91*, 148–156. [CrossRef] [PubMed]

55. Yoshida, K.; Terai, Y.; Mizoiri, S.; Aibara, M.; Nishihara, H.; Watanabe, M.; Kuroiwa, A.; Hirai, H.; Hirai, Y.; Matsuda, Y.; et al. B chromosomes have a functional effect on female sex determination in Lake Victoria cichlid fishes. *PLoS Genet.* **2011**, *7*, e1002203. [CrossRef] [PubMed]

© 2018 by the authors. Licensee MDPI, Basel, Switzerland. This article is an open access article distributed under the terms and conditions of the Creative Commons Attribution (CC BY) license (http://creativecommons.org/licenses/by/4.0/).

GCAT
TACG
GCAT
genes

MDPI

Review

L Chromosome Behaviour and Chromosomal Imprinting in *Sciara Coprophila*

Prim B. Singh [1,2,*] and Stepan N. Belyakin [2,3]

[1] Nazarbayev University School of Medicine, 5/1 Kerei, Zhanibek Khandar Street,
    Astana Z05K4F4, Kazakhstan
[2] Epigenetics Laboratory, Department of Natural Sciences, Novosibirsk State University, Pirogov str. 2,
    Novosibirsk 630090, Russia; belyakin@mcb.nsc.ru
[3] Genomics laboratory, Institute of Molecular and Cellular Biology SB RAS, Lavrentyev ave, 8/2,
    Novosibirsk 630090, Russia
* Correspondence: prim.singh@nu.edu.kz; Tel.: +7-7172-694-706

Received: 19 July 2018; Accepted: 29 August 2018; Published: 3 September 2018

Abstract: The retention of supernumerary chromosomes in the germ-line of *Sciara coprophila* is part of a highly-intricate pattern of chromosome behaviours that have fascinated cytogeneticists for over 80 years. Germ-line limited (termed L or "limited") chromosomes are cytologically heterochromatic and late-replicating, with more recent studies confirming they possess epigenetic hallmarks characteristic of constitutive heterochromatin. Little is known about their genetic constitution although they have been found to undergo cycles of condensation and de-condensation at different stages of development. Unlike most supernumeraries, the L chromosomes in *S. coprophila* are thought to be indispensable, although in two closely related species *Sciara ocellaris* and *Sciara reynoldsi* the L chromosomes, have been lost during evolution. Here, we review what we know about L chromosomes in *Sciara coprophila*. We end by discussing how study of the L chromosome condensation cycle has provided insight into the site and timing of both the erasure of parental "imprints" and also the placement of a putative "imprint" that might be carried by the sperm into the egg.

Keywords: supernumerary chromosomes; heterochromatin; parent-of-origin effects; paternal X chromosome; maternal X chromosome; controlling element

1. Introduction

Sciara coprophila possesses a complicated, sometimes bizarre, pattern of chromosomal behaviours that involve both the regular chromosomes and the supernumerary germ-line limited or L chromosomes [1–3]. The cycles of both kinds of chromosome are intimately associated. Because of this, L chromosome behaviour is best described as part of the whole. Accordingly, we begin with the newly-fertilized egg in which the well-known X-X' device in mothers has conditioned the ooplasm to determine the sex of the embryo [4]. Looking at the chromosomal complements provided by each parent, the female pro-nucleus provides 5 chromosomes (3 autosomes, 1 X or X' chromosome and 1 L chromosome); sperm that forms the male pro-nucleus delivers 6–7 chromosomes (3 autosomes, 2 identical X chromosomes and 1–2 L chromosomes). After syngamy, the zygotic nucleus contains 11–12 chromosomes.

1.1. L Chromosome Elimination in the Soma

Cleavage divisions take place within a multinucleate syncytium and it is around the fifth to sixth division that the L chromosomes are eliminated from somatic nuclei of both males and females (Figure 1) [5]. Elimination results from a failure of L chromosomes to move pole-ward like the ordinary chromosomes. They are arrested in their anaphase separation and left behind at the equatorial plate to be found only later, discarded as amorphous clumps of chromatin at the periphery of the multinucleate cell. By contrast, L chromosomes are retained in the 2–4 primordial germ cells found at the posterior of the coenocyte after the fifth division. In this way, L chromosomes become limited to the germ line.

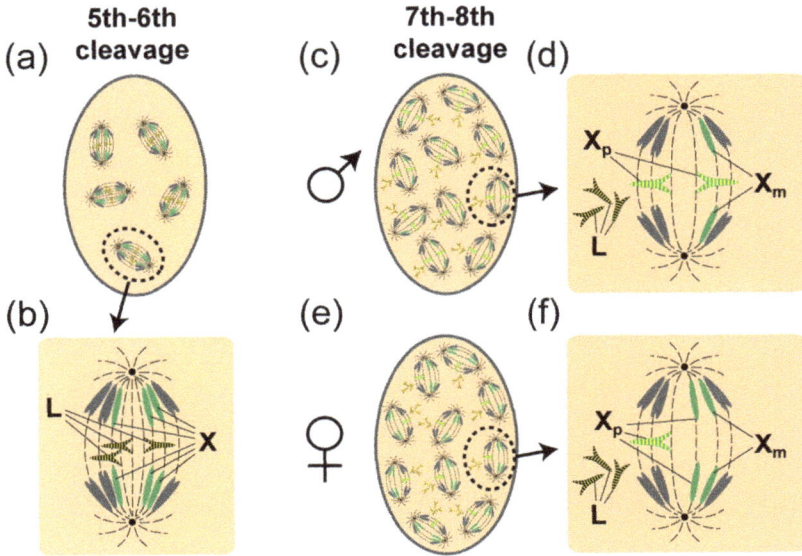

Figure 1. Elimination of L and paternal X chromosomes (Xp) in the embryonic soma. (**a**) Several mitoses are depicted in an embryo at the 5–6th embryonic cleavage where the L chromosomes are in the process of being eliminated. (**b**) One is magnified and shows that that the L chromosomes are left behind at the equatorial plate and fail to separate. The arms of the L chromosomes retain high levels of phosphorylation of histone H3 serine 10 (H3S10P) (shown as yellow stripes). (**c**) In males at the 7–8th cleavage two Xps are eliminated. (**d**) As with the eliminated L chromosomes, the arms of the eliminated Xps fail to separate and likewise retain high levels of H3S10P (shown as yellow stripes). The Xps remain at the equatorial plate and are eliminated giving a somatic male (XO) chromosome constitution. In (**d**) phosphorylated L chromosomes that have already been eliminated are shown. (**e**) In females, a single Xp is eliminated at the 7th–8th cleavage division. (**f**) The single Xp remains at the equatorial plate and retains high levels of H3S10P (shown as yellow stripes). Note that by contrast to the eliminated chromosomes the chromosomes that move away to the poles have lost H3S10P. Eliminations normally take place at the periphery of the multinucleate cell but because the diagram is in 2 dimensions mitoses are seen to fill the cell. The L chromosomes are black, the autosomes are grey and the Xps are green. Phosphorylation is depicted as yellow stripes.

After the somatic elimination of the L chromosomes there are nine chromosomes (6 autosomes and 3 X chromosomes) in somatic cells; the germ line contains 11–12 since the L chromosomes are retained. Strikingly, the eliminations in the soma do not end there. Another elimination event is programmed to take place soon after, which is regulated by the X-X′ mechanism that determines the sex of the embryo. At the seventh or eighth cleavage division, there is an elimination of X chromosomes, which are exclusively paternal in origin [1]. The somatic eliminations are complete at around 4–6 h post oviposition (see Table 16 in [6]). Thereafter the number of chromosomes is reduced to seven in the male soma (XO), because two paternal X chromosomes (Xps) are lost along with all the L chromosomes and to eight chromosomes in the female soma (XX or X′X), where one Xps is lost along with all L chromosomes (Figure 1d,e).

After the eliminations in the soma, germ cells undergo another division that is followed by a long resting stage during which the germ cells migrate to the site of the presumptive gonad. Soon after reaching the gonad another round of eliminations takes place.

1.2. L Chromosome Elimination in the Germ-Line

Between 24–48 h post-oviposition the 30 or so resting germ cells are in the process of migrating to the site of the presumptive gonad [6]. Each germ cell possesses three X chromosomes (two paternal and one maternal), the autosomes and all of the L chromosomes. Just prior to or concurrently with elimination, the regular chromosomes in both sexes form fairly discrete bodies calledpro-chromosomes [7], which are differentiated. The four paternal chromosomes are more diffuse and lightly-staining while the maternal homologues are more condensed and darkly-stained [6]. The fifth paternal chromosome, which is the extra X chromosome, remains condensed and it is this chromosome that is eliminated at 60–72 h post-oviposition from the germ cells in both sexes. Along with the paternal X chromosome all but two L chromosomes are eliminated [6]. Notably, the L chromosomes, which are typically heterochromatic and darkly-stained, become diffuse and lightly-strained at the same time as the paternal chromosome set [6]. After the eliminations two L chromosomes and eight regular chromosomes remain (Figure 2).

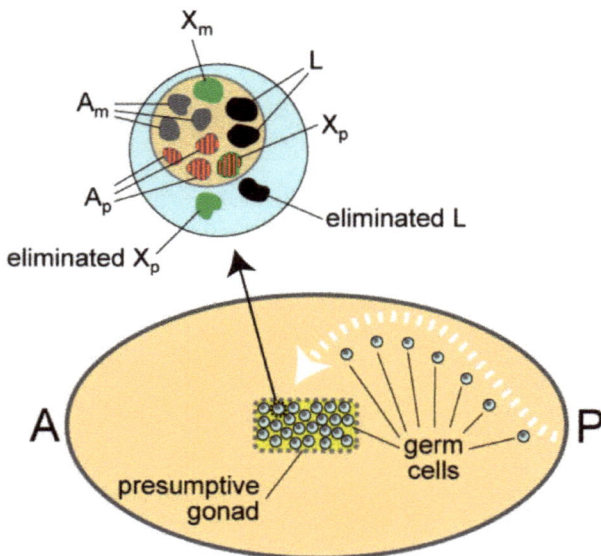

Figure 2. Elimination of L and Xp chromosomes in resting germ cells of the presumptive gonad. After rounds of cell division at the posterior of the embryo, germ cells enter a quiescent state and migrate to the presumptive gonad. Around 30 arrive in the gonad. In the gonad of both sexes, the chromosomes in the resting germ cells form distinct entities called pro-chromosomes, as shown in the magnified germ cell above the embryo. All but two L chromosomes are eliminated and one Xp is eliminated. The eliminated chromosomes lie in the cytoplasm. In the nucleus, the three paternal autosomes and the retained Xp are acetylated (acetylation is shown as red stripes); these four paternal chromosomes are also cytologically more diffuse than all the other chromosomes. Notably, the eliminated Xp is under-acetylated, which is also the case for all the other chromosomes. The L chromosomes are black, the autosomes are grey and the X chromosomes are green. Acetylation is shown as red stripes. Am are the maternal autosomes and Ap are the paternal autosomes. Xm is the maternal X chromosome and Xp is the paternal X chromosome. A is anterior and P is posterior.

The mechanism for elimination in germ cells is very different to that which takes place in soma of the early embryo (Figure 1). The germ cells have an intact nuclear membrane and elimination involves passage of whole chromosomes through the nuclear membrane into the cytoplasm where they become very smooth and round in contour and remain darkly-stained until they disappear. How passage through the membrane is achieved is not known, although the close apposition of the Xp to be eliminated to the nuclear membrane has led to the suggestion that membrane-spanning proteins, which interact with silenced chromatin—such as the lamin B receptor [8]—might be involved [9]. Elimination of the X chromosome may precede or follow L chromosome elimination in germ cells, unlike the strict order of elimination for the two kinds of chromosomes in the soma [6].

1.3. L Chromosome Behaviour During Male Meiosis

Meiosis is initiated during pupation some 23 days after oviposition [6]. In females, it is orthodox. The L chromosomes also pair and undergo the typical reduction divisions. In stark contrast, in males, both meiotic divisions are unequal, the first to a striking degree [1–3,10]. The first meiotic division is monocentric and brings about the selective elimination of the regular paternal homologues.

Accordingly, the four maternally-derived regular chromosomes move to the single pole and notably, along with them, move all L chromosomes of which there are usually two in number (Figure 3). Unlike the ordinary chromosomes, the parental origin of the L chromosomes is immaterial: The L chromosomes retained in the germ-line can be derived from either parent [6,11]. Thus, with regard to the L chromosomes, the first meiotic division in males is not a reduction division and they are included in unreduced number in the secondary spermatocyte and thereafter in the sperm [6]. The reason for the germ-line eliminations described above (Figure 2) becomes plain; in their absence, the number of L chromosomes would increase over successive generations.

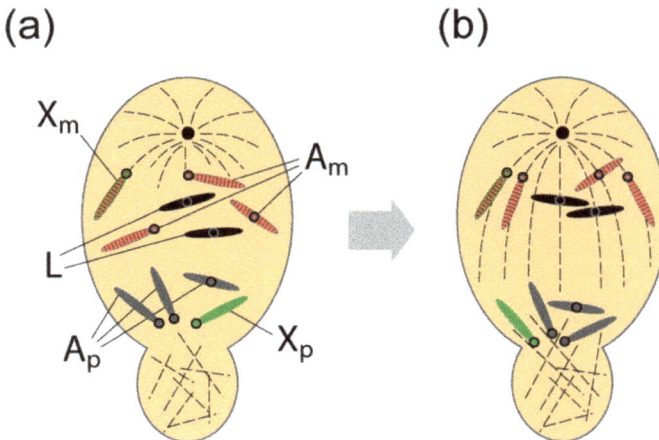

Figure 3. Retention of L and regular maternal homologues during first meiotic division in primary spermatocytes. (**a**) At meiotic prophase, the L chromosomes and regular maternal chromosomes display a similar yet separate localisation to the paternal homologues within the nucleus. A monopolar spindle is formed and non-spindle microtubules are generated in the cytoplasmic bud region. The maternal chromosomes are acetylated (acetylation is shown as red stripes). (**b**) The chromosomes move directly to an anaphase like stage where the L chromosomes and regular maternal chromosomes move towards the single pole while paternal homologues segregate into the bud. The maternal chromosomes are acetylated. Circles represent the positions of the centromeres. The L chromosomes are black, the autosomes are grey and the X chromosomes are green. Acetylation is shown as red stripes.

The second meiotic division is essentially orthodox except that the X-chromosome undergoes non-disjunction and is found precociously at the monopole (Figure 4). The X-dyad then passes into the sperm nucleus while the chromosome group devoid of X chromosomes is extruded and degenerates. At the end of the asymmetric meiosis the spermatocyte gives rise to one sperm cell and a bud containing the eliminated chromosomes.

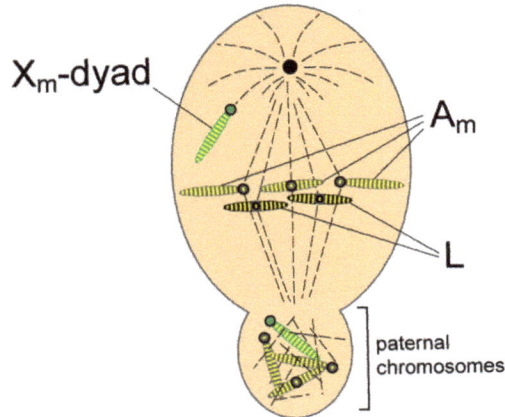

Figure 4. Lack of phosphorylation of the X centromere region at meiotic metaphase II. The Xm-dyad is seen precociously at the monopole. The arms of Xm retain histone histone 3 (H3) phosphorylation (given as yellow stripes), while the centromere region (the ribosomal DNA (rDNA)-chromosomal tip) is devoid of phosphorylation (no stripe in the X centromere). The regular maternal homologues and the L chromosomes lie on the metaphase plate and exhibit phosphorylation of histone H3 both along the arms and the centromere regions. The regular paternal chromosomes found in the bud are also phosphorylated except for the centromere region of Xp. Circles indicate centromere position on the chromosomes. The L chromosomes are black, the autosomes are grey and the X chromosomes are green. Phosphorylation is shown as yellow stripes.

Studies on the sciarid species, *Trichosia pubescens*, have shown that in spermatocytes, L chromosomes become de-condensed and indistinguishable from ordinary chromosomes during the interphase between the meiotic divisions [12]. Based on this observation it was suggested that the de-condensation represents gene activity likely required for proper spermatogenesis, a notion supported by the unpublished results of Crouse (cited in [6]) who observed that spermatocytes having no L chromosomes were smaller, develop more slowly and do not undergo typical (albeit unequal) meiosis. However, this conclusion must be tempered by the observation that both *Sciara ocellaris* and *Sciara reynoldsi* do not possess L chromosomes [6,11]. It has been argued that genes on L chromosomes required for spermatogenesis may have been transferred onto the autosomes in these closely-related species during evolution rendering the former dispensable [6].

2. The Epigenotype of L Chromosomes

A series of detailed studies by Goday and co-workers has led to a description of the localisation of epigenetic modifications (histone and DNA) and the non-histone chromosomal protein, heterochromatin protein 1 (HP1), on ordinary and L chromosomes [9,13–16]. The epigenetic modifications associated with L chromosomes are given in Table 1. L chromosomes possess modifications that are characteristic of constitutive heterochromatin (for reviews see [17,18]). There is a lack of acetylation on histones histone 3 (H3) and H4, a modification associated with gene activity, while there is an over-representation of repressive histone marks such as methylation of lysines 9 on

H3 and 20 on H4 as well as an enrichment of 5-methyl-cytosine in chromosomal DNA. Also enriched are two HP1 homologues, ScoHET1 and ScoHET2; HP1 is a hallmark of heterochromatin [19].

Table 1. Epigenetic modifications associated with the supernumerary L chromosomes at different stages of development. L chromosomes have epigenetic profile characteristic of constitutive heterochromatin. Apart from the first somatic mitosis, the L chromosomes are de-acetylated. Where investigated, L chromosomes also share the repressive epigenetic modifications and non-histone heterochromatin protein 1 (HP1) chromosomal proteins. Analyses of phosphorylation show that during the somatic eliminations the arms of the L chromosomes retain phosphorylation with H3S10P. During meiosis L chromosomes show a complex pattern of phosphorylation (see text for details). Data for acetylation is taken from [9]. The distribution of repressive epigenetic modifications and HP1 protein was taken from Figures 6 and 7 of [13] in "pre-meiotic" and "young germ nuclei" that are likely to be in or around larval mitosis I. For description of phosphorylation during the meiotic divisions, the data was taken from [15]. For phosphorylation of L chromosomes in the embryonic soma the data was taken from [16].

Stage	Epigenetic Modifications
First somatic division	L chromosomes are positive for H4K8Ac [9]
Syncytial embryo before elimination	L chromosomes are positive for H3S10P at pro-metaphase [16]. Proceeding to metaphase they become positive for H3S28P [16]. The centromeres are specifically stained with H3T3P antibodies at metaphase [16].
Syncytial embryo during elimination	L chromosome arms are stained with H3S10P at metaphase/anaphase [16]. By contrast, the regulars have lost their H3S10P staining at anaphase.
Resting stage germ cells before elimination of Xp	L chromosomes are negative for H4K8Ac, H4K12Ac and PanH3Ac [9].
Resting stage germ cells after elimination of Xp	L chromosomes are negative for H4K8Ac, H4K12Ac and PanH3Ac [9].
Germ cells undergoing larval mitosis I (9th day after oviposition)	L chromosomes are negative for H4K8Ac [9]. They are positive for H3K9me2, H3K9me3, H4K20me3, 5MeC and the two HP1-like proteins ScoHET1 and ScoHET2 [13].
Germ cells at end of mitosis III in third instar larvae (12th day after oviposition)	L chromosomes are negative for H4K8Ac [9].
Prophase of Meiosis I	L chromosomes are negative for H3S10P, H3S28P and H3T3P but positive for H3T11P [15]
Meiosis I	L chromosomes are negative for H3S10P, H3S28P and H3T3P but positive for H3T11P [15]. They are also negative for H4K8Ac [9].
Meiosis II	L chromosomes are positive for H3S10P, H3S28P and H3T3P and H3T11P [15]. They are also positive for ScoHET1 [13].

Study of histone phosphorylation has shown an association with L chromosomes eliminated in the soma. Localisation of phosphorylation on serine 10 of histone H3 (H3S10P) on L chromosomes undergoing elimination revealed that the levels remain high along the chromatid arms ([16]; Table 1). By comparison, the levels on the ordinary chromosomes have fallen [16]. The retention of H3S10P levels are also observed along the arms of the Xp chromosomes during their elimination from the soma [16]. This has led to the suggestion that inhibition of de-phosphorylation of the chromatid arms is part of the mechanism that orchestrates the programmed elimination of both kinds of chromosome from the soma during the embryonic cleavages [16].

Histone phosphorylation of L chromosomes in spermatocytes during the meiotic divisions is almost identical to that found for the regular maternal chromosomes ([15]; Table 1), indicating that similar mechanisms might operate to retain both kinds of chromosomes in the face of the selective elimination of regular paternal chromosomes at meiosis I. However, the mechanisms involved cannot be identical. This is because the regular chromosomes retained in spermatocytes during the first meiotic division are exclusively maternal in origin, while the L chromosomes that move en bloc to the same monopole can be derived from either parent [6,11]. How the paternal L chromosomes are stopped from moving backward along with the four ordinary paternal homologues is not known.

The parent-of-origin behaviours of the regular chromosomes in *Sciara* have been of great interest for many decades (for reviews see [1–3]). Notably it has been suggested that the study of L chromosome behaviour may provide insight into the mechanisms of so-called chromosomal imprinting [20] of the regular chromosomes in *Sciara*.

3. L Chromosomes and Chromosomal Imprinting of the Regular Chromosomes

Chromosomal imprinting was a term coined by Helen Crouse [20] to describe the reversible identification of homologues that exhibit the parent-of-origin-specific behaviour seen in spermatocytes, where the four regular paternal homologues are selectively eliminated during meiosis I (Figure 3). Imprinting is reversible because the maternal homologues that are retained and enter the sperm become the paternal chromosomes that behave in precisely the opposite manner at meiosis I in sons. The selective elimination of Xp chromosomes [1] in soma and germ-line provides another example of chromosome imprinting. Notably, the X-X′ device operating in the mother determines the sex of the soma by ensuring the correct number of Xp chromosomes are eliminated [21]. The mechanism of elimination itself requires a chromosomal segment embedded in the constitutive heterochromatin adjacent to the X centromere called the controlling element (CE; [20]).

Strikingly, while L chromosomes also undergo elimination in both soma and germ-line (Figures 1 and 2), they are not subject to chromosomal imprinting, which led Crouse et al. [11] to conclude that: " ... in Sciara there are two independent systems of chromosome identification: one which distinguishes supernumerary from regular; the other which identifies regular as paternal or maternal."

Based on the condensation cycle of both L and regular chromosomes Crouse et al. [6,11] attempted to identify specific germ-line stages where the identification (imprinting) process might be regulated. It is worth revisiting those analyses.

3.1. Erasure of Parental Imprints and Selective Retention of L Chromosomes and Regular Maternal Chromosomes at the First Spermatocyte Division

The de-condensation of the L-chromosomes, along with the regular paternal chromosomes, in or around the time of the germ line eliminations (Figure 2) was suggested to represent the time at which the parent-of-origin imprint on the regular chromosomes is erased [6]. Erasure ensures no difference between the regular chromosomes thus enabling meiosis to proceed in an orthodox manner in females. This timing is supported by studies on histone acetylation that have shown the de-condensed paternal chromosomes (except for the Xp to be eliminated) are initially hyper-acetylated [9] and thereafter, upon entry into the gonal mitotic divisions, both parental sets become equally acetylated [9]. Notably, the de-condensed L chromosomes remain under-acetylated (Table 1). That the L chromosomes are out-of-step with the regular chromosomes is most likely because L chromosomes are not imprinted.

Given that erasure might take place in resting spermatocytes, how are the chromosomes identified later at the first spermatocyte division? A mechanism for identification must exist because paternal homologues are selectively eliminated while the maternal homologues and L chromosomes are retained. How this might be achieved comes from the observation that elimination does not require homologue pairing and metaphase alignment–instead the chromosomes proceed directly to an anaphase-like stage [2,3]. There is an intrinsic separation of the chromosomes that follow different fates. This is so even in early germ cells where there is non-random clustering of the ordinary maternal homologues and L chromosomes in a nuclear compartment distinct to that occupied by the paternal homologues [9,22,23]. It remains to be proven but it may be that this compartmentalisation presages and of itself determines the later selective elimination of the regular paternal homologues.

In primary spermatocytes, there is also a reversal in the acetylation status of the ordinary chromosomes where the maternal homologues retain their high levels of acetylation while the paternal homologues become under-acetylated [9]. As explained, loss of acetylation is a feature of the Xp eliminated in the germ-line (Figure 2), indicating that the loss of acetylation is a characteristic of the parent-of-origin specific elimination of the ordinary paternal homologues. Interestingly, the non-imprinted L chromosomes are under-acetylated and thus out-of-step with the imprinted homologues that accompany them to the monopole [9].

In secondary spermatocytes, the L chromosomes exhibit a high degree of H3 histone phosphorylation (Table 1; [15]). This is also true for the maternal homologues except for the centromeric ribosomal DNA (rDNA) region of the X-dyad, which includes the CE, where there is a clear deficiency in H3 phosphorylation (Figure 4; [15]).

3.2. L Chromosomes and Imprinting of the Controlling Element

The CE regulates two different events. First, the meiotic non-disjunction of the precocious maternal X-dyad in secondary spermatocytes by centromere inactivation (Figure 4) and, second, the elimination of the now paternal X chromosomes in the embryo and germ lines (Figures 1 and 2). X-autosome translocations have shown that the CE represents the site of imprinting that enables the Xp to be distinguished from the Xm in soma and germ cells. The act of imprinting itself requires that the imprint is placed on a chromosome at the time when the parental genomes are separate, which can be in the respective germ-lines or in the brief period when the parental genomes lie separately within the pro-nuclei of the newly-fertilized egg. It has been suggested that if the maternal X chromosomes in the sperm carry a paternal imprint into the egg it is likely that the imprinting event takes place at the end of anaphase in meiosis II, when L chromosomes reach their greatest condensation–and therefore most likely to be resistant to imprinting–and the regulars are de-condensed and diffuse [see discussion in 11]. Alternatively, imprinting may take place in the ooplasm. This has been suggested on two grounds, first, because the X-X′ device in the mother conditions the ooplasm to eliminate the appropriate number of Xp chromosomes [4] and, second, by analogy with the situation in Coccids where imprinting is regulated by the mother with no contribution from the father [24].

Recently, a molecular model of how a CE can cause the elimination of an Xp during the early cleavage divisions has been posited (see Figure 2B in [25]). According to the model, the CE encodes a non-coding RNA (ncRNA) that acts upon the Xp's chromosome arms, leading to their failure to separate at the anaphase of the 7th to 8th cleavage division and subsequent elimination. Research in *Sciara coprophila* is entering an exciting phase for it may now be possible with the advent of molecular techniques to dissect the imprinting process and, in addition, to elucidate the mechanism(s) by which the L chromosomes avoid it.

Author Contributions: P.B.S. wrote the first draft. S.N.B. drew the figures.

Funding: This work was supported by a grant from the Ministry of Education and Science of Russian Federation #14.Y26.31.0024; P.B.S. was supported by Nazarbayev University grant 090118FD5311. S.N.B. was supported by the Russian Fundamental Scientific Research Project (0310-2018-0009).

Acknowledgments: In this section you can acknowledge any support given which is not covered by the author contribution or funding sections. This may include administrative and technical support, or donations in kind (e.g., materials used for experiments).

Conflicts of Interest: The authors declare no conflict of interest.

References

1. Metz, C.W.C. Chromosome behaviour, inheritance and sex determination in *Sciara*. *Am. Nat.* **1938**, *72*, 485–520. [CrossRef]
2. Gerbi, S.A. Unusual chromosome movements in sciarid flies. *Results Probl. Cell Differ.* **1986**, *13*, 71–104. [PubMed]
3. Goday, C.; Esteban, M.R. Chromosome elimination in sciarid flies. *Bioessays* **2001**, *23*, 242–250. [CrossRef]
4. Crouse, H.V. The nature of the influence of X-translocations on sex of progeny in *Sciara coprophila*. *Chromosoma* **1960**, *11*, 146–166. [CrossRef] [PubMed]

5.   DuBois, A.M. Chromosome behaviour during cleavage in the eggs of *Sciara coprophila* (*Diptera*) in relation to the problem of sex determination. *Z Zellf mikr Anat* **1933**, *19*, 595–614.

6.   Rieffel, S.M.; Crouse, H.V. The elimination and differentiation of chromosomes in the germ line of *Sciara*. *Chromosoma* **1966**, *19*, 231–276. [CrossRef] [PubMed]

7.   Berry, R.O. Chromosome behaviour in the germ cells and development of the gonads of *Sciara ocellaris*. *J. Morph.* **1941**, *68*, 547–583. [CrossRef]

8.   Singh, P.B.; Georgatos, S.D. HP1: Facts, open questions and speculation. *J. Struct. Biol.* **2002**, *140*, 10–16. [CrossRef]

9.   Goday, C.; Ruiz, M.F. Differential acetylation of histones H3 and H4 in paternal and maternal germline chromosomes during development of sciarid flies. *J. Cell Sci.* **2002**, *115*, 4765–4775. [CrossRef] [PubMed]

10. Metz, C.W. An apparent case of monocentric mitosis in *Sciara* (*Diptera*). *Science* **1926**, *63*, 190–191. [CrossRef] [PubMed]

11. Crouse, H.V.; Brown, A.; Mumford, B.C. L-chromosome inheritance and the problem of chromosome "imprinting" in *Sciara* (*Sciaridae, Diptera*). *Chromosoma* **1971**, *34*, 324–339. [CrossRef] [PubMed]

12. Amabis, J.M.; Reinach, F.C.; Andrews, N. Spermatogenesis in *Trichosia pubescens* (Diptera:Sciaridae). *J. Cell Sci.* **1979**, *36*, 199–213. [PubMed]

13. Greciano, P.G.; Goday, C. Methylation of histone H3 at Lys4 differs between paternal and maternal chromosomes in *Sciara ocellaris* germline development. *J. Cell Sci.* **2006**, *119*, 4667–4677. [CrossRef] [PubMed]

14. Greciano, P.G.; Ruiz, M.F.; Kremer, L.; Goday, C. Two new chromodomain-containing proteins that associate with heterochromatin in *Sciara coprophila* chromosomes. *Chromosoma* **2009**, *118*, 361–376. [CrossRef] [PubMed]

15. Escriba, M.C.; Giardini, M.C.; Goday, C. Histone H3 phosphorylation and non-disjunction of the maternal X chromosome during male meiosis in sciarid flies. *J. Cell Sci.* **2011**, *124*, 1715–1725. [CrossRef] [PubMed]

16. Escriba, M.C.; Goday, C. Histone X3 phosphorylation and elimination of paternal X chromosomes at early cleavages in sciarid flies. *J. Cell Sci.* **2013**, *126*, 3214–3222. [CrossRef] [PubMed]

17. Bannister, A.J.; Kouzarides, T. Regulation of chromatin by histone modifications. *Cell Res.* **2011**, *21*, 381–395. [CrossRef] [PubMed]

18. Jaenisch, R.; Bird, A. Epigenetic regulation of gene expression: How the genome integrates intrinsic and environmental signals. *Nat. Genet.* **2003**, *33*, 245–254. [CrossRef] [PubMed]

19. Saksouk, N.; Simboeck, E.; Dejardin, J. Constitutive heterochromatin formation and transcription in mammals. *Epigenetics Chromatin* **2015**, *8*, 3. [CrossRef] [PubMed]

20. Crouse, H.V. The controlling element in sex chromosome behaviour in *Sciara*. *Genetics* **1960**, *45*, 1429–1443. [PubMed]

21. Metz, C.W. Factors influencing chromosome movements in mitosis. *Cytologia* **1936**, *7*, 219–231. [CrossRef]

22. Kubai, D.F. Meiosis in *Sciara coprophila*: Structure of the spindle and chromosome behaviour during the first meiotic division. *J. Cell Biol.* **1982**, *93*, 655–669. [CrossRef] [PubMed]

23. Kubai, D.F. Nonrandom chromosome arrangements in germ line nuclei of *Sciara coprophila* males: The basis for nonrandom chromosome segregation on the meiosis I spindle. *J. Cell Biol.* **1987**, *105*, 2433–2446. [CrossRef] [PubMed]

24. Chandra, H.S.; Brown, S.W. Chromosome imprinting and the mammalian X chromosome. *Nature* **1975**, *253*, 165–168. [CrossRef] [PubMed]

25. Singh, P.B. Heterochromatin and the molecular mechanisms of "parent-of-origin" effects in animals. *J. Biosci.* **2016**, *41*, 759–786. [CrossRef] [PubMed]

© 2018 by the authors. Licensee MDPI, Basel, Switzerland. This article is an open access article distributed under the terms and conditions of the Creative Commons Attribution (CC BY) license (http://creativecommons.org/licenses/by/4.0/).

Article

Sequencing of Supernumerary Chromosomes of Red Fox and Raccoon Dog Confirms a Non-Random Gene Acquisition by B Chromosomes

Alexey I. Makunin [1,*], Svetlana A. Romanenko [1,2], Violetta R. Beklemisheva [1],
Polina L. Perelman [1,2], Anna S. Druzhkova [1,2], Kristina O. Petrova [2], Dmitry Yu. Prokopov [1,2],
Ekaterina N. Chernyaeva [3], Jennifer L. Johnson [4], Anna V. Kukekova [4], Fengtang Yang [5],
Malcolm A. Ferguson-Smith [6], Alexander S. Graphodatsky [1,2] and Vladimir A. Trifonov [1,2]

[1] Institute of Molecular and Cellular Biology Siberian Branch of the Russian Academy of Sciences, 630090 Novosibirsk, Russia; rosa@mcb.nsc.ru (S.A.R.); bekl@mcb.nsc.ru (V.R.B.); polina.perelman@gmail.com (P.L.P.); rada@mcb.nsc.ru (A.S.D.); dprokopov@mcb.nsc.ru (D.Y.P.); graf@mcb.nsc.ru (A.S.G.); vlad@mcb.nsc.ru (V.A.T.)
[2] Novosibirsk State University, 630090 Novosibirsk, Russia; petrova@mcb.nsc.ru
[3] Theodosius Dobzhansky Center for Genome Bioinformatics, Saint-Petersburg State University, 199004 Saint-Petersburg, Russia; echernya@gmail.com
[4] Department of Animal Sciences, University of Illinois at Urbana-Champaign, Urbana, IL 61801, USA; jjohnso@illinois.edu (J.L.J.); avk@illinois.edu (A.V.K.)
[5] Wellcome Sanger Institute, Wellcome Genome Campus, Hinxton, Cambridge CB10 1SA, UK; fy1@sanger.ac.uk
[6] Cambridge Resource Centre for Comparative Genomics, Department of Veterinary Medicine, Cambridge University, Cambridge CB3 0ES, UK; maf12@cam.ac.uk
* Correspondence: alex@mcb.nsc.ru

Received: 6 July 2018; Accepted: 7 August 2018; Published: 10 August 2018

Abstract: B chromosomes (Bs) represent a variable addition to the main karyotype in some lineages of animals and plants. Bs accumulate through non-Mendelian inheritance and become widespread in populations. Despite the presence of multiple genes, most Bs lack specific phenotypic effects, although their influence on host genome epigenetic status and gene expression are recorded. Previously, using sequencing of isolated Bs of ruminants and rodents, we demonstrated that Bs originate as segmental duplications of specific genomic regions, and subsequently experience pseudogenization and repeat accumulation. Here, we used a similar approach to characterize Bs of the red fox (*Vulpes vulpes* L.) and the Chinese raccoon dog (*Nyctereutes procyonoides procyonoides* Gray). We confirm the previous findings of the *KIT* gene on Bs of both species, but demonstrate an independent origin of Bs in these species, with two reused regions. Comparison of gene ensembles in Bs of canids, ruminants, and rodents once again indicates enrichment with cell-cycle genes, development-related genes, and genes functioning in the neuron synapse. The presence of B-chromosomal copies of genes involved in cell-cycle regulation and tissue differentiation may indicate importance of these genes for B chromosome establishment.

Keywords: supernumerary chromosomes; karyotype evolution; genome instability

---

1. Introduction

B chromosomes (or Bs) were first described over a century ago [1] and since then they have been found in a multitude of animal and plant species. The name reflects the fact that the Bs represent a variable addition to the normal (or A) chromosome set. Unlike other types of supernumerary chromosomes, such as marker chromosomes [2], double minutes [3], and circular extrachromosomal

DNA [4], Bs are widespread in populations and generally lack specific phenotypic effects. They often demonstrate a higher than Mendelian inheritance rate, resulting from non-random behavior during cell division: Accumulation in gametes relative to somatic cells or in ovaries relative to polar bodies. This phenomenon is known as the B chromosome drive [5,6].

In most species studied so far, Bs originated through ectopic segmental duplications of main genome chromosomal fragments, although cases involving interspecific hybridization [7] or polyploidization [8] are described. Subsequently, Bs undergo internal amplification, accumulation of additional DNA sequences—both unique and repetitive. Eventually, Bs degrade and disappear from the genome. Currently there are no documented cases of Bs older than several million years [9–11].

Several studies identified genes on Bs of plants, insects, and vertebrates, and for all of these lineages records of the transcribed B-chromosomal gene copies exist (reviewed in Reference [12–14]). B chromosomes were found to contain genes associated with cell division and tissue differentiation, suggesting their role in the Bs drive [10,15–17].

The first report of a protein coding gene on Bs, namely protooncogene *KIT*, was made for the red fox and raccoon dog [18]. Subsequent study refined the duplication breakpoints [19]. As a result of high-density bacterial artificial chromosome (BAC)-clone mapping of fox and raccoon dog genomes, the number of B-chromosomal regions increased to seven in the red fox and to five in the raccoon dog [20]. Since then, development of high-throughput sequencing and associated analytical methods allowed us to directly access the genetic content of Bs in mammals using samples of chromosomes isolated by flow sorting and microdissection [21,22].

Here, we utilize this approach to access the genetic content of Bs in the red fox and Chinese raccoon dog based on flow-sorted and microdissected chromosome sequencing. For the red fox, we supplement Bs sequencing data with copy number estimation based on the whole-genome sequencing data. We also combine Bs gene content for six species of mammals analyzed so far [21,22] to identify unifying features of genes involved in Bs formation.

2. Materials and Methods

Chromosomes were isolated from established primary fibroblast cell cultures using chromosome flow sorting [23,24] of several hundred chromosome copies, or microdissection [18,25,26] of single chromosome copies. For both methods, whole-chromosome amplification was performed either with degenerate oligonucleotide-primed polymerase chain reaction (DOP-PCR) [27] using 6-MW primer (5′-CCGACTCGAGNNNNNNATGTGG-3′) or with whole-genome amplification WGA1 kit (Sigma-Aldrich, Saint Louis, MO, USA) depending on the sample (Table S1). For newly acquired microdissected samples, the specificity of amplified chromosome DNA was checked with fluorescence in situ hybridization to metaphase chromosomes of the corresponding species (Figures S1 and S2) [18,23–25].

Samples were prepared for sequencing with Nextera or TruSeq v.3 library preparation kits (Illumina, San Diego, CA, USA). Paired-end sequencing was performed using Illumina MiSeq (Table S1). Sequencing data are available from NCBI short read archive (SRA) under accession PRJNA477942.

Genomic regions of chromosomes were identified using dopseq_pipeline of DOPseq_analyzer v.1.0 [21], which includes primer and adapter trimming with cutadapt v.1.8.3 [28], alignment to the dog (CanFam3.1) genome with BWA-MEM v.0.7.15 [29], removal of human contamination by comparative alignment to human (hg19) genome, filtering of the alignment (min_mapq = 20, min_len = 20), and genome classification based on mean distance between consecutive merged mapped read positions, which is expected to be lower for regions present on the sampled chromosome (see Table S2 for statistics). Only chromosomes of the dog genome assembly were considered at this point, while unassigned scaffolds were discarded. Automated target region detection was complemented by visual inspection of rainfall-style plots representing the distribution of distances between mapped reads along the reference chromosomes (Supplementary File S2).

Copy-number analysis was performed using genomic reads (PRJNA378561) generated during the fox genome assembly project [30]. We established a fibroblast cell culture from this individual and estimated the Bs number to be stable and equal to three. We microdissected all three Bs from a single metaphase plate. The resulting DNA samples were amplified, sequenced and used in this study as VVUB3, 4, and 6. Fox genomic reads were aligned to CanFam3.1 genome with BWA-MEM, and copy number estimation was performed with CNVnator v.3.3.0 [31]. The changes in read coverage corresponding to duplications and deletions were estimated with a sliding window size of 1000 bp, as a lower window size (200 bp) revealed a high number of false positive deletions resulting from uneven coverage for between-species read mapping. B chromosomal copy number was estimated from normalized read depth reported by CNVnator (Table S3).

As a part of comparative gene content study, we re-analyzed Siberian roe deer (*Capreolus pygargus* Pall.) and grey brocket deer (*Mazama gouazoubira* G. Fischer) Bs sequencing data [21] with dopseq_pipeline, using cattle reference genome UMD3.1 instead of Baylor Btau_4.6.1 and BWA-MEM instead of Bowtie2 with "very-sensitive-local" profile, as in the original publication [21]. Other parameters were set as described above for canid chromosomes.

Sets of genes found in Bs of six mammalian species were obtained from the Ensembl Genes 92 [32] database using R biomaRt package [33]. First, B-chromosomal gene sets were obtained for the original reference genomes: dog for red fox and Chinese raccoon dog, cattle for Siberian roe deer and grey brocket deer, mouse for field mice *Apodemus flavicollis* Melchior and *Apodemus peninsulae* Thomas. Then, information on gene homology was added, so that every gene was supplemented (if possible) with identifiers in human, mouse, dog, and cattle genomes (Supplementary File S3). Functional enrichment analysis was performed using DAVID GO v.6.8 [34,35] for the selections of genes representing all or one-to-one (i.e., single copy) orthologs in the human genome (Supplementary File S4). Functional clustering was performed only for categories that annotated at least 80% of genes in the gene sets. Default background datasets of human genes were used for each reference.

3. Results

3.1. Sequencing of Fox and Raccoon Dog B Chromosomes and Autosomes

To ascertain the adequacy of our approach for chromosomal region detection we included samples of autosomes: flow-sorted Chinese raccoon dog (*Nyctereutes procyonoides procyonoides* Gray, NPP) chromosome 6 (NPP6) and microdissected red fox (*Vulpes vulpes* L., VVU) chromosome 3 (VVU3). In both cases, detected regions are in good agreement with comparative cytogenetics data [20,23,24]: NPP6 is homologous to the entire dog (*Canis lupus familiaris* L., CFA) chromosome 3 (CFA3) and the distal portion of CFA13; VVU3 is homologous to entire chromosomes CFA6, CFA34, and CFA36. For NPP6, an additional 160 kbp region of CFA16 was found, suggesting a putative duplication or translocation. This result indicates the higher resolution of our method compared to comparative cytogenetics, but requires further validation. Slight depletion of reads in some genomic regions was observed on autosomes of both species: In a region of NPP6 corresponding to CFA3:56.4–62.8 Mbp and in a region of VVU3 corresponding to CFA6:55.8 Mbp up to chromosome end for VVU3 (Supplementary File S2). These changes were insufficient to be treated as deletions, but reflected a trend of under-representation of certain regions in isolated chromosome sequencing data [21].

Four samples of fox Bs were sequenced, including one sorted (VVUB2) [23] and three microdissected from a single metaphase plate (VVUB3, 5 and 6). Comparison of sequenced Bs to the dog genome sequence revealed 14 regions comprising 7.7 Mbp (Table 1). Three samples, including both sorted and microdissected Bs (VVUB2, 3 and 5), were in perfect agreement. A reduced set of regions was identified in the sample VVUB6. The seven regions previously detected in fox Bs by BAC clone mapping [20] were recovered successfully. Significant deletions lacking read coverage were observed in two regions: CFA13:34 Mbp and CFA22:24-25 Mbp.

Table 1. Genomic regions identified in red fox (*Vulpes vulpes* L., VVU) B chromosomes (Bs). Region coordinates are given according to the dog (CanFam3.1, CFA) genome assembly. VVUB2—flow sorted B sample, VVUB3, 5 and 6—three Bs microdissected from a single metaphase plate, BAC—bacterial artificial chromosome (BAC) clone mapping data from Reference [20].

Region	VVUB2	VVUB3	VVUB5	VVUB6	BAC
CFA5:70796855-70973839	+	+	+		
CFA6:75583707-76038617	+	+	+		
CFA10:18252019-18487716	+	+	+		
CFA12:48851593-49019024	+	+	+		+
CFA13:34034756-34423363 [1]	+	+	+	+	+
CFA13:47122582-47327423	+	+	+	+	+
CFA15:53805540-54125636	+	+	+	+	+
CFA19:41511154-44072972	+	+	+	+	+
CFA22:7086237-7370105	+	+	+	+	
CFA22:24782790-25362284 [2]	+	+	+		
CFA31:2880206-4129393	+	+	+	+	+
CFA32:14687852-15257168	+	+	+		
CFA34:2522638-2767316	+	+	+	+	+
CFA34:15179149-15450638	+	+	+		

[1] Deletions at CFA13:34121352-34138888 and CFA13:34202633-34373142; [2] Deletion at CFA22:24939999-25103044.

For the Chinese raccoon dog, a total of eight samples of Bs were analyzed: One sorted (NPPB1), and others microdissected (NPPB2-8) (Table 2). Here, the highest number of regions (27, total size 8.5 Mbp) was recovered from the sorted sample only, while microdissected samples demonstrated only a subset of these regions—in contrast to fox Bs, where both chromosome isolation methods resulted in similar sets of regions. We recovered only three of five regions previously identified on Chinese raccoon dog Bs by BAC clone mapping [20]. The region corresponding to CFA13:34 Mbp claimed to be present on both fox and raccoon dog Bs was discovered only in fox Bs, while the region on CFA29:4 Mbp (specific to Chinese raccoon dog) was not identified in any samples. Due to the larger size of raccoon dog Bs, in addition to whole-chromosome probes (NPPB2 and 3), we were able to dissect proximal (NPPB5 and 6), middle (NPPB7), and distal (NPPB4 and 8) portions of the B, hoping to recover gene order on the B. However, this approach had only limited success: Differences between same-portion samples (NPPB5 vs. 6, 4 vs. 8, see Table 2) were higher than between samples of different chromosomal portions from the same experiment (NPPB4 vs. 5, 6 vs. 7 vs. 8), and many regions were present throughout the whole B. This finding is in line with previous BAC-clone mapping results, which yielded repetitive locations of several genes throughout the NPPB [20].

We found two overlapping regions on Bs of the fox and raccoon dog: CFA13:47 Mbp with *KIT* protooncogene (Figure 1), which was the first gene to be discovered in Bs of mammals [18], and CFA32:13-15 Mbp, with no genes in the subregion present in both species. A similar region reuse pattern involving regions without any genes was previously observed for Bs in two field mouse species [22]. Several regions identified on Bs of the red fox and raccoon dog are located in close proximity in the dog genome, e.g., CFA15:53-54 Mbp in fox Bs and CFA15:58 Mbp in raccoon dog Bs, CFA19:41-44 Mbp (VVUB) and CFA19:38-39 Mbp (NPPB), CFA29:23 Mbp (VVUB) and CFA29:28-29 Mbp (NPPB).

Table 2. Genomic regions of Chinese raccoon dog (*Nyctereutes procyonoides procyonoides* Gray, NPP) Bs. Region coordinates are given according to dog (canFam3) genome assembly. NPPB1—flow-sorted Bs, NPPB2 and 3—whole microdissected Bs, NPPB4 and 8—microdissected distal part of Bs, NPPB5 and 6—microdissected proximal part of Bs, NPPB7—microdissected middle part of B, BAC—BAC clone mapping data from [20]. +—regions with >5 read positions (recovered automatically). ~—regions with <5 read positions (recovered manually).

Region	NPPB1	NPPB2	NPPB3	NPPB4	NPPB5	NPPB6	NPPB7	NPPB8	BAC
CFA1:207953-434482	+								
CFA3:7233009-7672854	+	+	+	+	~	~	~	+	
CFA3:30949630-31192570	+	+	+					+	
CFA5:5213259-5517316	+					+			
CFA5:60890426-60976094	+								
CFA5:84586827-84980563	+							+	
CFA13:47079087-47478846	+							+	+
CFA14:58162796-58517967	+								
CFA15:58671318-58919401	+	+	+	~	+	+	+		
CFA16:5042848-5239664	+	~	~	~		+		+	
CFA16:50655536-50916841	+								
CFA16:51357728-51710145	+								
CFA19:28836248-29004558	+	+	+	+	~	+	+	+	
CFA19:38810088-39213802	+	+	+			+	~	+	
CFA20:24724662-25132678	+	+	+			~		+	+
CFA21:36289982-36390083	+	+	~			+	+		
CFA23:47315786-47535793	+	+	+	~	+	+	+		
CFA24:11660722-11768215	~					+			
CFA24:43488879-43732408	~	~	~			+		~	
CFA26:18769412-18964210	+								
CFA27:31691325-31793171	+					+			
CFA27:37050854-37410557	+	+	+	+	+	+	+	+	
CFA28:3855680-4240133	+	+	+			+	+		+
CFA28:7498907-7656661	+	+	+	~		+	+	+	
CFA29:28913684-29259138	+							+	
CFA32:13104843-14881357	+					+			
CFA38:1062270-1094480	+					+			

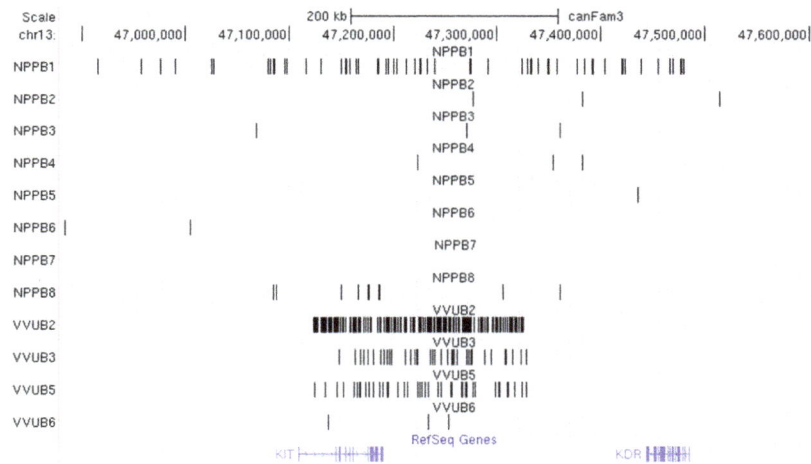

Figure 1. The region encompassing protooncogene *KIT* is present on B chromosomes (Bs) of the fox (VVUB2, 3, 5 and 6) and raccoon dog (NPPB1-8) visualized in UCSC genome browser (http://genome.ucsc.edu/). Coordinates are given for the dog (CanFam3.1) genome.

3.2. Fox Whole-Genome Sequencing Data

Whole-genome sequencing (WGS) allows the discovery of B-chromosomal regions, using coverage-based copy number analysis of individuals with B-chromosomes (B+) and comparison to individuals without Bs (B−) [10]. The regions with increased copy number only in the B+ genome are treated as present on Bs.

Here, we took advantage of the WGS data available for the fox individual whose fibroblast cell cultures were used for isolation of Bs VVUB3, 5 and 6. We aligned its reads to the dog genome and called copy number variants, anticipating finding a higher coverage in B-specific regions. We were able to avoid the WGS for the B− individual by focusing on the regions predicted with isolated B chromosome sequencing. The overall agreement between whole-genome copy number and isolated chromosome sequencing results was striking—all 14 regions were consistently recovered by both methods (Table S3). The median difference between breakpoint coordinates predicted by the two methods was about 700 bp, or less than one window size (1000 bp)—the limit of copy number calling resolution. Compared to the isolated chromosome sequencing, the copy-number dataset included multiple internal deletions, which presumably reflect unevenly reduced mapping efficiency for the cross-species read alignment. As a result, the total size of the B chromosomal regions detected by copy number analysis (6.2 Mbp, Table S3) was still significantly lower than for isolated chromosome sequencing (7.4 Mbp).

Estimates of normalized read depth allowed us to calculate the copy number for each of the 14 regions present on Bs and compare it to the number of Bs of the sampled individual ($n = 3$, Figure 2). In agreement with BAC clone mapping results [18–20], some of the regions were amplified, e.g., six copies of *KIT* region (CFA13:47 Mbp) were present on Bs, which implies two copies per B chromosome. The copy number varied not only between the regions but also within some regions, suggesting a partial amplification of these regions in Bs. The most striking example was observed in the region corresponding to CFA31:2–4 Mbp; the number of additional copies for different parts of this region varied from zero (putative deletion) to 20 (presumably seven copies per B chromosome). Existence of copy numbers lower than three together with the absence of these regions from chromosome VVUB6 suggest that B heterogeneity within an individual exists (Table 1, Table S3). Taken together, these observations suggest that a complex mixture of duplications and deletions occurred since the origin of Bs in red fox, which resulted in significant heterogeneity among Bs—even within a single individual.

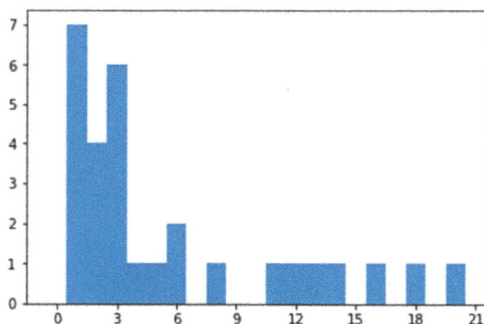

Figure 2. Numbers of additional copies for B chromosomal regions (identified by isolated chromosome sequencing) estimated based on whole-genome sequencing of the red fox individual with three Bs. Region parts with different copy numbers counted separately. Regions with copy number below three were lost from some of Bs, while regions with copy number above three were amplified within Bs. X—number of additional copies, Y—counts of regions.

3.3. Genetic Content of B Chromosomes in Deer Revisited

B chromosomes of two deer species, Siberian roe deer (*C. pygargus* Pall., CPYB) and grey brocket deer (*M. gouazoubira* G. Fischer, MGOB) were previously sequenced and analyzed using an earlier version of the chromosome region identification pipeline [21]. Here, we repeatedly identified target region using another version of the cattle (*Bos taurus* L., BTA) reference genome assembly (UMD3.1) and a more sensitive alignment algorithm. For both samples, this resulted in a significant increase in B region numbers.

In roe deer Bs, five regions were identified in addition to the two discovered previously (Table S4). The largest added region corresponded to BTAX:148 Mbp (292 kbp in size). It is interesting that the alignment to UMD3.1 assembly with Bowtie2 also revealed this region. Thus, this region became detectable due to the significant improvement of the sex chromosome assembly, rather than due to increased aligner sensitivity. The total size of B chromosomal regions increased from 1.96 Mbp to 2.36 Mbp.

In grey brocket deer, four regions were detected in addition to 25 identified previously (Table S5), and the size was increased from 9.31 Mbp to 10.46 Mbp. Among the newly discovered regions was the retrogene, *RASA1*, for which only exons were located in Bs. Similarly processed retrogenes were previously reported for Bs of fish [10].

In both species, false positive signals arose at stretches of telomeric (BTA20:72 Mbp) and centromeric (BTSAT4 at BTA19:0.4 Mbp, BTA:61 Mbp) repeats. The telomeric repeats are universal among mammals, and the observation of the same centromeric repeat in deer and cattle is in line with the previous report [36].

3.4. Comparison of B Chromosome Gene Content in Six Species of Mammals

To identify common features of Bs in various lineages of mammals, we compared genetic content of Bs across six species: red fox (VVUB), Chinese raccoon dog (NPPB), Siberian roe deer (CPYB), grey brocket deer (MGOB), and two field mouse species, yellow-necked mouse (*A. flavicollis* Melchior, AFLB) and Korean field mouse (*A. peninsulae* Thomas, APEB) [22]. For each species, we extracted Ensembl gene predictions overlapping the B chromosomal regions. The number of genes identified with such a method was higher than in both original studies, as other gene types, apart from protein-coding ones, were included in the analysis. Data on homologous human genes were also added from the Ensembl database (Table 3, Supplementary File S3).

Table 3. B chromosome genes statistics. Genes retrieved from the Ensembl 92 database. HSA hom—number of genes with homologs in human (*Homo sapiens* L.), HSA 1-to-1—number of genes with one-to-one (i.e., single copy) orthologs in human.

Sample	Reference	Region Size, bp	Genes	HSA hom	HSA 1-to-1
VVUB	CanFam3.1	7,708,416	49	17	9
NPPB	CanFam3.1	8,510,228	100	44	36
CPYB	UMD3.1	2,355,879	9	6	4
MGOB	UMD3.1	10,456,241	107	113	81
AFLB	mm10	3,421,582	101	37	25
APEB	mm10	12,641,228	152	49	41

Next, we ran functional annotation and clustering for human one-to-one homologs using the DAVID GO web interface (Table 4, Supplementary File S4). Unexpectedly, we found the top cluster to be associated with neuron synapses and cell junctions (enrichment score 2.15). Genes from this cluster were present on Bs of all species, except for the Siberian roe deer.

Table 4. Selected B chromosome gene functions. Neuron synapse and cell cycle proteins retrieved from DAVID Gene ontology (GO) clustering results, while differentiation and proliferation proteins are listed based on keyword match in the GO database.

	Neuron Synapse, Cell Junction	Cell Division, Microtubules	Differentiation, Proliferation
VVUB	CTNND2	CENPN	KIT
NPPB	LRRC7, CXCR4, KDR, ARHGAP32		AICDA, APOBEC1, ARNTL, BARX2, BTBD10, COL4A3BP, CXCR4, ENPP1, GDF3, GNAS, HMGCR, JAG1, KDR, KIT, MDM4
CPYB			TNNI3K
MGOB	SDK1, SDK2, GABRA4, GABRB1, PALLD, LPP, SHANK2	CC2D2A, EVI5, CHFR, CCND2, TRIM67, PALLD, CDC42EP4	ACVR2B, BCL6, BST1, CCND2, CD38, DHCR7, DLEC1, EOMES, EVI5, FBXL5, FGFBP1, FNIP1, GABRB1, GFI1, HPSE, KIT, MYD88, PLCD1, SDK2, SERPINB9, SSBP3, SST, SSTR2, TXK, ZNF268
AFLB	CADPS	DYNC1I2, MAPRE1, MAP7, RPS6, CENPE, HAUS6, SAXO1	ACER2, KDM6A, MAPRE1, NCK2, RPS6
APEB	GRID2, UNC13A, ADGRL3	PIK3C3, MYO9B, HAUS8, KIF23, MVB12A	BST2, DDA1, GRID2, JAK3, TESPA1

The cluster of cell division machinery and cell cycle control functions (enrichment score 1.26) included genes found in Bs of the grey brocket deer, both mouse species, and the red fox. A related cluster was associated with microtubules and centrosomes (enrichment score 0.94).

The B chromosomal gene set was also highly enriched with genes bound by developmental transcription factors (enrichment score 2.09). Still, genes involved in cell differentiation and proliferation did not form a significant cluster, although genes involved in these processes were found on Bs in all species studied. Although not automatically classified, *TNNI3K*, the first gene identified in Siberian roe deer Bs and the only confirmed example of transcribed B chromosomal gene in mammals [37], was included in this list based on the report of its role in cardiomyocyte differentiation from embryonic stem cells [38].

Multiple clusters were associated with specific intracellular processes: A cluster of genes found in all sampled Bs was associated with protein phosphorylation (enrichment score 1.14). The clusters with related protein domains included PH-like or Pleckstrin homology-like (phosphatidylinositol binding, enrichment score 0.97), SH or Src homology (phosphotyrosine binding, enrichment score 0.83), C2 domain (potentially phospholipid binding, found in Ca^{2+}-dependent channels, enrichment score 1.57), and immunoglobulin-like domains (enrichment score 1.59).

In line with previous observations [18,20,39], genes involved in cancer development were represented on Bs of four studied species: red fox (*KIT*), Chinese raccoon dog (*ENPP1, GNAS, HMGCR, KDR, RET*), brocket deer (*BCL6, RASA1, RET*), and Korean field mouse (*JAK3*). Still, the enrichment with this function was not significant (score 0.09, *p*-value > 0.05).

4. Discussion

Canidae is the only family of carnivores with species bearing Bs. Bs of the red fox (*V. vulpes*) and raccoon dog (*N. procyonoides*) are the most studied, while only brief descriptions were made for Bs of the maned wolf (*Chrysocyon brachyurus*) [40], short-eared dog (*Atelocynus microtis*) [41], Bengal fox (*V. bengalensis*) [42], and pale fox (*V. pallida*) [43].

Dot-like Bs of red fox were one of the first to be described among mammals [44,45]. During the famous experiment on fox domestication [46], Bs were found to be replicating late [47] and somatically mosaic with accumulation in gametes [48]. In meiosis, Bs form uni-, bi- and trivalents and occasionally associate (but do not conjugate) with sex chromosomes [49]. Furthermore, meiotic recombination between Bs was hypothesized based on mismatch repair protein visualization [50]. Fox Bs are characterized by lower DNA methylation [51] and central location in interphase nuclei [52].

The revolutionary finding of the first unique protein-coding gene, namely protooncogene *KIT*, on Bs of mammals was made with the fox [18]. Further studies provided additional data on breakpoint margins [19] and identified six additional regions apart from the *KIT* region on red fox Bs [20]. The current data entirely confirmed these findings and allowed the identification of seven additional regions. Using a combination of isolated chromosome and whole-genome sequencing data, we demonstrated a complex pattern of DNA amplification on Bs: Many regions are present in single copy, while some are duplicated, amplified to a higher copy number, or deleted—partially or completely. Both methods suggest that one of three Bs in the studied individual (VVUB6) was partially degenerated in comparison to the other two (VVUB3 and 5).

In raccoon dogs, Bs were described in three subspecies: Chinese (*N. p. procyonoides*, NPP) [53], Japanese (*N. p. viverrinus*, NPV) [54], and Korean (*N. p. koreensis*) [55]. Multiple studies were made for Chinese and Japanese subspecies, which differ in the number of Robertsonian translocations and by B morphology, and supposedly represent independent species [56]. In contrast to the fox Bs, which are the smallest karyotype elements comparable only to Y chromosomes, the raccoon dog Bs are similar in size to medium (NPP) or small (NPV) autosomes [24,56]. In both subspecies, somatic mosaicism and variable B chromosome morphology were recorded [57–60]. NPP Bs form bi- and multivalents in meiosis [61], have higher level of DNA methylation [51] and are located at the periphery of the interphase nucleus [52]. Interstitial telomeric sites are found along the arms of Bs of both subspecies [62]. 28S rRNA gene clusters were found on NPP Bs [59].

Protooncogene *KIT* was discovered in both NPP and NPV Bs [18]. In NPP, B chromosomal copies of *KIT* were found to be enriched with sequence variants, but not transcribed [63]. Using PCR mapping it was shown that the *KIT* region in NPP Bs also includes the 5′ part of neighboring *KDR* gene [19]. Further high-density BAC clone mapping experiment identified four additional regions on NPP Bs [20], but the current data did not detect two of these regions: CFA13:34 Mbp and CFA29:41 Mbp. Both of these demonstrated only one pair of BAC clone hybridization signals per three Bs in the individual studied, and thus might represent either variable additional regions, as described in *A. flavicollis* Bs [22], or mapping artifacts, e.g., due to specific repeat accumulation. For the remaining three regions, the current dataset agrees well with BAC clone mapping. For example, *KIT* region location in the distal subarm is confirmed by both methods. Extensive amplification of genes found on NPP Bs is supported by three facts: recurrent identification of the same regions in different B subarms; significant increase of B total size (~50–100 Mbp as suggested by similar autosome sizes) in comparison to 8.5 Mbp unique chromosomal regions identified; earlier evidence of repetitive BAC clone localizations [18–20].

A relationship between Bs of raccoon dog subspecies remains unknown. The onset of Bs in the common ancestor could have predated the set of Robertsonian translocations observed in the Japanese raccoon dog. However, the only unique gene shown to be present in NPV Bs is *KIT*. This is the most frequently reused B chromosomal gene, which makes an independent origin of Bs also possible. Sequencing of the Japanese raccoon dog Bs could help elucidate a common or independent origin of Bs in NPP and NPV.

Genetic content of Bs is far from random. Previous studies suggested that Bs often harbor genes involved in cell division and early development. Two common B drive mechanisms found in animals imply interventions into these two processes: Directed (non)disjunction towards ovaries in female meiosis and mitotic accumulation in the germ line, as opposed to somatic tissues. Genes related to cell cycle and tissue differentiation were reported in non-mammalian lineages: morphogene *ihhb* (Indian hedgehog b) on Bs of cichlid *Lithochromis rubripinnis* [15]; several cell division associated genes in another cichlid *Astatotilapia latifasciata* [10]; a set of expressed pseudogenes including kinesins, argonaute and other regulatory genes in rye (*Secale cereale*) [16,64]; five genes related to cell division (*CIP2A*, *KIF20A*, *CKAP2*, *CAP-G* and *MYCB2*) in a grasshopper *Eyprepocnemis plorans* [17].

Similar tendencies were observed in mammals, both in terms of gene function enrichment and reuses (genes encoding cell-cycle related signaling protein kinases—*KIT*, *RET* and *VRK1*, as well as splicing regulatory *KHDRBS3*). The only species with Bs seemingly lacking genes related to cell cycle or early development is the Siberian roe deer, in which the first gene described and shown to be transcribed, *TNNI3K*, [37] has been recently associated with cardiovascular cell differentiation [38].

Neuron synapse and signaling represents another interesting function highly enriched in mammalian Bs. These genes can be found on Bs of all species studied, except for the roe deer. Examples of genes from this category also occur in Bs of cichlid *Astatotilapia latifiscata* [10]. This functional category remains quite unexplained by the current model of B chromosome evolution.

Five single copy genes were found to be reused in the Bs of different species. Protooncogene *KIT* was the first gene identified on the Bs in mammals, and it is currently found in Bs of three species: red fox, Chinese raccoon dog, and grey brocket deer. It encodes a protein kinase controlling the cell cycle, and plays an important role in development. Another protooncogene encoding protein kinase, *RET*, is found on Bs of Chinese raccoon dog and grey brocket deer. A third previously known gene reuse example is *VRK1*, a gene which also encodes a protein kinase involved in cell cycle control. This gene is found in Bs of two field mice species, which most likely acquired the supernumerary chromosomes independently [22].

Two novel examples of gene reuse were identified in this study:

- *KHDRBS3* (as known as *T-STAR* and *SLM-2*) is found on Bs of the grey brocket deer and the Korean field mouse. It encodes an RNA-binding signal transduction protein involved in alternative splicing regulation expressed in the brain and gonads. Mutations in this gene are associated with a neurological disorder (child absence epilepsy [65]), and affect the progression of various cancers.
- *MYOF* was found on Bs in the Chinese raccoon dog and the yellow-necked mouse. It encodes myoferin, a plasma-associated protein involved in Ca^{2+}-channel formation, important in myoblast functioning [66] and involved in EGF-induced cell migration in breast cancer [67].

Two factors seemingly affect the onset of Bs. Genomic instability context, that is, propensity to chromosome rearrangements, is important, as it provides the source material for B chromosome formation. However, the fixation of Bs within a population may in fact depend on their own genes. Here, we revealed patterns of functional preferences and gene reuse in six mammalian species. These trends are, in general, compatible with the B chromosomal gene findings in other lineages. Further studies are needed to understand the exact molecular mechanisms that Bs used to "hack" cell division and differentiation processes while escaping from selection pressure.

Supplementary Materials: The following are available online at http://www.mdpi.com/2073-4425/9/8/405/s1: Supplementary Tables and Figures: Figures S1 and S2: FISH confirmation of chromosome specificity, Table S1: Sequencing statistics, Table S2: Alignment statistics, Table S3: Copy number variant calling for fox, Table S4: Regions of Siberian roe deer B chromosomes, Table S5: Regions of grey brocket deer B chromosomes; Supplementary File S2: Rainfall plots of distances between mapped read positions; Supplementary File S3: Homology data for genes found in B chromosomes of six mammal species; Supplementary File S4: Functional clustering for human one-to-one orthologs of B chromosome genes.

Author Contributions: Conceptualization, A.S.G. and V.A.T.; Methodology, A.I.M., S.A.R., V.R.B., P.L.P., A.S.D., E.N.C., J.L.J., A.V.K., F.Y., M.A.F.-S., V.A.T.; Formal Analysis, A.I.M., K.O.P., J.L.J.; Investigation, A.I.M., S.A.R., A.S.D., K.O.P., D.Y.P., E.N.C., J.L.J., A.V.K., V.A.T.; Resources, V.R.B., P.L.P., J.L.J., A.V.K., F.Y., M.A.F.-S., A.S.G., V.A.T.; Data Curation, A.L.M., V.R.B., P.L.P., A.V.K., V.A.T.; Writing-Original Draft Preparation, A.I.M.; Writing-Review & Editing, A.I.M., S.A.R., V.R.B., P.L.P., A.S.D., K.O.P., D.Y.P., A.V.K., F.Y., M.A.F.-S., A.S.G., V.A.T.; Visualization, A.I.M., S.A.R., K.O.P., D.Y.P., P.L.P., V.R.B., V.A.T.; Supervision, V.A.T., A.S.G.; Project Administration, V.A.T., A.S.G.; Funding Acquisition, A.S.G.

Funding: The work was supported by the Russian Science Foundation (RSF) grant number 16-14-10009.

Acknowledgments: We would like to thank Fedor Goncharov (Institute of Molecular and Cellular Biology SB RAS) for IMCB Galaxy server maintenance. DNA sequencing was performed by the "Molecular and cellular biology" facility at IMCB SB RAS and in Dobzhansky Center for Genome Bioinformatics SPbSU.

Conflicts of Interest: The authors declare no conflict of interest.

References

1. Wilson, E.B. The supernumerary chromosomes of Hemiptera. *Science* **1907**, *26*, 870–871.
2. Liehr, T.; Claussen, U.; Starke, H. Small supernumerary marker chromosomes (sSMC) in humans. *Cytogenet. Genome Res.* **2004**, *107*, 55–67. [CrossRef] [PubMed]

3. Storlazzi, C.T.; Lonoce, A.; Guastadisegni, M.C.; Trombetta, D.; D'Addabbo, P.; Daniele, G.; L'Abbate, A.; Macchia, G.; Surace, C.; Kok, K.; et al. Gene amplification as double minutes or homogeneously staining regions in solid tumors: Origin and structure. *Genome Res.* **2010**, *20*, 1198–1206. [CrossRef] [PubMed]

4. Cohen, S.; Segal, D. Extrachromosomal circular DNA in eukaryotes: Possible involvement in the plasticity of tandem repeats. *Cytogenet. Genome Res.* **2009**, *124*, 327–338. [CrossRef] [PubMed]

5. Jones, R.N. B-chromosome drive. *Am. Nat.* **1991**, *137*, 430–442. [CrossRef]

6. Houben, A. B chromosomes—A matter of chromosome drive. *Front. Plant Sci.* **2017**, *8*, 210. [CrossRef] [PubMed]

7. Schartl, M.; Nanda, I.; Schlupp, I.; Wilde, B.; Epplen, J.T.; Schmid, M.; Parzefall, J. Incorporation of subgenomic amounts of DNA as compensation for mutational load in a gynogenetic fish. *Nature* **1995**, *373*, 68–71. [CrossRef]

8. Dhar, M.K.; Friebe, B.; Koul, A.K.; Gill, B.S. Origin of an apparent B chromosome by mutation, chromosome fragmentation and specific DNA sequence amplification. *Chromosoma* **2002**, *111*, 332–340. [CrossRef] [PubMed]

9. Martis, M.M.; Klemme, S.; Banaei-Moghaddam, A.M.; Blattner, F.R.; Macas, J.; Schmutzer, T.; Scholz, U.; Gundlach, H.; Wicker, T.; Šimková, H.; et al. Selfish supernumerary chromosome reveals its origin as a mosaic of host genome and organellar sequences. *Proc. Natl. Acad. Sci. USA* **2012**, *109*, 13343–13346. [CrossRef] [PubMed]

10. Valente, G.T.; Conte, M.A.; Fantinatti, B.E.A.; Cabral-de-Mello, D.C.; Carvalho, R.F.; Vicari, M.R.; Kocher, T.D.; Martins, C. Origin and evolution of B chromosomes in the cichlid fish *Astatotilapia latifasciata* based on integrated genomic analyses. *Mol. Biol. Evol.* **2014**, *31*, 2061–2072. [CrossRef] [PubMed]

11. Lamb, J.C.; Riddle, N.C.; Cheng, Y.-M.; Theuri, J.; Birchler, J.A. Localization and transcription of a retrotransposon-derived element on the maize B chromosome. *Chromosome Res.* **2007**, *15*, 383–398. [CrossRef] [PubMed]

12. Makunin, A.I.; Dementyeva, P.V.; Graphodatsky, A.S.; Volobouev, V.T.; Kukekova, A.V.; Trifonov, V.A. Genes on B chromosomes of vertebrates. *Mol. Cytogenet.* **2014**, *7*, 99. [CrossRef] [PubMed]

13. Houben, A.; Banaei-Moghaddam, A.M.; Klemme, S.; Timmis, J.N. Evolution and biology of supernumerary B chromosomes. *Cell. Mol. Life Sci.* **2014**, *71*, 467–478. [CrossRef] [PubMed]

14. Ruban, A.; Schmutzer, T.; Scholz, U.; Houben, A. How next-generation sequencing has aided our understanding of the sequence composition and origin of B chromosomes. *Genes* **2017**, *8*, 294. [CrossRef] [PubMed]

15. Yoshida, K.; Terai, Y.; Mizoiri, S.; Aibara, M.; Nishihara, H.; Watanabe, M.; Kuroiwa, A.; Hirai, H.; Hirai, Y.; Matsuda, Y.; et al. B chromosomes have a functional effect on female sex determination in Lake Victoria cichlid fishes. *PLoS Genet* **2011**, *7*, e1002203. [CrossRef] [PubMed]

16. Banaei-Moghaddam, A.M.; Meier, K.; Karimi-Ashtiyani, R.; Houben, A. Formation and expression of pseudogenes on the B chromosome of rye. *Plant Cell Online* **2013**, *25*, 2536–2544. [CrossRef] [PubMed]

17. Navarro-Domínguez, B.; Ruiz-Ruano, F.J.; Cabrero, J.; Corral, J.M.; López-León, M.D.; Sharbel, T.F.; Camacho, J.P.M. Protein-coding genes in B chromosomes of the grasshopper *Eyprepocnemis plorans*. *Sci. Rep.* **2017**, *7*, 45200. [CrossRef] [PubMed]

18. Graphodatsky, A.S.; Kukekova, A.V.; Yudkin, D.V.; Trifonov, V.A.; Vorobieva, N.V.; Beklemisheva, V.R.; Perelman, P.L.; Graphodatskaya, D.A.; Trut, L.N.; Yang, F.; et al. The proto-oncogene *C-KIT* maps to canid B-chromosomes. *Chromosome Res.* **2005**, *13*, 113–122. [CrossRef] [PubMed]

19. Yudkin, D.V.; Trifonov, V.A.; Kukekova, A.V.; Vorobieva, N.V.; Rubtsova, N.V.; Yang, F.; Acland, G.M.; Ferguson-Smith, M.A.; Graphodatsky, A.S. Mapping of *KIT* adjacent sequences on canid autosomes and B chromosomes. *Cytogenet. Genome Res.* **2007**, *116*, 100–103. [CrossRef] [PubMed]

20. Duke Becker, S.E.; Thomas, R.; Trifonov, V.A.; Wayne, R.K.; Graphodatsky, A.S.; Breen, M. Anchoring the dog to its relatives reveals new evolutionary breakpoints across 11 species of the Canidae and provides new clues for the role of B chromosomes. *Chromosome Res.* **2011**, *19*, 685–708. [CrossRef] [PubMed]

21. Makunin, A.I.; Kichigin, I.G.; Larkin, D.M.; O'Brien, P.C.M.; Ferguson-Smith, M.A.; Yang, F.; Proskuryakova, A.A.; Vorobieva, N.V.; Chernyaeva, E.N.; O'Brien, S.J.; et al. Contrasting origin of B chromosomes in two cervids (Siberian roe deer and grey brocket deer) unravelled by chromosome-specific DNA sequencing. *BMC Genom.* **2016**, *17*, 618. [CrossRef] [PubMed]

22. Makunin, A.I.; Rajičić, M.; Karamysheva, T.V.; Romanenko, S.A.; Druzhkova, A.S.; Blagojević, J.; Vujošević, M.; Rubtsov, N.B.; Graphodatsky, A.S.; Trifonov, V.A. Low-pass single-chromosome sequencing of human small supernumerary marker chromosomes (sSMCs) and *Apodemus* B chromosomes. *Chromosoma* **2018**. [CrossRef] [PubMed]

23. Yang, F.; O'Brien, P.C.M.; Milne, B.S.; Graphodatsky, A.S.; Solanky, N.; Trifonov, V.; Rens, W.; Sargan, D.; Ferguson-Smith, M.A. A complete comparative chromosome map for the dog, red fox, and human and its integration with canine genetic maps. *Genomics* **1999**, *62*, 189–202. [CrossRef] [PubMed]

24. Nie, W.; Wang, J.; Perelman, P.; Graphodatsky, A.S.; Yang, F. Comparative chromosome painting defines the karyotypic relationships among the domestic dog, Chinese raccoon dog and Japanese raccoon dog. *Chromosome Res.* **2003**, *11*, 735–740. [CrossRef] [PubMed]

25. Trifonov, V.A.; Perelman, P.L.; Kawada, S.-I.; Iwasa, M.A.; Oda, S.-I.; Graphodatsky, A.S. Complex structure of B-chromosomes in two mammalian species: *Apodemus peninsulae* (Rodentia) and *Nyctereutes procyonoides* (Carnivora). *Chromosome Res.* **2002**, *10*, 109–116. [CrossRef] [PubMed]

26. Yang, F.; Trifonov, V.; Ng, B.L.; Kosyakova, N.; Carter, N.P. Generation of paint probes from flow-sorted and microdissected chromosomes. In *Fluorescence in Situ Hybridization (FISH) Application Guide*; Liehr, T., Ed.; Springer Protocols Handbooks; Springer: Berlin/Heidelberg, Germany, 2017; pp. 63–79.

27. Telenius, H.; Carter, N.P.; Bebb, C.E.; Nordenskjöld, M.; Ponder, B.A.J.; Tunnacliffe, A. Degenerate oligonucleotide-primed PCR: General amplification of target DNA by a single degenerate primer. *Genomics* **1992**, *13*, 718–725. [CrossRef]

28. Martin, M. Cutadapt removes adapter sequences from high-throughput sequencing reads. *EMBnet. J.* **2011**, *17*, 10–12. [CrossRef]

29. Li, H. Aligning sequence reads, clone sequences and assembly contigs with BWA-MEM. *arXiv*, 2013.

30. Kukekova, A.; Johnson, J.; Xiang, X.; Feng, S.; Liu, S.; Rando, H.; Kharlamova, A.; Herbeck, Y.; Serdyukova, N.; Xiong, Z.; et al. Red fox genome assembly identifies genomic regions associated with tame and aggressive behaviors. *Nat. Ecol. Evol.* **2018**, in press. [CrossRef] [PubMed]

31. Abyzov, A.; Urban, A.E.; Snyder, M.; Gerstein, M. CNVnator: An approach to discover, genotype, and characterize typical and atypical CNVs from family and population genome sequencing. *Genome Res.* **2011**, *21*, 974–984. [CrossRef] [PubMed]

32. Zerbino, D.R.; Achuthan, P.; Akanni, W.; Amode, M.R.; Barrell, D.; Bhai, J.; Billis, K.; Cummins, C.; Gall, A.; Girón, C.G. Ensembl 2018. *Nucleic Acids Res.* **2017**, *46*, D754–D761. [CrossRef] [PubMed]

33. Durinck, S.; Spellman, P.T.; Birney, E.; Huber, W. Mapping identifiers for the integration of genomic datasets with the R/Bioconductor package biomaRt. *Nat. Protoc.* **2009**, *4*, 1184. [CrossRef] [PubMed]

34. Huang, D.W.; Sherman, B.T.; Lempicki, R.A. Bioinformatics enrichment tools: Paths toward the comprehensive functional analysis of large gene lists. *Nucleic Acids Res.* **2009**, *37*, 1–13. [CrossRef] [PubMed]

35. Huang, D.W.; Sherman, B.T.; Lempicki, R.A. Systematic and integrative analysis of large gene lists using DAVID bioinformatics resources. *Nat. Protoc.* **2009**, *4*, 44–57. [CrossRef] [PubMed]

36. Li, Y.; Lee, C.; Chang, W.; Li, S.-Y.; Lin, C. Isolation and identification of a novel satellite DNA family highly conserved in several Cervidae species. *Chromosoma* **2002**, *111*, 176–183. [CrossRef] [PubMed]

37. Trifonov, V.A.; Dementyeva, P.V.; Larkin, D.M.; O'Brien, P.C.; Perelman, P.L.; Yang, F.; Ferguson-Smith, M.A.; Graphodatsky, A.S. Transcription of a protein-coding gene on B chromosomes of the Siberian roe deer (*Capreolus pygargus*). *BMC Biol.* **2013**, *11*, 1–11. [CrossRef] [PubMed]

38. Wang, Y.; Wang, S.-Q.; Wang, L.-P.; Yao, Y.-H.; Ma, C.-Y.; Ding, J.-F.; Ye, J.; Meng, X.-M.; Li, J.-J.; Xu, R.-X. Overexpression of cardiac-specific kinase TNNI3K promotes mouse embryonic stem cells differentiation into cardiomyocytes. *Cell. Physiol. Biochem.* **2017**, *41*, 381–398. [CrossRef] [PubMed]

39. Lamatsch, D.K.; Trifonov, V.; Schories, S.; Epplen, J.T.; Schmid, M.; Schartl, M. Isolation of a cancer-associated microchromosome in the sperm-dependent parthenogen *Poecilia formosa*. *Cytogenet. Genome Res.* **2011**, *135*, 135–142. [CrossRef] [PubMed]

40. Pieńkowska-Schelling, A.; Schelling, C.; Zawada, M.; Yang, F.; Bugno, M.; Ferguson-Smith, M. Cytogenetic studies and karyotype nomenclature of three wild canid species: Maned wolf (*Chrysocyon brachyurus*), bat-eared fox (*Otocyon megalotis*) and fennec fox (*Fennecus zerda*). *Cytogenet. Genome Res.* **2008**, *121*, 25–34. [CrossRef] [PubMed]

41. Hsu, T.C.; Benirschke, K. *An Atlas of Mammalian Chromosomes*; Springer: New York, NY, USA, 1973; Volume 4, p. 178.

42. Bhatnagar, V.S. Microchromosomes in the somatic cells of *Vulpes bengalensis* Shaw. *Chromosome Inf. Serv.* **1973**, *15*, 32.

43. Chiarelli, A.B. The chromosomes of the Canidae. In *The Wild Canids, Their Systematics, Behavioral Ecology, and Evolution*; Van Nostrand Reinhold Co.: New York, NY, USA, 1975; pp. 40–53.

44. Gustavsson, I.; Sundt, C.O. Chromosome complex of the family Canidae. *Hereditas* **1965**, *54*, 249–254. [CrossRef] [PubMed]

45. Moore, W.; Elder, R.L. Chromosomes of the fox. *J. Hered.* **1965**, *56*, 142–143. [CrossRef]

46. Trut, L.N. Early Canid Domestication: The Farm-Fox Experiment: Foxes bred for tamability in a 40-year experiment exhibit remarkable transformations that suggest an interplay between behavioral genetics and development. *Am. Sci.* **1999**, *87*, 160–169. [CrossRef]

47. Volobujev, V.T.; Radzhabli, S.I.; Belyaeva, E.S. Investigation of the nature and the role of additional chromosomes in silver foxes. III. Replication pattern in additional chromosomes. *Genetika* **1976**, *12*, 30.

48. Radzhabli, S.I.; Isaenko, A.A.; Volobujev, V.T. Investigation of the nature and the role of additional chromosomes in silver fox. IV. B-chromosomes behaviour in meiosis. *Genetika* **1978**, *14*, 438–443. [PubMed]

49. Świtoński, M.; Gustavsson, I.; Höjer, K.; Plöen, L. Synaptonemal complex analysis of the B-chromosomes in spermatocytes of the silver fox (*Vulpes fulvus* Desm.). *Cytogenet. Genome Res.* **1987**, *45*, 84–92.

50. Basheva, E.A.; Torgasheva, A.A.; Sakaeva, G.R.; Bidau, C.; Borodin, P.M. A-and B-chromosome pairing and recombination in male meiosis of the silver fox (*Vulpes vulpes* L., 1758, Carnivora, Canidae). *Chromosome Res.* **2010**, *18*, 689–696. [CrossRef] [PubMed]

51. Bugno-Poniewierska, M.; Solek, P.; Wronski, M.; Potocki, L.; Jezewska-Witkowska, G.; Wnuk, M. Genome organization and DNA methylation patterns of B chromosomes in the red fox and Chinese raccoon dogs. *Hereditas* **2014**, *151*, 169–176. [CrossRef] [PubMed]

52. Kociucka, B.; Sosnowski, J.; Kubiak, A.; Nowak, A.; Pawlak, P.; Szczerbal, I. Three-dimensional positioning of B chromosomes in fibroblast nuclei of the red fox and the Chinese raccoon dog. *Cytogenet. Genome Res.* **2013**, *139*, 243–249. [CrossRef] [PubMed]

53. Mäkinen, A.; Fredga, K. Banding analyses of the somatic chromosomes of raccoon dogs, *Nyctereutes procyonoides*, from Finland. In Proceedings of the 4th European Colloquium on Cytogenetics of Domestic Animals, Uppsala, Sweden, 10–13 June 1980.

54. Yosida, T.H.; Wada, M.Y.; Ward, O.G. Karyotype of a Japanese raccoon dog with 40 chromosomes including two supernumeraries. *Proc. Jpn. Acad. Ser. B* **1983**, *59*, 267–270. [CrossRef]

55. Wada, M.Y.; Lim, Y.; Wurster-Hill, D.H. Banded karyotype of a wild-caught male Korean raccoon dog, *Nyctereutes procyonoides koreensis*. *Genome* **1991**, *34*, 302–306. [CrossRef]

56. Ward, O.G.; Wurster-Hill, D.H.; Ratty, F.J.; Song, Y. Comparative cytogenetics of Chinese and Japanese raccoon dogs, *Nyctereutes procyonoides*. *Cytogenet. Genome Res.* **1987**, *45*, 177–186. [CrossRef] [PubMed]

57. Yosida, T.H.; Wada, M.Y. Cytogenetical studies on the Japanese raccoon dog. VI. Distribution of B-chromosomes in 1372 cells from 13 specimens, with special note on the frequency of the Robertsonian fission. *Proc. Jpn. Acad. Ser. B Phys. Biol. Sci.* **1984**, *60*, 301–305. [CrossRef]

58. Yosida, T.H.; Wada, M.Y. Cytogenetical studies on the Japanese raccoon dog. VIII. B-chromosomes observed in the spermatogonial metaphase cells. *Proc. Jpn. Acad. Ser. B* **1985**, *61*, 375–378. [CrossRef]

59. Szczerbal, I.; Switonski, M. B chromosomes of the Chinese raccoon dog (*Nyctereutes procyonoides procyonoides* Gray) contain inactive NOR-like sequences. *Caryologia* **2003**, *56*, 213–216. [CrossRef]

60. Wurster-Hill, D.H.; Ward, O.G.; Kada, H.; Whittemore, S. Banded chromosome studies and B chromosomes in wild-caught raccoon dogs, *Nyctereutes procyonoides viverrinus*. *Cytogenet. Genome Res.* **1986**, *42*, 85–93. [CrossRef]

61. Shi, L.; Tang, L.; Ma, K.; Ma, C. Synaptonemal complex formation among supernumerary B chromosomes: An electron microscopic study on spermatocytes of Chinese raccoon dogs. *Chromosoma* **1988**, *97*, 178–183. [CrossRef]

62. Wurster-Hill, D.H.; Ward, O.G.; Davis, B.H.; Park, J.P.; Moyzis, R.K.; Meyne, J. Fragile sites, telomeric DNA sequences, B chromosomes, and DNA content in raccoon dogs, *Nyctereutes procyonoides*, with comparative notes on foxes, coyote, wolf, and raccoon. *Cytogenet. Genome Res.* **1988**, *49*, 278–281. [CrossRef] [PubMed]

63. Li, Y.M.; Zhang, Y.; Zhu, W.J.; Yan, S.Q.; Sun, J.H. Identification of polymorphisms and transcriptional activity of the proto-oncogene *KIT* located on both autosomal and B chromosomes of the Chinese raccoon dog. *Genet. Mol. Res.* **2016**, *15*, 26909958. [CrossRef] [PubMed]

64. Ma, W.; Gabriel, T.S.; Martis, M.M.; Gursinsky, T.; Schubert, V.; Vrána, J.; Doležel, J.; Grundlach, H.; Altschmied, L.; Scholz, U. Rye B chromosomes encode a functional Argonaute-like protein with in vitro slicer activities similar to its A chromosome paralog. *New Phytol.* **2017**, *213*, 916–928. [CrossRef] [PubMed]
65. Sugimoto, Y.; Morita, R.; Amano, K.; Shah, P.U.; Pascual-Castroviejo, I.; Khan, S.; Delgado-Escueta, A.V.; Yamakawa, K. T-STAR gene: Fine mapping in the candidate region for childhood absence epilepsy on 8q24 and mutational analysis in patients. *Epilepsy Res.* **2001**, *46*, 139–144. [CrossRef]
66. Posey, A.D.; Demonbreun, A.; McNally, E.M. Ferlin proteins in myoblast fusion and muscle growth. *Curr. Top. Dev. Biol.* **2011**, *96*, 203–230. [CrossRef] [PubMed]
67. Turtoi, A.; Blomme, A.; Bellahcène, A.; Gilles, C.; Hennequière, V.; Peixoto, P.; Bianchi, E.; Noel, A.; De Pauw, E.; Lifrange, E.; et al. Myoferlin is a key regulator of EGFR activity in breast cancer. *Cancer Res.* **2013**, *73*, 5438–5448. [CrossRef] [PubMed]

© 2018 by the authors. Licensee MDPI, Basel, Switzerland. This article is an open access article distributed under the terms and conditions of the Creative Commons Attribution (CC BY) license (http://creativecommons.org/licenses/by/4.0/).

genes

MDPI

Review

Transmission and Drive Involving Parasitic B Chromosomes

R.N. Jones

Institute of Biological, Environmental and Rural Sciences (IBERS), Aberystwyth University, Edward Llwyd Building, Penglais Campus, Aberystwyth SY23 3DA, UK; neil.rnj@gmail.com

Received: 25 June 2018; Accepted: 26 July 2018; Published: 31 July 2018

Abstract: B chromosomes (Bs) are enigmatic additional elements in the genomes of thousands of species of plants, animals, and fungi. How do these non-essential, harmful, and parasitic chromosomes maintain their presence in their hosts, making demands on all the essential functions of their host genomes? The answer seems to be that they have mechanisms of drive which enable them to enhance their transmission rates by various processes of non-mendelian inheritance. It is also becoming increasingly clear that the host genomes are developing their own mechanisms to resist the impact of the harmful effects of the Bs.

Keywords: B chromosomes; transmission; drive; host/parasite interaction

1. Introduction

New technologies are the driver for advances in our knowledge of genetics and cytogenetics, including some fundamental questions concerning supernumerary B chromosomes (Bs): namely their origin, widespread existence, phenotypic effects, molecular organization, modes of inheritance, and population equilibrium frequencies. Their main properties can be summarised as follows: (i) they occur in thousands of species, and in all known cases they are dispensable and individuals with none are always present; (ii) they never pair with the standard A chromosomes (As) of their hosts; (iii) their behaviour at meiosis is irregular and can lead to elimination; (iv) they are usually smaller than the As and often heterochromatic; (v) several cases they have recently been shown to carry genes; (vi) they are harmful to their host organisms when present in high numbers; (vii) they have various mechanisms of accumulation, including nondisjunction and meiotic drive [1]. The context and background to the story is recorded in a number of reviews, listed in Table 1.

Table 1. Selection of reviews on B chromosomes.

Author	Title	Reference
Jones, R.N.	B Chromosome Drive	[1]
Jones, R.N. et al.	B Chromosomes	[2]
Camacho, J.P.M.	B Chromosomes	[3]
Houben, A.; et al.	Biology and Evolution of B Chromosomes	[4]
Houben, A.; et al.	Evolution and Biology of Supernumerary B Chromosomes	[5]
Banaei-Moghaddam, A.M. et al.	Genes on B Chromosomes: Old Questions Revisited with New Tools	[6]
Valente, G.T. et al.	B Chromosomes: from Cytogenetics to Systems Biology	[7]
Houben, A.	B Chromosomes—A Matter of Chromosome Drive	[8]
Ruban, A. et al.	How Next-Generation Sequencing Has Aided Our Understanding of the Sequence Composition and Origin of B Chromosomes	[9]
Coan, R.L.B. et al.	Landscape of Transposable Elements Focusing on the B Chromosome of the Cichlid Fish *Astatotilapia latifasciata*	[10]

These reviews are augmented by the papers appearing in this special edition of Genes. The present review considers one aspect of this story, namely the modes of inheritance of Bs, with particular reference to transmission and drive which enables them to maintain their presence in populations against a gradient of harmful effects. It is convenient to deal with plants (Table 2) and animals (Table 3) separately.

2. Plants

There are several mechanisms of B chromosome accumulation in plants, as summarised in Table 2, of which post-meiotic nondisjunction in microspores and megaspores are the best known and are considered first. The model species, and virtually the only ones for investigating this process are rye and maize.

2.1. Rye (Secale Cereal, 2n = 2x = 14 + Bs)

Directed nondisjunction at the first pollen grain mitosis in rye was discovered, described and represented diagrammatically by Hasegawa in 1934 [11]. A photograph later captured an undivided single rye B at the equator of the unequal spindle at anaphase (Figure 1).

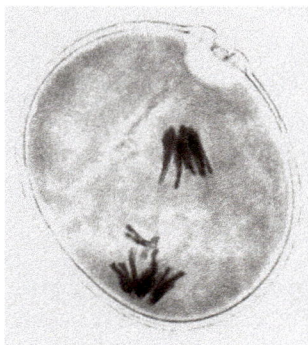

Figure 1. Nondisjunction of a rye B at first pollen grain mitosis (photo by author).

Hasegawa's diagrams are reproduced in Houben [8], together with a full account of the latest information on the cellular and molecular components of the nondisjunction process. Suffice it to say here that there is a controlling element at the end of the long arm of the B, and that a deleted B missing this region can only undergo nondisjunction when the standard B is also present and acts in trans to provide the essential function. Houben has discussed the action of this essential region in detail, together with the variation in centromere structure between that of the B and the standard A chromosomes. The sticking sites on either side of the B centromere are also considered in detail, and these comprehensive studies will not be repeated here. The autonomy of the rye B nondisjunction was established when it became known that it behaves in the same way in hexaploid wheat as it does in rye [12]; although the addition line cannot be easily maintained due to the low pairing level of the rye Bs in the wheat background [13]. This pairing observation raises the interesting point that the transmission level of the rye Bs are subject to interaction with the host background genotype, be it in a related species or in different strains of rye, and it seems that the B itself controls its own transmission properties. The output of Bs through meiosis will also clearly impact on the level at which nondisjunction can operate, determined by the number of Bs passing through to the gametophytes. Transmission data for several varieties of rye is summarized and reviewed in [14,15], and reveal a wide range of variation. Japanese populations have a B frequency of up to 90%, with most individuals having 2Bs [16], while in the Swedish variety Östgöta Gråråg it is down to as low as a few percent [17]. Müntzing also compared transmission frequencies in Östgöta Gråråg with the variety Vasa II, and found a wide variation based on meiotic pairing, but not in the nondisjunction rates.

He also drew attention to the structural variation in the Bs themselves [18,19], particularly at the end of the long arm where the nondisjunction controlling element was later located. The standard B of Vasa II is larger than that of Östgöta Gråråg. Matthews and Jones [15] also concluded that the pairing rate of Bs at meiosis is the main factor determining their equilibrium frequencies in natural populations, and this pairing is a property of the Bs themselves. Jiménez et al. [20] and Puertas et al. [21], Figure 2 also report, from studies on high and low transmission lines in Korean rye, that the transmission frequency is a property of the Bs themselves, in terms of their capacity to form bivalents rather than univalents at meiosis, and does not depend directly on the level of nondisjunction in the pollen grain.

Metaphase of meiosis in 2B plants

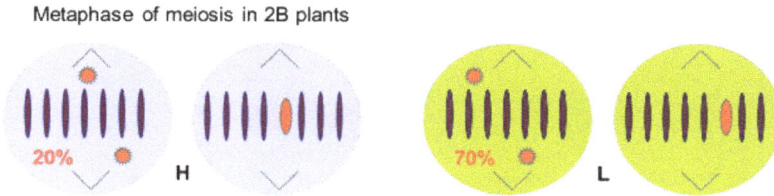

Figure 2. Metaphase of meiosis in 2B plants of two transmission genotypes of Korean rye, based on Jiménez et al [20]. Genotypes were selected homozygous for A chromosome 'genes' controlling HIGH (H) and LOW (L) transmission rates of B chromosomes. Transmission rates reflect resistance of the A background to the Bs.

These observations lead us to consider the possibility of long-term host/parasite co-evolution, and to the idea that the parasitic effects of rye Bs might be beneficial in the long term [22]. We should also remember a classic paper by Östergren [23] on the parasitic nature of extra fragment chromosomes. This idea of the co-evolution of Bs and their hosts is a recurring theme in both plants and animals and is the basis by which the frequency of Bs in natural populations is determined.

2.2. Maize (Zea Mays, 2n = 2x = 20 + Bs)

The mechanism of nondisjunction in maize was discovered by Roman in 1947 [24], using a translocation between the B and A chromosome 4, known as the A-B interchange TB-4a, and a marker gene to track the process. The behaviour of the B does not depend on it remaining intact. This approach was taken since the small size of the maize B in the pollen grains did not allow its behaviour to be followed cytologically. The methodology of Roman's genetic analysis is covered in Jones and Ruban [25]. It turns out that the nondisjunction of the B occurs at the second division of the male gametophyte, which in itself does not lead to an increase in the number of Bs in subsequent generations, and it occurs in 50–100% of pollen grains [26]. The drive depends on the sperm nuclei carrying the B preferentially fertilising the egg nucleus about two thirds of the time, although how this happens is not fully understood [26]. Inheritance through the female side is normal. There are a number of studies using A-B translocations that have identified the regions of the B enable the centromere to undergo nondisjunction (Figure 3, based on Carlson [26], Jones and Ruban [25]).

Figure 3. Structure of the maize B chromosome, based on Carlson [26]. See text for details of the numbering of the segments.

These sites, and the basis of their actions, have been extensively described [26], and are summarised briefly here. Region 1 is the short distal euchromatic tip of the B, region 2 is the longer proximal euchromatin and region 3 the proximal centromeric heterochromatin. Region 4, comprising the centromere and B short arm modifies the rate of nondisjunction. In relation to region 4, it is noteworthy that the maize B has a high level of autonomy and control over its own inheritance. This was demonstrated by Rosato et al. [27] using crosses between genotypes for high and low transmission rates in races of Pisingallo maize from Argentina. Carlson and Roseman [28] had already shown that genotypes of maize Bs could control their own transmission rates by the suppression of meiotic loss. Drive mechanisms based on pollen grain mitosis are well known in several other species, many of which listed in Table 2 with references. There are other mechanisms of drive that are not based on pollen grain mitosis (Table 2), and some of these are worthier of more detailed description.

Table 2. Summary of various mechanisms of B accumulation in plants.

Nondisjunction at First Pollen Grain Mitosis
Aegilops speltoides, Alopecurus pratensis, Anthoxanthum aristatum, Brachycome lineariloba, Briza media. Dactylis glomerata, Deschampsia bottnica, Deschampsia caespitosa, Deschampsia wibeliana, Festuca arundinacea, Festuca pratensis, Haplopappus gracilis, Holcus lanatus, Phleum phleoides, [2] *Panicum maximum* [29] *Aegilops mutica* [30]
Pollen Grain Mitosis of Extra Divisions
Sorghum-purpureo-sericium [31]
Somatic Non-Disjunction in the Developing Inflorescences
Crepis capillaris [32]
Female Meiotic Drive
Lilium callosum [33], *Phleum nodosum* [34] *Plantago serraria* [35] *Trillium grandiflorum* [36]
Female Meiotic Drive and Male Meiotic Drag
Picea sitchensis [37] *Hypochoeris maculata* [38]
Male Drive
Haplopappus validus, Clarkia elegans, Iseilema laxum [2] *Briza humilis* B^L [39]
Somatic Nondisjunction Coincident with Flower Initiation
Crepis capillaris [40]
No Apparent Mechanism
Allium schoenoprasum [41] *Xanthisma texanum* [42] *Centauria scabiosa, Poa alpina, Ranunculus acris, Ranunculus ficaria* [2].

Table 3. Summary of various mechanisms of B-chromosome accumulation in animals.

No Drive
Eyprepocnemis plorans (grasshopper) [43]   *Prochilodus lineatus* (fish) [44]   *Metagagrella tenuipes* (Arachnida, Japanese harvestman) [45]
Mechanism (?) to Boost B-number in Males
Euphydryas colon (Lepidoptera). [46]
Female Meiotic Drive
Pseudococcus obscurus (mealy bug) [47]   *Myrmeleotettix maculatus* (grasshopper) [48]   *Melanoplus femur-rubrum* (grasshopper) [49]   *Heteracris littoralis* (grasshopper) [50]   *Omocestus burri* (grasshopper) [51]
Male Drive
Rattus fuscipes (Australian bushrat) [52]
Male Meiotic Drive and the Opposite of Drive, i.e., Female Drag
Chortoicetes terminifera (Australian plague locust) [53]
PSR Enhances Transmission by Losing Paternal Chromosomes Except Itself
Nasonia vitripennis (parasitic wasp) [54]
Female Meiotic Drive and Male Meiotic Drag
Myrmeleotettix maculatus (grasshopper) [55]   *Locusta migratoria* [56]
B Elimination during Spermiogenesis
Eumigus monticola, Eyprepocnemis plorans [57]

Kimura and Kayano [33] were first to describe a mathematical theory to analyse the mechanism of distribution of Bs in a natural plant population. Their theory also suggested a mechanism by which the deleterious effect of Bs could be reduced in the course of evolution. Female meiotic drive is also noted in three other cases (Table 2). In *Crepis capillaris*, there is somatic nondisjunction coincident with flower initiation [40]. It is noteworthy that there is a number species with widespread B polymorphisms where no mechanism of drive could be determined (*Allium schoenoprasum, Xanthisma texanum, Centauria scabiosa*), and numerous others where we have no specific information, or species (most of them) where transmission has not yet been determined. There is a very narrow basis on which our current view of drive and B/A co-evolution is built.

In *Lilium callosum* the transmission of Bs through the pollen is normal, but there is drive through the female line which occurs at meiosis [33]. The spindle is asymmetrical at metaphase I and single unpaired Bs tend to be located 80% of the time in the larger area of the spindle which will give rise to the egg cell, and only 20% of the time in the area which will give rise to the embryo. An interesting feature of this process is that it depends on an asymmetric spindle, like the pollen mitosis in rye and other Gramineae.

3. Animals

In most of the animal kingdom gametes are produced directly by meiosis. There is no gametophyte phase to the life cycle and female or male drive is mainly based on B-behaviour at meiosis. *Eyprepocnemis plorans* [43] is the most intensively studied, and best understood in terms of population dynamics, of the species lacking drive. The authors explain that in a newly invaded population, a particular form of the B has substantial drive, and that co-evolution of drive between the nascent B and suppressor genes in the As gradually reduces the level of B-drive. The B eventually becomes neutral and stochastic loss then eliminates it from the population. When a new variant of the B appears with drive it can displace the original variant, and a new cycle of drive suppression and drift to extinction occurs. The time it takes for extinction to occur is related to population size [58]. In *Eyprepocnemis plorans* [43] the transmission pattern of three variants of the Bs was studied in detail at meiosis in both male and female animals and in controlled crosses. The Bs appear to be inherited in a regular manner with no tendency to either accumulation or loss, which begs the question of their status in the population. The idea that they are neutral, based on some historic perspective, or that they have an accumulation mechanism unrelated to meiotic drive (mating preferences or sperm precedence), or natural selection can all be advanced, but at the end of the day there is no definitive evidence to answer the questions. No net accumulations of Bs was found in the fish species *Prochilodus lineatus* [44], and the argument was again advanced that the Bs were possibly driven in the past but this was neutralised by drive-suppressor genes in the A genome. Nur and Brett [47] reported on genotypes suppressing meiotic drive in *Pseudococcus obscurus*. Female meiotic drive is the most common mode of transmission and is best known in grasshoppers and in the mealy bug *Pseudococcus obscuruus*. Preferential segregation of univalent Bs at meiosis is one of the mechanisms of B-drive in animals, and this has usually been inferred from breeding experiments, as in *Melanoplus femur-rubrum* [49], although there is only one case in which this has been demonstrated at cytological level and that is in *Myrmeleotettix maculatus*, Figure 4 [48].

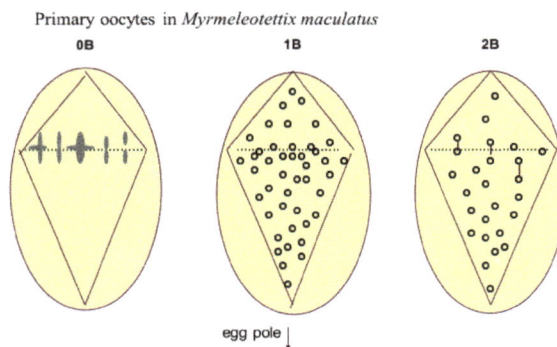

Figure 4. Meiotic drive in primary oocytes of *Myrmeleotettix maculatus*, based on Hewitt [48]. The 0B nucleus shows the seven A chromosome bivalents lined up at the equator of the asymmetrical spindle at metaphase I. In the 1B and 2B nuclei there are more Bs in the region of the nucleus which will form the egg cells. 2B bivalents are shown as joined dots. This preferential chromosome distribution matches very closely the level of preferential transmission determined from breeding experiments.

Mechanisms of male drive and of male drive and female drag have also been reported (Table 2), as well as specialised systems -in s, *Eumigus monticola*, and *Eyprepocnemis plorans*. In the wasp *Nasonia vitripennis* the males are haploid and develop from unfertilized eggs, while the diploid females develop from fertilized eggs. Some individuals in this species carry a genetic element, termed PSR (paternal sex ratio). This element is transmitted through sperm where it causes condensation and subsequent loss of paternal chromosome set in fertilized eggs. Diploid females are thereby converted into haploid males. The authors have shown that PSR has the properties of a supernumerary chromosome B with a number of B-specific repetitive sequences. The PSR appears to produce a trans-acting product which causes condensation of the paternal chromosome set, but is not itself affected. The fascinating effect is that the PSR enhances its own transmission by eliminating the rest of the genome, and can thus considered to be the ultimate 'selfish' genetic element.

4. Conclusions

Among the relatively small number of species that have been investigated, relative to the total number of known species, there are a variety of mechanisms of B chromosome drive, including those where no drive has been identified. In plants, the most common process takes place post-meiotically whereas in animals it mainly involves meiosis. In several cases of both plants and animals, the population equilibrium B frequency is a result of conflict between the As and Bs: Bs providing the drive and As acting in various ways to suppress the increase in B numbers. Even within populations there are differences in B equilibrium frequencies between different populations. The most instructive case is that of the grasshopper *Eyprepocnemis plorans*, where suppressor genes in the As have neutralised one of the B-types, and over time stochastic events can eliminate that B from the population. Eventually, a new B can arise with its drive restored and the cycle can be repeated over again. In the plant world, there is also variation in B transmission in certain populations, particularly in maize and in rye, and evidence of suppression of B-drive by the A genomes. However, there are no cases known as yet where the As have completely suppressed and neutralised any B. The view that Bs are nuclear parasites, albeit involved in host/parasite interactions, seems to hold true with the balance in favour of the parasites rather than the host. There is much more to learn about this aspect of the story of Bs, involving the species which have been investigated as well as the many others about which we have no knowledge.

Funding: This research received no external funding.

Conflicts of Interest: The author declares no conflict of interest.

References

1. Jones, R.N. B chromosome drive. *Am. Nat.* **1991**, *137*, 430–442. [CrossRef]
2. Jones, R.N.; Rees, H. *B Chromosomes*; Academic Press: New York, NY, USA, 1982.
3. Camacho, J.P.M. B chromosomes. In *The Evolution of the Genome*; Gregory, T.R., Ed.; Elsevier: Amsterdam, The Netherlands, 2012.
4. Houben, A.; Banaei-Moghaddam, A.M.; Klemme, S. Biology and evolution of B chromosomes. In *Plant Genome Diversity Volume 2*; Greilhuber, J., Dolezel, J., Wendel, J.F., Eds.; Springer: Vienna, Austria, 2013; pp. 149–165.
5. Houben, A.; Banaei-Moghaddam, A.M.; Klemme, S.; Timmis, J.N. Evolution and biology of supernumerary B chromosomes. *Cell. Mol. Life Sci.* **2014**, *71*, 467–478. [CrossRef] [PubMed]
6. Banaei-Moghaddam, A.M.; Martis, M.M.; Macas, J.; Gundlach, H.; Himmelbach, A.; Altschmied, L.; Mayer, K.F.; Houben, A. Genes on B chromosomes: Old questions revisited with new tools. *Biochim. Biophys. Acta* **2015**, *1849*, 64–70. [CrossRef] [PubMed]
7. Valente, G.T.; Nakajima, R.T.; Fantinatti, B.E.; Marques, D.F.; Almeida, R.O.; Simoes, R.P.; Martins, C. B chromosomes: From cytogenetics to systems biology. *Chromosoma* **2017**, *126*, 73–81. [CrossRef] [PubMed]
8. Houben, A. B chromosomes—A matter of Chromosome Drive. *Front. Plant Sci.* **2017**, *8*, 210. [CrossRef] [PubMed]
9. Ruban, A.; Schmutzer, T.; Scholz, U.; Houben, A. How Next-Generation sequencing has aided our understanding of the sequence composition and origin of B chromosomes. *Genes* **2017**, *8*, 294. [CrossRef] [PubMed]
10. Coan, R.L.B.; Martins, C. Landscape of transposable elements focusing on the B chromosome of the cichlid fish *Astatotilapia latifasciata*. *Genes* **2018**, *9*, 269. [CrossRef] [PubMed]
11. Hasegawa, N. A cytological study on 8-chromosome rye. *Cytologia* **1934**, *6*, 68–77. [CrossRef]
12. Lindström, J. Transfer to wheat of accessory chromosomes from rye. *Hereditas* **1965**, *54*, 149–155. [CrossRef]
13. Müntzing, A. Chromosomal variation in the Lindström strain of wheat carrying accessory chromosomes of rye. *Hereditas* **1970**, *66*, 279–286. [CrossRef]
14. Matthews, R.B.; Jones, R.N. Dynamics of the B chromosome polymorphism in rye. I. Simulated populations. *Heredity* **1982**, *48*, 345–369. [CrossRef]
15. Matthews, R.B.; Jones, R.N. Dynamics of the B chromosome polymorphism in rye. II. Estimates of parameters. *Heredity* **1983**, *50*, 119–137. [CrossRef]

16. Kishikawa, H. Cytogenetic studies of B chromosomes in rye, *Secale cereale* L. in Japan. *Agric. Bull. Saga Univ.* **1965**, *21*, 1–81.
17. Müntzing, A. Cytological studies of extra fragment chromosomes in rye II. Transmission and multiplication of standard fragments and iso-fragments. *Hereditas* **1945**, *31*, 457–477. [CrossRef] [PubMed]
18. Lima-De-Faria, A. B chromosomes of rye at pachytene. *Port. Acta Biol.* **1948**, *2*, 167–174.
19. Lima-De-Faria, A. The evolution of the structural pattern in a rye B chromosome. *Evolution* **1963**, *17*, 289–295. [CrossRef]
20. Jiménez, M.M.; Romero, F.; Puertas, M.J. B Chromosomes in inbred lines of rye (*Secale cereal* L.) I. Vigour and fertility. *Genetica* **1994**, *92*, 149–154. [CrossRef]
21. Puertas, M.J.; Jiménez, G.; Manzanero, S.; Chiavarino, A.M.; Rosato, M.; Naranjo, C.A.; Poggio, L. Genetic control of B chromosome transmission in maize and rye. *Chromosom. Today* **2000**, *13*, 79–82.
22. Gonzáez-Sánchez, M.; Chiavarino, M.; Jiménez, G.; Manzanero, S.; Rosato, M.; Puertas, M.J. The parasitic effects of rye B chromosomes might be beneficial in the long term. *Cytogenet. Genome Res.* **2004**, *106*, 386–393. [CrossRef] [PubMed]
23. Östergren, G. Parasitic nature of extra fragment chromosomes. *Bot. Not.* **1945**, *2*, 157–163.
24. Roman, H. Mitotic nondisjunction in the case of interchanges involving the B-type chromosome in maize. *Genetics* **1947**, *32*, 391–409. [PubMed]
25. Jones, R.N.; Ruban, A. Are B chromosomes useful? In *Plants, People, Planet*; 2018; submitted for publication.
26. Carlson, W.R. The B chromosome of maize. *Crit. Rev. Plant Sci.* **1986**, *3*, 201–226. [CrossRef]
27. Rosato, M.; Chiavarino, A.M.; Naranjo, C.; Puertas, M.; Poggio, L. Genetic control of B chromosome transmission rate in *Zea mays ssp. mays* (Poaceae). *Am. J. Bot.* **1996**, *83*, 1107–1112. [CrossRef]
28. Carlson, W.R.; Roseman, R. A new property of the maize B chromosome. *Genetics* **1992**, *131*, 211–223. [PubMed]
29. Komatsu, T.; Nakajima, K. B chromosomes in diploid Guineagrass (*Panicum maximum* JACQ). *Jpn. J. Breed.* **1988**, *38*, 151–157. [CrossRef]
30. Ohta, S. Mechanisms of B-chromosome ac in *Aegilops mutica* Boiss. *Genes Genet. Syst.* **1996**, *71*, 23–29. [CrossRef]
31. Darlington, C.D.; Thomas, P.T. Morbid mitosis and the activity of inert chromosomes in *Sorghum*. *Proc. R. Soc. B* **1941**, *130*, 127–150. [CrossRef]
32. Parker, J.S.; Jones, G.H.; Edgar, L.; Whitehouse, C. The population cytogenetics of *Crepis capillaris*. II. The stability and inheritance of B-chromosomes. *Heredity* **1989**, *63*, 19–27. [CrossRef]
33. Kimura, M.; Kayano, H. The maintenance of supernumerary chromosomes in wild populations of *Lilium callosum* by preferential segregation. *Genetics* **1961**, *46*, 1699–1712. [PubMed]
34. Fröst, S. The inheritance of accessory chromosomes in plants, especially in *Ranunculus acris* and *Phleum nodosum*. *Hereditas* **1969**, *61*, 317–326. [CrossRef]
35. Fröst, S. The cytological be and mode of transmission of accessory chromosomes in *Plantago serraria*. *Hereditas* **1959**, *45*, 191–210. [CrossRef]
36. Rutishauser, A. Genetics of fragment chromosomes in *Trillium grandiflorum*. *Heredity* **1956**, *10*, 195–204. [CrossRef]
37. Kean, V.M.; Fox, D.P.; Faulkner, R. The Ac mechanism of the supernumerary (B-) chromosome in *Picea sitchensis* (Bong.) Carr. and the effect of this on male and female flowering. *Silvae Genet.* **1982**, *31*, 126–131.
38. Parker, J.S.; Taylor, S.; Ainsworth, C.C. The B-Chromosome system of *Hypochoeris maculata* III. Variation in B-Chromosome transmission rates. *Chromosoma* **1982**, *85*, 299–310. [CrossRef]
39. Murray, B.G. The structure, meiotic behaviour and effects of B chromosomes in *Briza humilis* Bieb. (Gramineae). *Genetica* **1984**, *63*, 213–219. [CrossRef]
40. Rutishauser, A.; Roethlisberger, E. Boosting of B chromosomes in *Crepis capillaris*. *Chromosom. Today* **1966**, *1*, 28–30.
41. Bougourd, S.M.; Plowman, A.B.; Ponsford, N.R.; Elias, M.L.; Holmes, D.S.; Taylor, S. The case for unselfish B-chromosomes: Evidence from *Allium schoenoprasum*. In Proceedings of the Kew Chromosome Conference IV, Kew, UK, 30 August–2 September 1994; Royal Botanic Gardens: Kew, UK, 1995; pp. 21–34.
42. Berger, C.A.; Feely, E.J.; Witkus, E.R. The cytology of *Xanthisma texanum* D.C. IV. Megasporogenesis and embryo formation, pollen mitosis and embryo formation. *Bull. Torrey Bot. Club* **1956**, *83*, 428–434. [CrossRef]

43. Lopez-Leon, M.D.; Cabrero, J.; Camacho, J.P.M.; Cano, M.I.; Santos, J.L. A widespread B chromosome polymorphism maintained without apparent drive. *Evolution* **1992**, *46*, 529–539. [CrossRef] [PubMed]

44. Oliveira, C.; Saboya, S.M.R.; Foresti, F.; Senhorini, J.O.; Bernardino, G. Increased B chromosome frequency and absence of drive in the fish *Prochiodus lineatus*. *Heredity* **1997**, *79*, 473–476. [CrossRef]

45. Gorlov, I.P.; Tsurusaki, N. Morphology and meiotic/mitotic behavior of B Chromosomes in a Japanese harvestman, *Metagagrella tenuipes* (Arachnida: Opiliones): No evidence for B accumulation mechanisms. *Zool. Sci.* **2000**, *17*, 349–355. [PubMed]

46. Pearse, F.K.; Ehrlich, P.R. B Chromosome variation in *Euphydryas colon* (Lepidoptera: Nymphalidae). *Chromosoma* **1979**, *73*, 263–274. [CrossRef]

47. Nur, U.; Brett, B.L.H. Genotypes suppressing meiotic drive in a B chromosome in the mealy bug, *Pseudococcus obscurus*. *Genetics* **1985**, *110*, 73–92. [PubMed]

48. Hewitt, G.M. Meiotic drive for a B-chromosomes in the primary oocytes of *Myrmeleotettix maculatus* (Orthoptera: Acrididae). *Chromosoma* **1976**, *56*, 381–391. [CrossRef] [PubMed]

49. Nur, U. Maintenance of a "parasitic" B chromosome in the grasshopper *Melanoplus femur-rubrum*. *Genetics* **1977**, *87*, 499–512. [PubMed]

50. Cano, M.I.; Santos, J.L. Cytological basis of the B chromosome accumulation mechanism in the grasshopper *Heteracris littoralis* (Ramb). *Heredity* **1989**, *62*, 91–95. [CrossRef] [PubMed]

51. Santos, J.L.; Del Cerro, A.L.; Fernandez, A.; Diez, M. Meiotic behaviour of B chromosomes in the grasshopper *Omocestus burri*: A case of drive in females. *Hereditas* **1993**, *118*, 139–143. [CrossRef]

52. Thompson, R.L.; Westerman, M.; Murray, D. B chromosomes in *Rattus fuscipes* I. Mitotic and meiotic chromosomes and the effects of B chromosomes on chiasma frequency. *Heredity* **1984**, *52*, 355–362. [CrossRef]

53. Gregg, P.C.; Webb, G.C.; Adena, M.A. The dynamics of B chromosomes in populations of the Australian plague locust, *Chortoicetes terminifera* (Walker). *Can. J. Genet. Cytol.* **1984**, *26*, 194–208. [CrossRef]

54. Nur, U.; Werren, J.H.; Eickbush, D.G.; Burke, W.D.; Eickbush, T.H. A "selfish" B chromosome that enhances its transmission by eliminating the paternal genome. *Science* **1988**, *240*, 512–514. [CrossRef] [PubMed]

55. Shaw, M.W.; Hewitt, G.M. The effect of temperature on meiotic transmission rates of the B chromosome of *Myrmeleotettix maculatus* (Orthoptea: Acrididiae). *Heredity* **1984**, *53*, 259–268. [CrossRef]

56. Pardo, M.C.; Lopez-Leon, M.D.; Cabrerro, J.; Camacho, J.P.M. Transmission analysis of mitotically unstable B chromosomes in *Locusta migratoria*. *Genome* **1994**, *37*, 1027–1034. [CrossRef] [PubMed]

57. Cabrero, J.; Martin-Pecina, M.; Ruiz-Ruano, F.J.; Gomez, R.; Camacho, J.P.M. Post-meiotic B chromosome expulsion, during spermiogenesis, in two grasshopper species. *Chromosoma* **2017**, *26*, 633–644. [CrossRef] [PubMed]

58. Camacho, J.P.; Cabrero, J.; Lopez-Leon, M.D.; Shaw, M.W. Evolution of a near-neutral B chromosome. *Chromosom. Today* **1997**, *12*, 301–318.

© 2018 by the author. Licensee MDPI, Basel, Switzerland. This article is an open access article distributed under the terms and conditions of the Creative Commons Attribution (CC BY) license (http://creativecommons.org/licenses/by/4.0/).

![genes logo] *genes*

MDPI

Article

Landscape of Transposable Elements Focusing on the B Chromosome of the Cichlid Fish *Astatotilapia latifasciata*

Rafael L. B. Coan and Cesar Martins *

Department of Morphology, Institute of Biosciences, São Paulo State University (UNESP), 18618-689 Botucatu, SP, Brazil; rafaelbcoan@gmail.com
* Correspondence: cmartins@ibb.unesp.br; Tel.: +55-14-3880-0462

Received: 17 March 2018; Accepted: 17 May 2018; Published: 23 May 2018

Abstract: B chromosomes (Bs) are supernumerary elements found in many taxonomic groups. Most B chromosomes are rich in heterochromatin and composed of abundant repetitive sequences, especially transposable elements (TEs). B origin is generally linked to the A-chromosome complement (A). The first report of a B chromosome in African cichlids was in *Astatotilapia latifasciata*, which can harbor 0, 1, or 2 Bs Classical cytogenetic studies found high a TE content on this B chromosome. In this study, we aimed to understand TE composition and expression in the *A. latifasciata* genome and its relation to the B chromosome. We used bioinformatics analysis to explore the genomic organization of TEs and their composition on the B chromosome. The bioinformatics findings were validated by fluorescent in situ hybridization (FISH) and real-time PCR (qPCR). *A. latifasciata* has a TE content similar to that of other cichlid fishes and several expanded elements on its B chromosome. With RNA sequencing data (RNA-seq), we showed that all major TE classes are transcribed in the brain, muscle, and male and female gonads. An evaluation of TE transcription levels between B- and B+ individuals showed that few elements are differentially expressed between these groups and that the expanded B elements are not highly transcribed. Putative silencing mechanisms may act on the B chromosome of *A. latifasciata* to prevent the adverse consequences of repeat transcription and mobilization in the genome.

Keywords: repetitive elements; RNA-Seq; genomics; evolution; cytogenetics; supernumerary elements; extra chromosomes

---

1. Introduction

B chromosomes (Bs) are supernumerary elements in addition to autosomal (A) chromosomes and have been observed in several species of animals, plants, and fungi. B chromosomes have an evolutionary pathway distinct from that of A chromosomes and a non-Mendelian form of inheritance [1]. Diverse publications have shown that Bs can have neutral, deleterious, or beneficial effects on their hosts: the presence of an additional B chromosome is correlated to sex determination in the cichlid fish *Lithochromis rubripinnis*, V-shaped phenotype in the common frog *Rana temporaria* and antibiotic resistance in the fungus *Nectria haematococca* [1–4].

B chromosomes are usually heterochromatic due to the abundance of repetitive elements in their composition [5]. Regardless of the predominance of repetitive content on Bs, they can carry genes with different levels of integrity, including fully transcribed copies. Gene copies found on B chromosomes can modulate the gene expression of A-complement genes and even influence metabolic pathways [6–8]. B chromosomes usually consist of a mosaic of sequences from the A complement; B origin is generally linked to genomic instability in the A chromosomes and involves the formation of a proto-B and later expansion of repetitive elements [9–11]. Repetitive elements, including transposable elements (TEs—DNA transposons and retrotransposons) [12–15], represent a large portion of most eukaryotic genomes and are also a major component of B chromosome constitution [3,13]. Furthermore, gene clusters such as U2 small nuclear RNA (snRNA) [16], 18S ribosomal RNA (rRNA), and H1 [11], H3, and H4 histones [17] have also been found on Bs in many species. Classical cytogenetic analysis is used to locate B chromosome origins on different A-complement pairs with the use of probes from multigenic families and TEs [11,18–21]. The TE content on B chromosomes is related to the TE content of the A complement, and repeat expansion on Bs is a common characteristic of these supernumerary elements, caused by lack of recombination and low selective pressure [3]. For example, on the B chromosome of rye, accumulation of satellite DNA, Ty1/Copia, a long terminal repeat (LTR) retrotransposon, and a few unclassified sequences are present [22]. On the B chromosome of the fish *Alburnus alburnus*, an expanded Ty3/Gypsy sequence showed similarity to the reverse transcriptase coding gene [23].

In addition to the accumulation of repetitive elements on Bs, evidence also exists that those elements can be transcribed. The StarkB element specific to the maize B chromosome has variable expression among individuals and loci [24]. In a different study, this element was found to be expressed with the Gypsy and Copia TEs. In fact, StarkB expression showed evidence of dose dependence; its expression increased as the number of Bs increased. This result demonstrates the influence of the number of B chromosomes in the transcription of their sequences [25]. Another recent study found differentially expressed repetitive sequences between 0B and 4B individuals in rye. The same study also noted differences in transcription among the different tissues [26].

Among African cichlids, B chromosomes have been observed in twenty species [2,19,27–29], being first described in *Astatotilapia latifasciata*, which may have 0, 1, or 2 Bs [27]. Cichlid fish from the Great Lakes of East Africa have experienced a rapid adaptive radiation and the B chromosome occurrence represents a new enigma to be investigated in the focus of evolutionary biology [30–32]. The Bs of *A. latifasciata* are among the largest B chromosomes investigated to date. They are heterochromatic and contain abundant repetitive elements, as evidenced by classical cytogenetic mappings of 18S ribosomal DNA (rDNA) and the Rex1 and Rex3 elements [19,27]. A recent next-generation sequencing (NGS) analysis also revealed an accumulation of diverse classes of repetitive sequences on the B chromosome of this species [10]. Thus, our study aims to understand TE content, distribution, and transcription on the *A. latifasciata* genome and its impact on the B chromosome. We used a combination of classical molecular cytogenetics and NGS to evaluate the TE landscape of the species and find representative sequences on the B chromosome. We analyzed the transcription levels of the repeats and their possible relations to B-enriched sequences. Understanding the repeat content of the A complement and its relation to the B chromosome will help elucidate the constitution and perpetuation of the B chromosome in *A. latifasciata*, as well as its influence on the cell biology of the species.

2. Materials and Methods

2.1. Samples

All animal samples were obtained from the fish facility of the Integrative Genomics Laboratory, São Paulo State University, Botucatu, Brazil. We complied with the ethical principles adopted by the Brazilian College of Animal Experimentation, with approval from the Institute of Biosciences/UNESP São Paulo State University ethics committee (protocol no 486-2013). We used samples from males and females with 1 and 2 B chromosomes (B+) or without Bs (B-). Samples were genotyped as B- and B+ by PCR with specific primers for B chromosome presence or absence [33].

All datasets for bioinformatics analysis were previously sequenced. DNA sequencing was performed in four male (M1-0B, M2-1B, M3-1, and M4-2B) and two female samples (F1-0B and F2-1B) from two previous studies [10,34]. The sequencing data included 0B, 1B, and 2B samples. RNA sequencing (RNA-seq) was performed for three tissues: brain, muscle, and gonads. Within each tissue, six samples were 1B and six 0B. Each B condition (1B or 0B) had three males and three females, with 36 total RNA-seq samples. All data are available in SaciBase (sacibase.ibb.unesp.br/). For fluorescent in situ hybridization (FISH), we used a 1B sample confirmed though molecular PCR and cytogenetic analysis. For real-time PCR (qPCR), we used B- (sample ID 04) and B+ (sample IDs 907, 908, 910, 913, 918, and 919) samples confirmed by PCR.

With the exception of M4-2B, a B+ samples with 2B chromosomes, all genomic B+ samples in this study were 1B. We prioritize the B- (0B) and B+ (1B or 2B) nomenclature though the text, but when necessary we will indicate the number of B chromosomes under analysis and discussion (0B, 1B, or 2B). All B+ samples used for expression analysis were 1B. All investigated samples are summarized in Supplementary Table S1.

2.2. Pipelines for Repeat Identification and Repeat Landscape Construction

To estimate the repetitive content in the genome of *A. latifasciata*, we used an assembled B- genome containing male and female reads as reference [34]. We first created a custom repeat library with RepeatModeler 1.0.8 [35] according to the instructions and using the default parameters. Any ID issue in the created *fasta* file was manually checked. We merged the custom repeat library with Repbase Update 20150807 [36] in order to obtain a comprehensive repeat library for input to RepeatMasker. Despite the manual curation, some sequences maintained the "Unknown" status. In the second step of the repeat identification, the merged library was used as input for RepeatMasker 4.0.5 [37] to search for repeat copy number and organization in the assembled genome. RepeatMasker was run with the "slow (-s)", "align (-a)", and "library (-lib)" parameters. To summarize the RepeatMasker results, we used "buildSummary.pl", a Perl script from the RepeatMasker package. The output files from RepeatMasker were also used as input for the "createrepeatlandscape.pl" and "calcdivergencefromalign.pl" scripts to calculate the Kimura divergence values and plot the repeat landscape. Both are helper Perl scripts from the RepeatMasker package.

2.3. Comparative Analysis Pipeline

For the first search of expanded elements on the B chromosome of *A. latifasciata*, we used RepeatExplorer [38], which performs a graph-based clustering of raw Illumina reads. RepeatExplorer uses a small subset of sequenced reads (0.1–0.5× coverage) [39] as input, providing a fast and accurate way to compare two or more datasets.

Raw Illumina datasets from B- (0B) and B+ (2B) genomes [10], available at SaciBase, were used as input for the RepeatExplorer pipeline. RepeatExplorer provides a set of helper scripts to prepare the data for clustering. Reads were quality filtered based on default parameters of "paired_fastq_filtering.R" and a random sample of five million reads for each genome comprising approximately 0.5× genome coverage was selected. Finally, graph-based clustering was applied for de novo repeat identification and comparative analysis [38,39] using the RepeatExplorer pipeline with the default parameters and developer recommendations [40]. Clusters from RepeatExplorer accumulate reads that come from the two datasets and represent a common element. Thus, each cluster has an element (or similar elements) with a specific number of associated reads proportion. The results from clustering were visually inspected with respect to their graphic composition, which indicates the type of repeat and proportion of reads from each genome. We manually chose clusters with the highest content of B+ (2B) reads compared to B- (0B) reads for the later steps. The longest contigs from the selected clusters were annotated with RepeatMasker and Basic Local Alignment Search Tool (BLAST) using the command line and default parameters. To solve the eventual ambiguities in the annotation, priority was given to RepeatMasker with a joint search for conversed domains via BLAST- Conserved Domains Database (CDD); the identification of TE-related proteins was key to the classification of the assembled contigs. We used the annotated contigs found by RepeatExplorer to design primers for probe construction for fluorescence in situ hybridization and qPCR validation (see next sections).

To further characterize the repetitive content of the *A. latifasciata* B chromosome, we performed a coverage ratio analysis on alignments of the six sequenced samples (Supplementary Table S2).

With this methodology, we could analyze the TE expansion on the B chromosome at individual loci. A similar approach was previously used in order to find B chromosome blocks [8,10]. The coverage ratio analysis was done as follows. Raw Illumina reads were filtered to eliminate adapters and bacterial contaminants. We created a library of adapters and bacterial genomes and aligned the reads against this library. We used Bowtie2 2.1 [41] with the "–very-fast-local" parameter, thus excluding reads similar to the sequences in the contaminant/adapter library. This protocol created a contaminant-free dataset. We checked the read quality distribution with fastQC 0.10.1 [42] and performed filtering with the FASTX-toolkit 0.0.13 [43]. We used a Phred score of 28 over 80% of the read as a quality cut-off. We then applied Pairfq 0.11 [44] to restore paired-end reads. Filtered reads were aligned with Bowtie2 against the *A. latifasciata* reference genome (see the Repeat identification and landscape pipeline section) using the "–very-sensitive" parameter. Alignment statistics from BAM files were extracted with Qualimap 2.2 [45].

Binary Alignment Map (BAM) files were used to extract coverage values. We used previously generated RepeatMasker results (see the Repeat identification and landscape pipeline section) to extract coverage for only the TE regions. Since our analysis was focused on TEs, repetitive elements with the "Simple_repeat" and "Low_complexity" traits were excluded from the annotation files. To extract coverage information from BAM files, we used bedtools 2.25 [46]. The single-copy gene hypoxanthine phosphoribosyltransferase (*HPRT*) was used as a cut-off for low-coverage regions (see details in Supplementary Methods). The average coverage for each TE interval was calculated using custom bash and python scripts. We used the average coverage of each interval to calculate the coverage ratios between the different genomes, always using the M1-0B sample as reference (Supplementary Table S3; Supplementary Methods). The ratios between *HPRT* coverages (Supplementary Table S4) were used to normalize the different read coverages among the samples.

2.4. Transcriptome Analysis

We had access to 36 previously sequenced messenger RNA (mRNA) libraries (available at SaciBase) from the muscle, gonads, and brain of B- (0B) and B+ (1B) males and females [47]. They were sequenced on an Illumina HiSeq 2000 (Illumina, San Diego, CA, USA) and include approximately 30 million reads each. The data are freely available at SaciBase. Reads were filtered to remove adapters and contaminants, similarly to the genomic preprocessing. They were quality filtered to maintain reads with at least a Phred score of 28 over 80% of the read. Filtered reads were submitted to RepEnrich 1.2 [48], which determines the enrichment of transcripts in the assembled genome using repeat coordinates as reference. RepEnrich returns a count table from each element and evaluates the repetitive nature of these elements; the results can be used for expression estimation within a tissue or for differential expression. We followed the developer's tutorial to perform these analyses [49]. To find TE expression values, we used the Bioconductor package edgeR 3.4.2, and the count table was produced by RepEnrich. We calculated TE expression within each specific B- tissue to evaluate whether TE expression was present. To find the expression in a particular tissue, we used edgeR to calculate the RPKM (reads per kilobase per million mapped reads) values of each B- tissue. We separated the male and female gonads in the analysis, as they are morphologically and functionally different tissues. Later, the differential expression levels between B- and B+ samples across tissues were evaluated based on the recommendations of the RepEnrich and edgeR developers, using a generalized linear model (GLM) statistical model. We calculated the log2 fold change and false discovery rate (FDR) for each tissue and compared B- and B+ values within tissues. Male and female gonads were analyzed separately. To consider an element as differentially expressed, we used a log2 fold-change cut-off of 1.2 and an FDR cut-off of 0.05.

2.5. Experimental Validation of TEs

Total DNA was extracted with the Qiagen DNeasy® Blood & Tissue Kit (Qiagen, Hilden, Germany) using the fish caudal fin. Primers were designed based on annotated contigs from selected RepeatExplorer clusters (Supplementary Tables S5 and S6) using Oligo Explorer® 1.2 [50], Primer-BLAST [51], and PCR Primer Stats [52]. Amplicons produced by conventional PCR were sequenced using the Sanger method to confirm the sequences found by RepeatExplorer. After confirmation, sets of FISH probes were PCR-labeled with biotin-dUTP.

Mitotic chromosome preparations were performed according to [53] with modifications. Kidney tissue was dilacerated with forceps and a syringe, placed in a Potassium chloride (KCL) 0.075 M solution and incubated for 30 min. The cell suspension was fixed with a 3:1 methanol:acetic acid mixture following three washes and 900 rpm centrifugations (10 min each), discarding the supernatant at the end of each centrifugation. The slides with the chromosomes were stained with Giemsa, and the metaphases were visualized under a light microscope.

Fluorescence in situ hybridization (FISH) was performed on the chromosome squash preparations according to the protocol of [54] with modifications. Slides were dehydrated with 70% ethanol, pretreated with RNAse A (37 °C for 20 min) and pepsin/HCl (3 min), washed (3 times 2× saline sodium citrate (SSC) for 2 min each), dehydrated (2× 70% ethanol for 2 min; 2× 90% ethanol for 2 min; 1× 100% ethanol for 4 min) and dried (1 h at 60 °C). For denaturation, formamide (70%) was applied to slides for 37 s at 65 °C and immediately dehydrated (2× 70% ethanol for 2 min; 2× 90% ethanol for 2 min; 1× 100% ethanol for 4 min). Overnight hybridization was performed with the previously described PCR-labeled probes. The slides were washed (42 °C formamide (50%) for 5 min; 2 times 2× SSC for 5 min; 2 times 4× SSC/Tween for 5 min) and then detected with fluorescein isothiocyanate (FITC) (0.2 μL FITC in 100 μL 4× SSC/Tween at 37 °C for 30 min). A final wash was conducted (3 times 4× SSC/Tween for 3 min), and the slides were stained with 4′,6-diamidino-2-phenylindole (DAPI). Slides were visualized on a BX61 Olympus microscope (Olympus, Tokyo, Japan), and the hybridized metaphases were registered under an Olympus DP71 digital camera coupled to the microscope. The metaphase images were cropped using Adobe Photoshop CS2.

Quantitative real-time PCR (qPCR) of genomic DNA was performed on StepOnePlus™ Real-Time PCR equipment from Applied Biosystems, using the GoTaq® qPCR Master Mix kit (Promega, Madison, USA). Cycling conditions were 95 °C for 10 min; 40 cycles of denaturation at 95 °C for 15 s and annealing/extension at 60 °C for 60 s; and 1 final cycle of dissociation at 95 °C. The results were used to calculate the gene dosage ratio (GDR) using the $2^{-\Delta CT}$ method [55], by relative quantification with respect to the autosomal single copy gene *HPRT* on an Excel spreadsheet.

3. Results

3.1. DNA Transposable Elements Are the Most Represented Class in the A. latifasciata Genome

We first focused on the repetitive content in the B- genome and detected similar patterns of TE composition among *A. latifasciata* and other cichlids [51,52]. An estimation of TE copy number and percentage on the B- genome, here considered the canonical genome, was obtained with RepeatMasker (Figure 1a, Supplementary Table S7, and Supplementary file 1). The *A. latifasciata* genome contains 28.41% repetitive DNA, the majority of which consists of TEs. DNA transposons are the most represented category, followed by long interspersed nuclear element (LINE), LTR, and short interspersed nuclear element (SINE) retrotransposons. In total, retrotransposons (LTR, LINE, and SINE) represent 354,288 copies, while 464,043 copies of DNA transposons are present. Among the retroelements, LINEs had both higher copy number and more bases masked on the genome. SINEs are present in higher number than LTRs, but LTRs have more bases masked on the genome, probably due to their larger size. We found slightly over 7% unclassified elements in the genome.

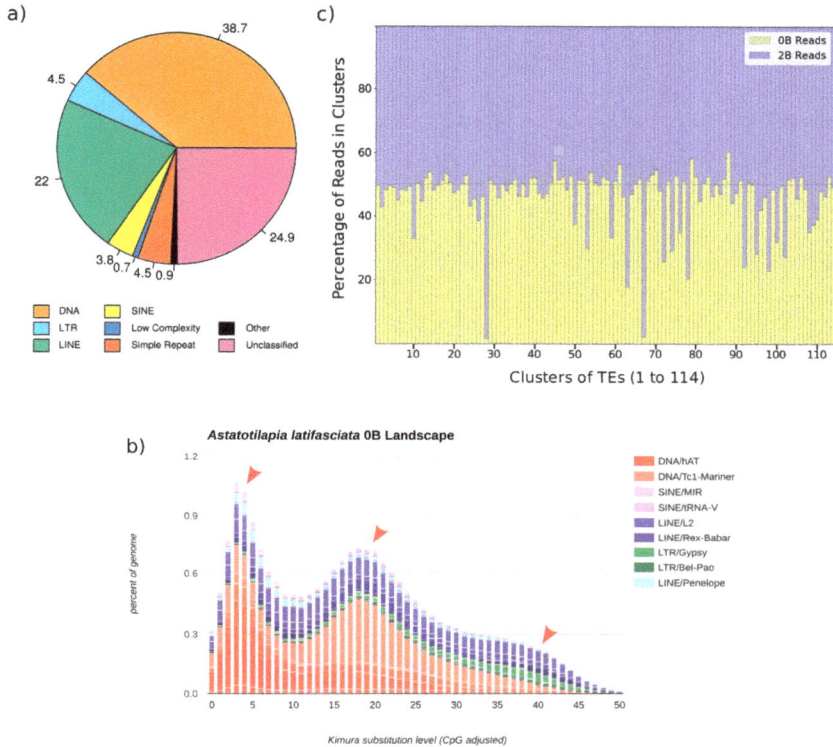

Figure 1. Genomic characteristics of repeated DNAs in the *Astatotilapia latifasciata* genome. (**a**) Percentage of each type of repetitive sequence in the total repetitive portion detected by RepeatMasker on the B- genome. Numbers in this figure represent the percentage of each repeat group compared to the total repetitive content, rather than their percentage in the genome. DNA transposons and long interspersed nuclear element (LINEs) dominate the repetitive content. Unclassified sequences comprise mostly multigenic families and gene clusters. (**b**) Repeat landscape of the *A. latifasciata* genome. For each element, the graph shows the sequence divergence from its consensus (*x*-axis) in relation to the number of copies on the genome (*y*-axis). Peaks represent insertion waves (red arrows) of elements into the genome. Elements with older insertion waves are shown on the right side of the graph, while newer insertions are depicted on the left side. Different colors show distinct element types, as described at the right. For a detailed and interactive version of the graph, please refer to Supplemental file 2. (**c**) Comparative analysis of repeats on the B- (0B) and B+ (2B) genomes via RepeatExplorer. Each column represents 100% of the reads in a cluster; the read proportions from the B- (0B) and B+ (2B) genomes are shown in yellow and blue, respectively. Clusters with higher proportions of B+ (2B) reads (in blue) are shown by an expansion of the blue color into the lower part of the graph (yellow). LINE: Long interspersed nuclear element; LTR: Long terminal repeat; SINE: Short interspersed nuclear element; TE: Transposable element.

Tc1-Mariner (4.37% of the genome) and hAT-Ac (1.77% of the genome) are the most highly represented DNA elements. Together, they represent over 50% of the DNA transposons in the *A. latifasciata* genome. A high diversity of other DNA elements is present, but each of them represents only a small percentage of the total. The predominant LINE is L2 (2.44%), but the element Rex (1.06% of the genome) is also represented. All major classes of LTR are found in the genome, with the

Gypsy (0.61%) and Pao (Bel/Pao, 0.21%) classes being the most abundant. Among SINEs, a high variation in element copy number is present, and families derived from transfer RNA (tRNA) are the most abundant.

These findings show that *A. latifasciata* has a similar transposable element composition to those of other available cichlid fish genomes (see Discussion section).

3.2. Three Main Bursts of DNA Element Insertion during Evolution Explain Their Genomic Abundance

Using the assembled B- genome and RepeatMasker results, a repeat landscape was constructed (Figure 1b, and Supplementary file 2 for an interactive version). It indicates the relative waves of insertion of specific families in the genome and can show bursts of insertion events during the evolution of the species, where a rapid increase in repeat copy number occurred. From these bursts, we can infer mobilization events in the evolution of the genome as well as putative TE activity represented by newer insertions [56]. Three main burst events were found in *A. latifasciata*'s evolution. In an older burst, a small accumulation of retrotransposons, both LTRs (Bel/Pao, Gypsy, and ERV) and LINEs (Rex-Babar and L2) occurred. The second burst of TEs had DNA elements as its major component. Tc1-Mariner showed the greatest increase in copy number during this period. The third and most recent burst of TE insertions was due to the accumulation of hAT DNA elements. Although hAT copy number began to increase in the second insertion wave, the third insertion wave involved predominantly hAT elements. As a trend, we see DNA elements dominating the landscape of TE insertions during *A. latifasciata* genome evolution. Retroelements show signs of constant mobilization; for example, L2 and Rex, which seem to have both younger and older copies, are represented by a constant appearance in the landscape.

3.3. Accumulation of TEs in the B Chromosome of A. latifasciata

Repetitive element accumulation is a hallmark of B chromosomes. Understanding the TE compositions of Bs can assist in the determination of their origins and possible effects in the cell. To establish whether an accumulation of TEs is present in the B chromosome of *A. latifasciata*, we performed a comparative analysis using RepeatExplorer and coverage ratios. The RepeatExplorer pipeline yielded 114 clusters (Supplementary file 3) with approximately 75% of the reads in clusters and 25% in singlets. For each cluster, we visually compared the number of reads from the B- (0B) and B+ (2B) datasets. Clusters having higher proportions of B+ (2B) reads (i.e., the number of 2B reads was much higher than the number of 0B reads) were selected for manual inspection and subsequent analysis (Figure 1c, Table 1, Supplementary file 3). From a total of 20 analyzed clusters, four were chosen for FISH probe construction (see next section for validation of results); clusters with better contig annotations and genomic relevance were selected. The clusters with higher proportions of 2B reads were clusters 28 and 67, annotated as Bel/Pao and Gypsy, respectively. This result indicates an expansion of those TEs on the extra chromosome. Other clusters selected for FISH were cluster 63 (hAT element) and cluster 74 (L2 element).

Table 1. Information of clusters with higher proportions of 2B reads selected for fluorescence in situ hybridization (FISH) probe construction.

Cluster	0B Reads	2B Reads	Annotation	Repeat Class
28	359	23,389	Bel/Pao	Retro/LTR
63	2939	13,563	DNA/hAT	DNA
67	351	15,603	Gypsy	Retro/LTR
74	4575	11,117	L2	Retro/LINE

Retro: Retrotransposon; DNA: DNA transposon; LTR: Long terminal repeat retrotransposon; LINE: Long interspersed nuclear element.

Although RepeatExplorer returned different expanded elements on the B chromosome, we decided to further analyze these elements to obtain a more detailed picture of B chromosome TE composition. In this way, we validated the RepeatExplorer results by bioinformatics and built a database of putative expanded elements with their respective loci. We performed a coverage ratio analysis on the several NGS datasets available (0B, 1B, and 2B, males and females; Supplementary Tables S1 and S2).

The coverage ratios showed that several elements were expanded on the B chromosome, consistent with the RepeatExplorer results (Table 2; Supplementary file 4). In general, the repeat class with the most accumulation on B was DNA transposons, followed by LINEs and LTRs. A large number of unclassified sequences was also found on B, showing the mobility of diverse sequences to the supernumerary element.

Table 2. Coverage ratio analysis of the six sequenced individuals. Values were normalized to the M1-0B genome (B- male). Values show the number of extra copies in a sample compared to that in the reference. Numbers are the sum of all loci expanded on the B chromosome.

Element	Class	Family	F1-0B/M1-0B	F2-1B/M1-0B	M2-1B/M1-0B	M3-1B/M1-0B	M4-2B/M1-0B
Gypsy-188_DR-I	LTR	Gypsy	0.00	357.90	478.75	505.95	977.98
AlRepB-26	LINE	L2	2.07	454.47	438.12	454.37	976.85
BEL32-I_DR	LTR	Pao	11.24	276.16	324.04	343.45	731.48
Gypsy-23_GA-I	LTR	Gypsy	0.00	311.83	336.77	360.23	709.62
Mariner-N13_DR	DNA	TcMar-Tc1	17.81	276.37	294.65	312.57	648.31
AlRepB-738	DNA	hAT-Ac	23.67	320.79	356.62	382.23	631.88
BEL32-LTR_DR	LTR	Pao	5.46	215.99	278.12	309.23	593.38
Maui	LINE	L2	777.82	916.34	626.59	665.77	312.38
AlRepC-927	DNA	hAT-Ac	253.98	250.69	96.81	92.91	0.00
AlRepB-157	Satellite	Satellite	545.89	537.41	39.69	23.75	44.40
TZSAT	Satellite	Satellite	181.04	177.25	4.73	4.52	6.86

M: Male; F: Female; 0B: Absence of B chromosomes; 1B: Presence of one B chromosome; 2B: Presence of two B chromosomes.

Examples of major TEs expanded on the B chromosome, as shown by coverage ratio analysis, include Gypsy-188_DR-I (Gypsy), AlRepB-26 (L2), Bel32-I_DR (Bel/Pao), Mariner-N13_DR (Tc1-Mariner), and AlRepB-738 (hAT-Ac) (Tables 1 and 2). RepeatExplorer also found elements from these families, and some of them were validated by qPCR, sequencing, and FISH (see next section). The copy number of these elements varies with the number of Bs in the genome; a B+ (2B) individual has accumulated more copies than a B+ (1B) individual has. Several other TEs also follow this pattern (Table 2).

Not only do our results point to the accumulation of TEs in the B chromosome, but the data also show different trends in the accumulation of elements by sex or by B presence and sex. For example, the Maui element (member of L2 family) is expanded in females and B chromosome carriers. Additionally, the element AlRepC-927 (hAT family) is absent in the B+ (2B) individual, which may indicate a recent mobilization or a population-related bias. Although an individual analysis of each element lies outside the scope of this work, such examples show the plasticity and mobility that TEs can achieve, especially considering the presence of a B chromosome.

3.4. B Chromosome TE Accumulation Is Validated by Cytogenetics and Molecular Techniques

Based on the bioinformatics analysis, we expected a higher number of TE copies in the B chromosome than in the A complement. Therefore, we sought to validate our bioinformatics findings concerning B chromosome TE composition. Sequences detected by RepeatExplorer were used to design a custom set of primers to validate the elements expanded in the B chromosome of *A. latifasciata* (Supplementary Tables S5 and S6). Custom probes were obtained by PCR and hybridized on mitotic chromosome preparations (Supplementary Table S5). Fluorescence in situ hybridization analysis revealed intense hybridization signals, mostly in the metacentric B chromosome (Figure 2a). As predicted by the bioinformatics pipeline, the signals from the custom probes were concentrated in the B chromosome, although they also had a diffuse presence in the A complement.

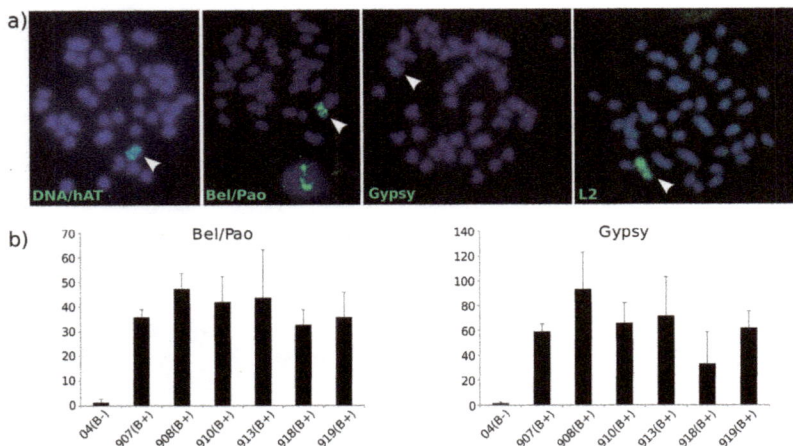

Figure 2. Distribution and copy number variation of repetitive DNAs in the *A. latifasciata* genome. (**a**) Fluorescent in situ hybridization mapping of four selected elements (DNA/hAT, Bel/Pao, Gypsy, and L2) on *A. latifasciata* metaphasic chromosomes, from a B+ specimen. Probes were PCR-labeled by biotin-dUTP, and the signal was detected by fluorescein isothiocyanate (FITC, green). Red arrowheads indicate the B chromosomes. The signal intensity is higher in the B chromosome than in the A complement. (**b**) qPCR of the Bel/Pao and Gypsy elements detected by the RepeatExplorer pipeline. Sample number 04 had no B chromosome (B-), while the others had at least one B (B+). The data show that B+ individuals had higher copy numbers than B- individuals, corroborating the previous results. GDR: Gene dosage ratio.

Quantitative real-time PCR was used to quantify the sequences in relation to the normalizer, the single-copy gene *HPRT*. The Bel/Pao and Gypsy elements were chosen due their abundance in the B chromosome (Supplementary Table S6). Relative quantification showed higher copy numbers of these elements in B+ individuals (Figure 2b). All B+ samples showed higher GDRs than the B- samples (sample 04). Thus, the results show that B+ individuals had higher element copy numbers than did B- individuals. The data from qPCR agreed with the results of RepeatExplorer and FISH, which demonstrated the accumulation of certain elements in the B chromosome (in this case, Bel/Pao and Gypsy).

3.5. TE Transcription across Tissues and Higher Expression in B Chromosome Samples

Based on the TE structural analysis of the B- and B+ genomes, we sought to establish the transcription levels of TEs in both conditions. According to the repeat landscape, some elements show signs of recent mobilization, and our first aim was to determine whether these elements are transcribed. The B- RPKM expression values show that many TEs are transcribed in various *A. latifasciata* tissues (Supplementary Figure S1; Supplementary Table S7; Supplementary file 5). All major families are expressed, with high variability among the individual elements. Although not suitable for comparisons between tissues, these RPKM values reveal that TE transcription occurs in the *A. latifasciata* genome. In fact, elements with recent insertion waves, which are thus putatively active, are expressed in different tissues. Tc1-Mariner and hAT have the highest expression levels among DNA transposons. L2, RTE-BovB, Rex, L1, Penelope, and MIR have high expression among retrotransposons. Such elements exemplify the relation between the repeat landscape and activity in the genome. The expression of each family tended to be constant among the tissues.

Next, we examined whether elements in the B chromosome had different expression levels than those in the rest of the genome. Such a difference would show the putative activity of elements present in the B chromosome compared to those in the A complement. All tissues showed differential expression of repetitive elements between the B+ (1B) and B- (0B) genomes (Figure 3; Table 3; Supplementary file 6). A high variability was found in the families expressed in different tissues, but DNA transposons showed differential expression in all of them. In brain, 15 upregulated and 5 downregulated elements were found. Retroelements dominated the differential expression; 14 out of 20 differentially expressed sequences were retroelements. In muscle, only three elements showed differential expression, and all were upregulated in the B+ tissue. Excluding an unclassified element, the other two were DNA transposons. In male gonads, we detected 14 differentially expressed elements, with only 3 upregulated in the B+ genome. In female gonads, only three elements showed differential expression, and all were upregulated in the B+ genome. Our results indicate that, with the exception of male gonads, a trend toward the upregulation of elements in B+ tissues is present.

Table 3. Repetitive elements with differential expression between B- and B+ individuals. B expansion represents how many more copies an individual had than the reference M1-0B. RC, rolling circle TE; srpRNA, signal recognition particle RNA.

Element	Superfamily	Class	Fold-Change	B Expansion
Brain				
AlRepD-1119	L2	LINE	1.4829	22/88/115/126/187
CR1-28_HM	CR1	LINE	3.2887	No
REX1-3_XT	Rex-Babar	LINE	1.7297	No
GYPSY2-I_CB	Gypsy	LTR	5.2102	No
AlRepC-299	Unknown	Unknown	1.5748	9/16/4/6/0
I-6_AAe	I	LINE	3.6624	No
L1-6_DR	L1	LINE	−3.8609	2/0/0/0/2
BovBa-1_EF	RTE-BovB	LINE	4.1560	No
AlRepD-1964	Unknown	Unknown	1.1420	24/117/121/120/160
Gypsy-34-I_DR	Gypsy	LTR	4.5448	No
AlRepD-520	Unknown	Unknown	1.8113	47/30/6/2/0
Mariner-1_SP	TcMar-Fot1	DNA	4.0150	0/0/2/5/12
ERV1-3N-EC_I-int	ERV1	LTR	2.4732	No
Gypsy-17-I_DR	Gypsy	LTR	−1.7877	2/2/2/4/16
Gypsy52-I_DR	Gypsy	LTR	5.4846	No
Tc1-2Eso	TcMar-Tc1	DNA	2.4478	No
Rex1-52_DR	Rex-Babar	LINE	−3.1220	No
Helitron-2_DR	Helitron	RC	−1.2385	No
RMER17C-int	ERVK	LTR	2.6684	No
Gypsy-20-I_DR	Gypsy	LTR	−2.1144	0/0/2/0/0
Muscle				
AlRepC-299	Unknown	Unknown	7.5228	9/16/4/6/0
AlRepB-358	hAT-Ac	DNA	3.1883	207/209/146/139/11
Mariner-1_SP	TcMar-Fot1	DNA	11.0098	0/0/2/5/12
Gonad Male				
AlRepD-4130	hAT-Ac	DNA	2.7431	37/87/51/56/31
HEROTn	R2-Hero	LINE	−6.4246	No
7SLRNA	srpRNA	srpRNA	3.4512	No
P-27_HM	P	DNA	−2.5097	No
AgaP15	P	DNA	−5.8314	No
Charlie16	hAT-Charlie	DNA	−5.7227	No
AlRepE-134	DNA	DNA	−2.9391	6/3/3/5/3
ERV-4_CPB-I	ERV1	LTR	−2.9007	No
Gypsy-10_GA-LTR	Gypsy	LTR	−2.6009	No
AlRepE-2243	Unknown	Unknown	3.6990	5/2/2/0/2
CR1-20_CQ	CR1	LINE	−2.9297	No
Dada-1_ON	Dada	DNA	−3.1755	No
Copia3-I_XT	Copia	LTR	−5.6443	No
AlRepD-3555	Unknown	Unknown	−4.5855	No
Gonad Female				
AlRepD-1636	Unknown	Unknown	5.5199	23/94/96/104/111
hAT-27_LCh	DNA	DNA	3.0980	No
AlRepD-4141	Unknown	Unknown	6.7706	2/0/0/0/2

Each number is the sum of all loci in the analyzed individuals (F1-0B/M1-0B, F2-1B/M1-0B, M2-1B/M1-0B, M3-1B/M1-0B, M4-2B/M1-0B).

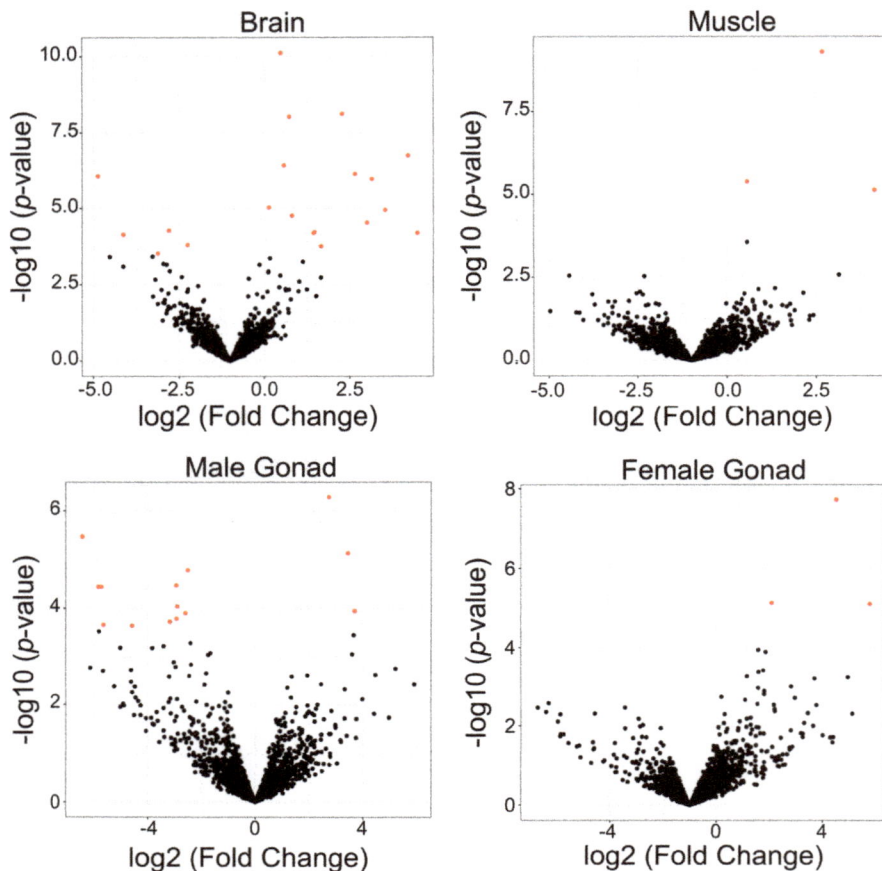

Figure 3. Volcano plots showing differential expression between B- and B+ individuals in brain, muscle, and male and female gonads. Red points indicate elements with differential expression above the threshold (log2 fold-change 1.2 and false discovery rate (FDR) <5%). Upregulated repeats in the B+ genomes are represented on the right side, and downregulated repeats in the B+ genomes are represented on the left side of each figure.

Most elements with differential expression between the B- and B+ genomes were not accumulated in the B chromosome. Exceptions existed, such as the element AlRepB-358 (hAT family), which had differential expression in B+ muscle and high copy number in 1B individuals, according to the coverage ratio analysis.

The only two elements with differential expression in two tissues (brain and muscle) were Mariner-1_SP and AlRepC-299. A BLAST of AlRepC-299 revealed 98% similarity with sequences from the *V2R* gene cluster of the cichlid *Haplochromis chilotes* (accession number AB780556.1) and 97% identity with sequences from *Oreochromis niloticus* (also a cichlid) in a region with no annotation but close to pseudogenes (accession number AB270897.1).

4. Discussion

4.1. Genomic Organization of TEs in A. latifasciata

The primary aim of our work was to characterize the repetitive content of the reference B- genome of *A. latifasciata* and show its relation to those of other cichlid fishes. DNA transposons were the major TE class found, while retroelements represented a smaller fraction. This composition is similar to those of other teleost fishes, including other African cichlids. In general, teleost fishes show one of the greatest diversities of TEs among vertebrates, with a high number of families populating the genome [57,58]. Among DNA transposons, Tc1-Mariner and hAT (class II TEs) are represented at higher percentages in the genome of *A. latifasciata*. This result is similar to those in other African cichlids, where class II elements are responsible for approximately 10% of repetitive content, a number also observed in *A. latifasciata*. L2 (a LINE retroelement) also predominates in the genome and, with Tc1-Mariner, has the highest genome percentage in both *A. latifasciata* and other cichlid fishes [57,59].

Next, we constructed a repeat landscape that shows several waves of TE insertion into the genome. We found that LTRs such as Gypsy and Bel/Pao entered the genome in its early evolutionary stages. DNA transposons such as Tc1-Mariner had later insertions and are accompanied by large increases in copy number. Similar landscapes are found in other cichlid fishes [59], and together with the TE content results, they support the idea of TE mobilization events in a common ancestor of the group. In general, TE families with high copy numbers in the genome of *A. latifasciata* show prominent insertion waves and expansions in the B chromosome. Elements with higher copy numbers seem to dominate B genome content, probably due to the low selective pressure on B. These findings demonstrate the ability of the B chromosome to act as a repository for sequences that can propagate over time.

The insertions of Penelope, Rex-Babar, L2, Tc1-Mariner, and hAT elements may indicate recent activity. In fact, these families have several elements with high expression in the analyzed tissues, and thus, some copies may still be active and populating new loci. Furthermore, some families are constant on the landscape, which indicates active mobilization during the evolution of the genome. Our data show a landmark of TEs: their activity is variable over time, with bursts following purification cycles [60]. Additionally, the correlation of transcription with the repeat landscape can help delimit elements for future studies of repeat mobilization and function.

Among the most highly transcribed elements, several have high copy numbers in the genome. Although transcription is not directly correlated to mobilization events, some expressed elements are related to recent insertion waves; some sequences even show evidence of open reading frames (ORFs) capable of expressing protein-coding sequences (data not shown). Considering that transposition events are a major cause of genomic instability and rearrangement [61], we should consider that, although transcribed, these copies might have lost the capacity to mobilize.

Environmental conditions can alter TE activity, change their copy number, and cause large genome alterations, especially through epigenetic silencing modifications. Posttranscriptional alterations such as small interfering (siRNA) can silence TEs and limit their copy number increase, abolishing the correlation between TE transcription and transposition events [60,62,63]. Even with several transcribed elements in the genome, posttranscriptional silencing can limit their transposition and copy number alteration. At this point, even with evidence of recent mobilization events and element transcription, we speculate that tight control is exerted on TEs.

4.2. Organization and Transcription of TEs in the B Chromosome

After examining the TE composition and expression in the B- genome, we were interested in searching for these characteristics in the B chromosome. For the first comparison between the B- and B+ genome datasets, we used RepeatExplorer, which is a de novo method of element identification and assembly [38,58] that can be used for comparative quantification between two or more datasets [64]. RepeatExplorer has the advantage of using raw Illumina reads with very low coverage; we selected this method for FISH probe construction and to test our hypothesis. Coverage information from the clusters obtained by RepeatExplorer is indicative of expanded elements in a particular dataset [39]. According to [38], the number of reads in a cluster is proportional to the quantity of that element on the genome. Here, we used a combination of methods (RepeatExplorer and coverage ratios) to give a broad picture of *A. latifasciata* B chromosome TE composition.

With RepeatExplorer, we could identify and validate several expanded elements on the B chromosome. All FISH-mapped elements showed similar patterns of hybridization, demonstrating a high number of copies on the B chromosome and thus corroborating the clustering data. We chose to perform a second validation with qPCR with the two elements most present in the B chromosome. Gypsy and Bel/Pao were selected for their abundance on the B chromosome compared to that on the A complement. Both analyses validated the B expansion of these specific TEs.

After our finding of expanded elements on the B chromosome of *A. latifasciata*, our aim was to specify the repeat families present and their loci; for that, we used a coverage ratio approach. The data from the coverage ratio analysis showed the diversity of TE composition on the B chromosome of the species and corroborated the results of RepeatExplorer. The elements detected by RepeatExplorer showed more coverage in B+ genomes. The results of the coverage ratios reflect the sum of all copies that are at least duplicated in the B chromosome. We selected this conservative threshold due to the inherent variation in repeat copy numbers. For some elements, 1B and 2B individuals had a two-times difference in coverage values, showing the effect of the number of B chromosomes in a species. Since this pattern was not present in all elements, it also indicates the variability of B sequences. Fluctuations in B block copy number are present even among siblings, as found by [29] in *Metriaclima lombardoi*, and our results corroborate these findings. Our sequencing data comes from different populations and is also prone to variation among individuals, which was evidenced by coverage analysis and qPCR.

Another characteristic detected by RepeatExplorer and by the coverage ratio analysis is the number of other types of sequences in the B chromosome. The *V2R* gene sequence was found in some contigs assembled by RepeatExplorer, along with *HOX* and *SOX* gene fragments. Considering the quantity of retroelements expanded on the B, these elements may have transferred a number of sequences during their activity cycles, as they are facilitators of sequence mobilization [65]. Retroelements such as LINEs have weak poly-A signals and are good candidates to facilitate the mobilization of adjacent sequences. Furthermore, enzymes can also incorporate non-TE mRNAs [66,67], leading to the inclusion of different types of sequences in the B chromosome. In general, TEs take advantage of the low selective pressure on the B chromosome for their insertion and perpetuation [3]. The combination of low selective pressure and drive (a B chromosome characteristic) allows repeats to expand and increase copy number in the B chromosome [9,10]. Such expansion or contraction of sequences in Bs is a hallmark of supernumeraries in diverse species [64], and is now presented here for *A. latifasciata*. Our results support the findings from other B chromosomes, which also show remarkable sequence expansions.

In addition to the analysis of TE expression in the B- genome, another goal of this study was to find signs of TE expression in the B chromosome of this species. Our data show evidence of the transcription of B chromosome sequences, although only a few TEs demonstrated differential expression between the B- and B+ genomes. This result points to a relatively stable B chromosome and is an important finding of our study. Moreover, differentially expressed elements expanded on the B chromosome usually had coverage differences smaller than those of highly expanded ones, such as Gypsy-188 and Bel32. Therefore, the majority of expanded elements in B+ genomes did not show differential expression between B- and B+ individuals. Expanded elements may have been present in the early stages of B chromosome formation, and we hypothesize that silencing mechanisms target them. If not silenced, elements with high copy numbers in the B chromosome could generate high genomic instability [60]. In general, increases in repeat copy number cause their transcriptional reduction through silencing mechanisms [68,69]. However, elements with high copy number can have high expression, as in the hAT family in maize [70]. Expression levels are also influenced by the number of B chromosomes in the cell; a higher number of Bs is associated with higher differential expression [25]. Our data are based on 1B samples, and therefore, small changes in differential expression from the B sequences may be difficult to detect. Our findings also point to a general trend of opposite TE expression in the gonads. Several elements are downregulated in males, while some are upregulated in females. We speculate that this DE among males and females could benefit the B chromosome drive exclusive of female meiosis [71].

The expression of repetitive elements on B chromosomes is highly dependent on the species and the location of the repeat in the supernumerary element. Expression is also related to the TE family. In rye, the elements expressed on the B chromosome are correlated to the presence of high copy number in the genome and even modulate the transcription of A-complement copies [6]. Differential expression among 0B and 4B rye individuals revealed a high quantity of expressed B repeats together with several gene fragments [26]. Gypsy and Mariner elements are expanded and expressed on the B chromosome of the grasshopper *Eyprepocnemis plorans*, with a predisposition to euchromatic regions [72]. Considering that methylation patterns vary on each B chromosome, their transcription levels may be altered. DNA methylation inactivates B chromosomes, reducing their effects on the cell [73]. It is plausible that the high-copy number elements in the B+ genome of *A. latifasciata* have their transcription regulated to avoid disruptive interference in the physiology of the cell. The B chromosome of *A. latifasciata* has the potential to generate noncoding RNA regulatory sequences [74], and therefore the regulation of this extra chromosome could help to stabilize the genome.

5. Conclusions

Research on TEs, especially those on B chromosomes, brings technical and analytical challenges, but understanding their organization in the genome can shed light on intrinsic mechanisms of gene and genome regulation. In this regard, this study is an important step to clarify B chromosome organization and structure. We found several expanded elements on the B chromosome of *A. latifasciata* and confirmed them through high-throughput sequencing, bioinformatics, FISH mapping, and qPCR quantification. We also described a general landscape of repeat copy number, relative insertion times in the genome, and transcription levels.

The influence of TEs on shaping the genome is clear, and studying their role in the formation and perpetuation of B chromosomes requires different approaches, such as the investigation of the transcription levels of messenger RNAs and noncoding RNAs. We found evidence that expansion of TEs on the B chromosome can be followed by the evolution of regulatory mechanisms that control TE activity. A deeper understanding of the epigenetic mechanisms acting on the accessory chromosome will also be necessary to determine the level of influence of the discovered elements on B.

Supplementary Materials: The following are available online at http://www.mdpi.com/2073-4425/9/6/269/s1. Figure S1: Repetitive element transcription levels quantified by RPKM, Table S1: Summary of most important information about the investigated samples, Table S2: Samples and statistics for the six read dataset alignments against the *A. latifasciata* reference genome, Table S3: Metrics of HPRT gene coverage in the six alignments, Table S4: Ratios between HPRT coverages in reference to the M1-0B dataset, Table S5: Primers used for the construction of custom FISH probes, Table S6: Primers used for qPCR, Table S7: Copy number estimates for several elements found in the *A. latifasciata* genome, File S1: Repeat copy numbers in the *A. latifasciata* 0B genome, File S2: *A. latifasciata* 0B landscape, File S3: Clusters from RepeatExplorer, File S4: Coverage ratios normalized to the M1-0B genome as reference, File S5: RPKM values of repetitive elements from brain, muscle and male and female gonads, File S6: Differential expression values between B- and B+ in brain, muscle and male and female gonads, Supplementary Methods: Determination of cut-off for coverage regions. References [75–77] are cited in the supplementary materials.

Author Contributions: R.L.B.C. and C.M. conceived and designed the experiments; R.L.B.C. performed the experiments; R.L.B.C. and C.M. analyzed the data; R.L.B.C. and C.M. wrote the paper.

Acknowledgments: This work was financially supported through grants from the São Paulo Research Foundation (FAPESP) (2013/04533-3; 2014/16763-6; 2015/16661-1) and the National Counsel of Technological and Scientific Development (CNPq) (134446/2014-3; 305321/2015-3).

Conflicts of Interest: The authors declare no conflict of interest.

References

1. Camacho, J.P.; Sharbel, T.F.; Beukeboom, L.W. B-chromosome evolution. *Philos. Trans. R. Soc. Lond. B Biol. Sci.* **2000**, *355*, 163–178. [CrossRef] [PubMed]
2. Yoshida, K.; Terai, Y.; Mizoiri, S.; Aibara, M.; Nishihara, H.; Watanabe, M.; Kuroiwa, A.; Hirai, H.; Hirai, Y.; Matsuda, Y.; et al. B chromosomes have a functional effect on female sex determination in Lake Victoria cichlid fishes. *PLoS Genet.* **2011**, *7*, e1002203. [CrossRef] [PubMed]
3. Houben, A.; Banaei-Moghaddam, A. Evolution and biology of supernumerary B chromosomes. *Cell. Mol. Life Sci.* **2014**, *71*, 467–478. [CrossRef] [PubMed]
4. Ploskaya-Chaibi, M.; Voitovich, A.M.; Novitsky, R.V.; Bouhadad, R. B-chromosome and V-shaped spot asymmetry in the common frog (*Rana temporaria* L.) populations. *Comptes Rendus Biol.* **2015**, *338*, 161–168. [CrossRef] [PubMed]
5. Beukeboom, L.W. Bewildering Bs: An impression of the 1st B-Chromosome conference. *Heredity* **1994**, *73*, 328–336. [CrossRef]
6. Carchilan, M.; Kumke, K.; Mikolajewski, S.; Houben, A. Rye B chromosomes are weakly transcribed and might alter the transcriptional activity of a chromosome sequences. *Chromosoma* **2009**, *118*, 607–616. [CrossRef] [PubMed]
7. Adnađević, T.; Jovanović, V.M.; Blagojević, J.; Budinski, I.; Cabrilo, B.; Bijelić-Čabrilo, O.; Vujošević, M. Possible influence of B Chromosomes on genes included in immune response and parasite burden in *Apodemus flavicollis. PLoS ONE* **2014**, *9*, e112260. [CrossRef] [PubMed]
8. Valente, G.T.; Nakajima, R.T.; Fantinatti, B.E.A.; Marques, D.F.; Almeida, R.O.; Simões, R.P.; Martins, C. B chromosomes: From cytogenetics to systems biology. *Chromosoma* **2017**, *126*, 73–81. [CrossRef] [PubMed]
9. Banaei-Moghaddam, A.M.; Martis, M.M.; Macas, J.; Gundlach, H.; Himmelbach, A.; Altschmied, L.; Mayer, K.F.X.; Houben, A. Genes on B chromosomes: Old questions revisited with new tools. *Biochim. Biophys. Acta* **2015**, *1849*, 64–70. [CrossRef] [PubMed]
10. Valente, G.T.; Conte, M.A.; Fantinatti, B.E.A.; Cabral-de-Mello, D.C.; Carvalho, R.F.; Vicari, M.R.; Kocher, T.D.; Martins, C. Origin and evolution of B chromosomes in the cichlid fish *Astatotilapia latifasciata* based on integrated genomic analyses. *Mol. Biol. Evol.* **2014**, *31*, 2061–2072. [CrossRef] [PubMed]
11. Silva, D.M.A.; Pansonato-Alves, J.C.; Utsunomia, R.; Araya-Jaime, C.; Ruiz-Ruano, F.J.; Daniel, S.N.; Hashimoto, D.T.; Oliveira, C.; Camacho, J.P.M.; Porto-Foresti, F.; et al. Delimiting the origin of a B chromosome by FISH mapping, chromosome painting and DNA sequence analysis in *Astyanax paranae* (Teleostei, Characiformes). *PLoS ONE* **2014**, *9*, e94896. [CrossRef] [PubMed]
12. Charlesworth, B.; Sniegowski, P.; Stephan, W. The evolutionary dynamics of repetitive DNA in eukaryotes. *Nature* **1994**, *371*, 215–220. [CrossRef] [PubMed]
13. Shapiro, J.A.; von Sternberg, R. Why repetitive DNA is essential to genome function. *Biol. Rev.* **2005**, *80*, 227–250. [CrossRef] [PubMed]
14. Wicker, T.; Sabot, F.; Hua-Van, A.; Bennetzen, J.L.; Capy, P.; Chalhoub, B.; Flavell, A.; Leroy, P.; Morgante, M.; Panaud, O.; et al. A unified classification system for eukaryotic transposable elements. *Nat. Rev. Genet.* **2007**, *8*, 973–982. [CrossRef] [PubMed]
15. Rebollo, R.; Romanish, M.T.; Mager, D.L. Transposable elements: An abundant and natural source of regulatory sequences for host genes. *Annu. Rev. Genet.* **2012**, *46*, 21–42. [CrossRef] [PubMed]
16. Bueno, D.; Palacios-Gimenez, O.M.; Cabral-de-Mello, D.C. Chromosomal mapping of repetitive DNAs in the grasshopper *Abracris flavolineata* reveal possible ancestry of the B chromosome and H3 histone spreading. *PLoS ONE* **2013**, *8*, e66532. [CrossRef] [PubMed]
17. Teruel, M.; Cabrero, J.; Perfectti, F.; Camacho, J.P.M. B chromosome ancestry revealed by histone genes in the migratory locust. *Chromosoma* **2010**, *119*, 217–225. [CrossRef] [PubMed]
18. Cabral-de-Mello, D.C.; Moura, R.C.; Martins, C. Chromosomal mapping of repetitive DNAs in the beetle *Dichotomius geminatus* provides the first evidence for an association of 5S rRNA and histone H3 genes in insects, and repetitive DNA similarity between the B chromosome and A complement. *Heredity* **2010**, *104*, 393–400. [CrossRef] [PubMed]
19. Fantinatti, B.E.; Mazzuchelli, J.; Valente, G.T.; Cabral-de-Mello, D.C.; Martins, C. Genomic content and new insights on the origin of the B chromosome of the cichlid fish *Astatotilapia latifasciata. Genetica* **2011**, *139*, 1273–1282. [CrossRef] [PubMed]

20. Utsunomia, R.; de Andrade Silva, D.M.Z.; Ruiz-Ruano, F.J.; Araya-Jaime, C.; Pansonato-Alves, J.C.; Scacchetti, P.C.; Hashimoto, D.T.; Oliveira, C.; Trifonov, V.A.; Porto-Foresti, F.; et al. Uncovering the ancestry of B chromosomes in *Moenkhausia sanctaefilomenae* (Teleostei, Characidae). *PLoS ONE* **2016**, *11*, e0150573. [CrossRef] [PubMed]

21. Silva, D.M.Z.; Daniel, S.N.; Camacho, J.P.M.; Utsunomia, R.; Ruiz-Ruano, F.J.; Penitente, M.; Pansonato-Alves, J.C.; Hashimoto, D.T.; et al. Origin of B chromosomes in the genus *Astyanax* (Characiformes, Characidae) and the limits of chromosome painting. *Mol. Genet. Genom.* **2016**, *291*, 1407–1418. [CrossRef] [PubMed]

22. Martis, M.M.; Klemme, S.; Banaei-Moghaddam, A.M.; Blattner, F.R.; Macas, J.; Schmutzer, T.; Scholz, U.; Gundlach, H.; Wicker, T.; Šimková, H.; et al. Selfish supernumerary chromosome reveals its origin as a mosaic of host genome and organellar sequences. *Proc. Natl. Acad. Sci. USA* **2012**, *109*, 13343–13346. [CrossRef] [PubMed]

23. Ziegler, C.G.; Lamatsch, D.K.; Steinlein, C.; Engel, W.; Schartl, M.; Schmid, M. The giant B chromosome of the cyprinid fish *Alburnus alburnus* harbours a retrotransposon-derived repetitive DNA sequence. *Chromosome Res.* **2003**, *11*, 23–35. [CrossRef] [PubMed]

24. Lamb, J.C.; Riddle, N.C.; Cheng, Y.M.; Theuri, J.; Birchler, J.A. Localization and transcription of a retrotransposon-derived element on the maize B chromosome. *Chromosome Res.* **2007**, *15*, 383–398. [CrossRef] [PubMed]

25. Huang, W.; Du, Y.; Zhao, X.; Jin, W. B chromosome contains active genes and impacts the transcription of A chromosomes in maize (*Zea mays* L.). *BMC Plant Biol.* **2016**, *16*, 88. [CrossRef] [PubMed]

26. Ma, W.; Gabriel, T.S.; Martis, M.M.; Gursinsky, T.; Schubert, V.; Vrána, J.; Doležel, J.; Grundlach, H.; Altschmied, L.; Scholz, U.; et al. Rye B chromosomes encode a functional Argonaute-like protein with in vitro slicer activities similar to its A chromosome paralog. *New Phytol.* **2017**, *213*, 916–928. [CrossRef] [PubMed]

27. Poletto, A.B.; Ferreira, I.A.; Martins, C. The B chromosomes of the African cichlid fish *Haplochromis obliquidens* harbour 18S rRNA gene copies. *BMC Genet.* **2010**, *11*, 1. [CrossRef] [PubMed]

28. Kuroiwa, A.; Terai, Y.; Kobayashi, N.; Yoshida, K.; Suzuki, M.; Nakanishi, A.; Matsuda, Y.; Watanabe, M.; Okada, N. Construction of chromosome markers from the Lake Victoria cichlid *Paralabidochromis chilotes* and their application to comparative mapping. *Cytogenet. Genome Res.* **2014**, *142*, 112–120. [CrossRef] [PubMed]

29. Clark, F.E.; Conte, M.A.; Ferreira-Bravo, I.A.; Poletto, A.B.; Martins, C.; Kocher, T.D. Dynamic sequence evolution of a sex-associated B chromosome in Lake Malawi cichlid fish. *J. Hered.* **2017**, *108*, 53–62. [CrossRef] [PubMed]

30. Turner, G.F.; Seehausen, O.; Knight, M.E.; Allender, C.J.; Robinson, R.L. How many species of cichlid fishes are there in African lakes? *Mol. Ecol.* **2001**, *10*, 793–806. [CrossRef] [PubMed]

31. Kocher, T.D. Adaptive evolution and explosive speciation: The cichlid fish model. *Nat. Rev. Genet.* **2004**, *5*, 288–298. [CrossRef] [PubMed]

32. Seehausen, O. African cichlid fish: A model system in adaptive radiation research. *Proc. Biol. Sci.* **2006**, *273*, 1987–1998. [CrossRef] [PubMed]

33. Fantinatti, B.E.A.; Martins, C. Development of chromosomal markers based on next-generation sequencing: The B chromosome of the cichlid fish *Astatotilapia latifasciata* as a model. *BMC Genet.* **2016**, *17*, 119. [CrossRef] [PubMed]

34. Jehangir, M. Genome Assembly of the Cichlid Fish *Astatotilapia latifasciata* with Focus in Population Genomics of B Chromosome Polymorphism. Available online: http://hdl.handle.net/11449/151740 (accessed on 12 March 2018).

35. Smit, A.; Hubley, R. RepeatModeler Open-1.0 2015. Available online: http://www.repeatmasker.org/RepeatModeler/ (accessed on 18 December 2015).

36. Bao, W.; Kojima, K.K.; Kohany, O. Repbase Update, a database of repetitive elements in eukaryotic genomes. *Mob. DNA* **2015**, *6*, 11. [CrossRef] [PubMed]

37. Smit, A.; Hubley, R.; Green, P. RepeatMasker Open-4.0 2013. Available online: http://www.repeatmasker.org/ (accessed on 5 July 2015).

38. Novak, P.; Neumann, P.; Pech, J.; Steinhaisl, J.; Macas, J. RepeatExplorer: A Galaxy-based web server for genome-wide characterization of eukaryotic repetitive elements from next-generation sequence reads. *Bioinformatics* **2013**, *29*, 792–793. [CrossRef] [PubMed]

39. Novák, P.; Neumann, P.; Macas, J. Graph-based clustering and characterization of repetitive sequences in next-generation sequencing data. *BMC Bioinform.* **2010**, *11*, 378. [CrossRef] [PubMed]

40. Novak, P.; (Institute of Plant Molecular Biology, České Budějovice, Czech Republic); Macas, J.; (Institute of Plant Molecular Biology, České Budějovice, Czech Republic). Personal Communication, 2014.

41. Langmead, B.; Salzberg, S.L. Fast gapped-read alignment with Bowtie 2. *Nat. Methods* **2012**, *9*, 357–359. [CrossRef] [PubMed]

42. Andrews, S. Fastqc 2012. Available online: https://www.bioinformatics.babraham.ac.uk/projects/fastqc/ (accessed on 15 January 2014).

43. Hannon FASTX-Toolkit 2010. Available online: http://hannonlab.cshl.edu/fastx_toolkit/index.html (accessed on 1 August 2014).

44. Staton, E. Pairfq 2014. Available online: https://github.com/sestaton/Pairfq (accessed 15 April 2014).

45. Garcia-Alcalde, F.; Okonechnikov, K.; Carbonell, J.; Cruz, L.M.; Gotz, S.; Tarazona, S.; Dopazo, J.; Meyer, T.F.; Conesa, A. Qualimap: Evaluating next-generation sequencing alignment data. *Bioinformatics* **2012**, *28*, 2678–2679. [CrossRef] [PubMed]

46. Quinlan, A.R. BEDTools: The swiss-army tool for genome feature analysis. In *Current Protocols in Bioinformatics*; Wiley Online Library: Hoboken, NJ, USA, 2014; ISBN 0471250937.

47. Marques, D.F. Functional analysis of B chromosome presence using cichlid *Astatotilapia latifasciata* as model. Available online: http://hdl.handle.net/11449/141953 (accessed on 12 March 2018).

48. Criscione, S.W.; Zhang, Y.; Thompson, W.; Sedivy, J.M.; Neretti, N. Transcriptional landscape of repetitive elements in normal and cancer human cells. *BMC Genom.* **2014**, *15*, 583. [CrossRef] [PubMed]

49. Criscione, S. RepEnrich Tutorial. Available online: https://github.com/nerettilab/RepEnrich (accessed on 17 April 2016).

50. Gene Link Oligo Explorer 1.2 2005. Available online: http://www.genelink.com/tools/gl-oe.asp (accessed on 14 November 2014).

51. Ye, J.; Coulouris, G.; Zaretskaya, I.; Cutcutache, I.; Rozen, S.; Madden, T.L. Primer-BLAST: A tool to design target-specific primers for polymerase chain reaction. *BMC Bioinform.* **2012**, *13*, 134. [CrossRef] [PubMed]

52. Stothard, P. The sequence manipulation suite: JavaScript programs for analyzing and formatting protein and DNA sequences. *Biotechniques* **2000**, *28*, 1102–1104. [CrossRef] [PubMed]

53. Bertollo, L.; Takahashi, C.; Moreira-Filho, O. Citotaxonomic consideration on *Hoplias lacerdae* (Pisces, Erythrinidae). *Braz. J. Genet.* **1978**, *1*, 103–120.

54. Pinkel, D.; Straume, T.; Gray, J.W. Cytogenetic analysis using quantitative, high-sensitivity, fluorescence hybridization. *Proc. Natl. Acad. Sci. USA* **1986**, *83*, 2934–2938. [CrossRef] [PubMed]

55. Bel, Y.; Ferré, J.; Escriche, B. Quantitative real-time PCR with SYBR Green detection to assess gene duplication in insects: Study of gene dosage in *Drosophila melanogaster* (Diptera) and in *Ostrinia nubilalis* (Lepidoptera). *BMC Res. Notes* **2011**, *4*, 84. [CrossRef] [PubMed]

56. Chalopin, D.; Fan, S.; Simakov, O.; Meyer, A.; Schartl, M.; Volff, J.N. Evolutionary active transposable elements in the genome of the coelacanth. *J. Exp. Zool. Part B Mol. Dev. Evol.* **2014**, *322*, 322–333. [CrossRef] [PubMed]

57. Chalopin, D.; Naville, M.; Plard, F.; Galiana, D.; Volff, J.N. Comparative analysis of transposable elements highlights mobilome diversity and evolution in vertebrates. *Genome Biol. Evol.* **2015**, *7*, 567–580. [CrossRef] [PubMed]

58. Sotero-Caio, C.G.; Platt, R.N.; Suh, A.; Ray, D.A. Evolution and Diversity of Transposable Elements in Vertebrate Genomes. *Genome Biol. Evol.* **2017**, *9*, 161–177. [CrossRef] [PubMed]

59. Brawand, D.; Wagner, C.E.; Li, Y.I.; Malinsky, M.; Keller, I.; Fan, S.; Simakov, O.; Ng, A.Y.; Lim, Z.W.; Bezault, E.; et al. The genomic substrate for adaptive radiation in African cichlid fish. *Nature* **2014**, *513*, 375–381. [CrossRef] [PubMed]

60. Belyayev, A. Bursts of transposable elements as an evolutionary driving force. *J. Evol. Biol.* **2014**, *27*, 2573–2584. [CrossRef] [PubMed]

61. Hedges, D.J.; Deininger, P.L. Inviting instability: Transposable elements, double-strand breaks, and the maintenance of genome integrity. *Mutat. Res. Fundam. Mol. Mech. Mutagen.* **2007**, *616*, 46–59. [CrossRef] [PubMed]

62. Feschotte, C.; Jiang, N.; Wessler, S.R. Plant transposable elements: Where genetics meets genomics. *Nat. Rev. Genet.* **2002**, *3*, 329–341. [CrossRef] [PubMed]

63. Slotkin, R.K.; Martienssen, R. Transposable elements and the epigenetic regulation of the genome. *Nat. Rev. Genet.* **2007**, *8*, 272–285. [CrossRef] [PubMed]

64. Klemme, S.; Banaei-Moghaddam, A.M.; Macas, J.; Wicker, T.; Novák, P.; Houben, A. High-copy sequences reveal distinct evolution of the rye B chromosome. *New Phytol.* **2013**, *199*, 550–558. [CrossRef] [PubMed]

65. Kaessmann, H. Origins, evolution, and phenotypic impact of new genes. *Genome Res.* **2010**, *20*, 1313–1326. [CrossRef] [PubMed]

66. Maestre, J.; Tchénio, T.; Dhellin, O.; Heidmann, T. mRNA retroposition in human cells: Processed pseudogene formation. *EMBO J.* **1995**, *14*, 6333–6338. [PubMed]

67. Moran, J.V.; DeBerardinis, R.J.; Kazazian, H.H. Exon shuffling by L1 retrotransposition. *Science* **1999**, *283*, 1530–1534. [CrossRef] [PubMed]

68. Meyers, B.C.; Tingey, S.V.; Morgante, M. Abundance, distribution, and transcriptional activity of repetitive elements in the maize genome. *Genome Res.* **2001**, *11*, 1660–1676. [CrossRef] [PubMed]

69. González, L.G.; Deyholos, M.K. Identification, characterization and distribution of transposable elements in the flax (*Linum usitatissimum* L.) genome. *BMC Genom.* **2012**, *13*, 644. [CrossRef] [PubMed]

70. Vicient, C.M. Transcriptional activity of transposable elements in maize. *BMC Genom.* **2010**, *11*, 601. [CrossRef] [PubMed]

71. Houben, A. B chromosomes—A matter of chromosome drive. *Front. Plant Sci.* **2017**, *8*, 210. [CrossRef] [PubMed]

72. Montiel, E.E.; Cabrero, J.; Camacho, J.P.M.; López-León, M.D. Gypsy, RTE and Mariner transposable elements populate *Eyprepocnemis plorans* genome. *Genetica* **2012**, *140*, 365–374. [CrossRef] [PubMed]

73. Bugno-Poniewierska, M.; Solek, P.; Wronski, M.; Potocki, L.; Jezewska-Witkowska, G.; Wnuk, M. Genome organization and DNA methylation patterns of B chromosomes in the red fox and Chinese raccoon dogs. *Hereditas* **2014**, *151*, 169–176. [CrossRef] [PubMed]

74. Ramos, É.; Cardoso, A.L.; Brown, J.; Marques, D.F.; Fantinatti, B.E.A.; Cabral-de-Mello, D.C.; Oliveira, R.A.; O'Neill, R.J.; Martins, C. The repetitive DNA element BncDNA, enriched in the B chromosome of the cichlid fish *Astatotilapia latifasciata*, transcribes a potentially noncoding RNA. *Chromosoma* **2017**, *126*, 313–323. [CrossRef] [PubMed]

75. Altschul, S.F.; Gish, W.; Miller, W.; Myers, E.W.; Lipman, D.J. Basic local alignment search tool. *J. Mol. Biol.* **1990**, *215*, 403–410. [CrossRef]

76. Haas, B.J.; Papanicolaou, A.; Yassour, M.; Grabherr, M.; Blood, P.D.; Bowden, J.; MacManes, M.D. De novo transcript sequence reconstruction from RNA-seq using the Trinity platform for reference generation and analysis. *Nat. Protoc.* **2013**, *8*, 1494–1512. [CrossRef] [PubMed]

77. Wu, T.D.; Watanabe, C.K. GMAP: A genomic mapping and alignment program for mRNA and EST sequences. *Bioinformatics* **2005**, *21*, 1859–1875. [CrossRef] [PubMed]

© 2018 by the authors. Licensee MDPI, Basel, Switzerland. This article is an open access article distributed under the terms and conditions of the Creative Commons Attribution (CC BY) license (http://creativecommons.org/licenses/by/4.0/).

MDPI

St. Alban-Anlage 66

4052 Basel

Switzerland

Tel. +41 61 683 77 34

Fax +41 61 302 89 18

www.mdpi.com

Genes Editorial Office

E-mail: genes@mdpi.com

www.mdpi.com/journal/genes

www.ingramcontent.com/pod-product-compliance
Lightning Source LLC
Chambersburg PA
CBHW051727210326
41597CB00032B/5634